SECOND EDITION

The Life of a Photograph

Archival Processing
Matting
Framing
Storage

LAURENCE E. KEEFE
DENNIS INCH

FOCAL PRESS
Boston London

Focal Press is an imprint of Butterworth Publishers.

Library of Congress Cataloging-in-Publication Data

Keefe, Laurence E.
The life of a photograph: archival processing, matting, framing, and storage / Laurence E. Keefe, Dennis Inch.
p. cm.
Includes bibliographical references.
ISBN 0-240-80024-9
1. Photographs — Conservation and restoration. 2. Photographs — Trimming, mounting, etc. I. Inch, Dennis. II. Title.
TR465.K44 1990
771′.46 — dc20 89-36459
CIP

British Library Cataloguing in Publication Data

Keefe, Laurence E.
The life of a photograph: archival processing, matting, framing, and storage. — 2nd ed.
1. Photographs. Conservation
I. Title II. Inch, Dennis
770′.28′6

ISBN 0-240-80024-9

Butterworth Publishers
80 Montvale Avenue
Stoneham, MA 02180

10 9 8 7 6 5 4 3 2 1

Printed in the United States of America

For our families

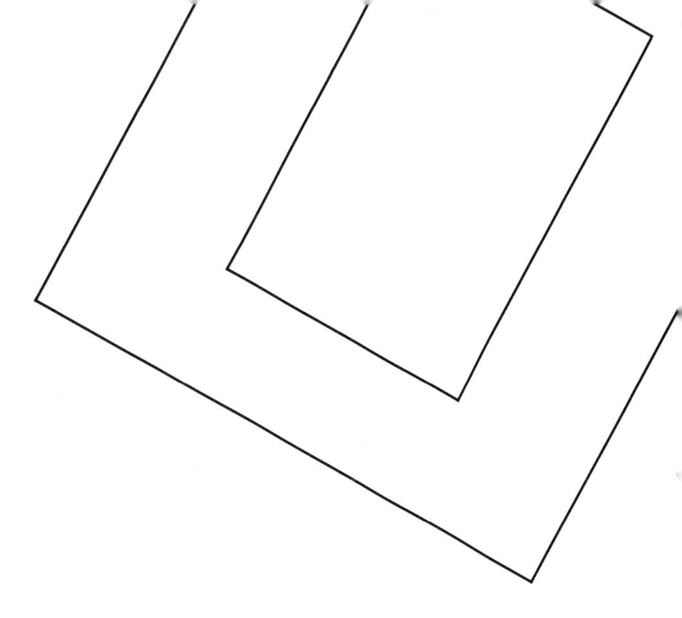

Contents

Acknowledgments

Now that the finished book lies before us and the work of it behind us, we are surprised as we look back over the years at how many people have contributed so much to what often seemed a frustratingly lonely task. Our gratitude is to all of them, while a few deserve special mention here.

Technical insights on specific topics were offered to us by Thomas T. Hill, James M. Reilly, and Guenther Cartwright, all of the Rochester Institute of Technology. Eugene Ostroff of the Smithsonian Institution has provided many illuminating comments on the field of conservation.

The following organizations were not responsible for specific contributions to this book. However, we would like to note that we have been strongly encouraged by the existence and activities of the American Institute for Conservation, the Society of American Archivists, the American Museum Association, the American Association for State and Local History, and the International Museum of Photography at George Eastman House. Also we thank Nathan Lyons of the Visual Studies Workshop for his influence as an educator.

And, having saved the best for last, we must especially thank Francis J. Crociata for his optimism, good spirits, and perseverance in pushing a pair of balking authors toward their final goal.

Despite all the help that we have received from these and other sources, we want to emphasize categorically that all errors of omission or commission that appear in this book remain the responsibility of its authors.

LEK
DI

The Life of a Photograph

Introduction

There is nothing permanent about the photograph. Whether in the silent sulfiding of silver emulsions or the fiery explosion of nitrate film, each image from the past carries within itself the seeds of its own destruction.

When we pull twenty-year-old slides from their boxes, we may find only faded ghosts from a past we had thought secured. The Cameron portraits of nineteenth-century Romantic poets grow dim and yellow, and before this century is over they may no longer exist save as copies. Well-kept collections like the National Geographic Society's invaluable files undergo relentless decay because the materials themselves are unstable and because the size of the collection makes copying prohibitively expensive. Acid rain in the northeastern United States causes hydrolysis of paper fibers, and humidity in the tropics breeds a proliferation of mold on the gelatin of slide and print emulsions.

The only comfort one can draw from all this must be the coldly philosophical reflection that not every image of an event need survive beyond its immediate use. What would we do if every single grip-and-grin check-passing publicity photo survived into all eternity? Some argue that we already suffer from a surfeit of this trite imagery.

The slow decay that eats away at our photographic roots must be nature's way of encouraging us to weed out the trivia. The stern inevitability of this pruning and thinning certainly clears away the deadwood effectively. But, because of its highly arbitrary character, we trust this natural process of removal at the peril of our own values. Fire, rust, and rot do not discriminate. The quality of the photographic record that will survive us lies in our own hands.

Responsibility for conservation is widespread. The central role played by photography in modern civilization means that we are all committed in some fashion to edit and preserve this medium's output. The result, it might be hoped, is that at least the best of our past and current work will survive, and along with it some of the other, less important but historically illustrative items that the camera produces in such abaundance. We need not only the brooding Lincoln portraits and ghastly Gettysburg battlefield scenes of a Mathew Brady but also the anonymous Civil War tintypes of green privates and callow lieutenants to flesh out the historic record. Future historians will find family picnic snapshots an essential counterpoint to professionally documented news events.

We — all of us — have two distinct options in preserving our collective photographic record. First, we can take what already exists in the form of prints and negatives, and then frame and store them in ways that best preserve them. In the case of old photographs, this is the only course possible.

Yet as we work toward this goal, we come upon a disturbing contradiction: some photogaphic processes have end products so unstable, so inherently volatile, that even with the best of care they destroy themselves. Until a few years ago this pertained mostly to color materials. This could be excused because mass-produced color photography is a relatively recent means of expression. Not until five or ten years ago did many fine art photographers start to work in color. Even then most worked in relatively stable media like dye transfer.

Now, instead of improving, the prospects for permanence grow worse. Starting around 1880, there existed many brands of silver-based photographic paper that were as stable as any printing technique known in the graphic arts. After the introduction of "safety film," the same held true for black and white film. But in recent years we see these materials supplanted by resin-coated papers and chromogenic black and white film. Even its manufacturers agree that the plastic paper is unstable, though they have done so only after much prodding. The new chromogenic film will be as susceptible to fading as color film.

This leads us to our second option, which is to insist, collectively, that manufacturers of photographic material make permanence one of their design parameters. As good businesspeople working in the free enterprise tradition, the photographic industrialists supply only what the market demands. You, as the consumer, can help the market speak for itself. For example, the next time you buy photographic paper and film, or a pack of instant print film, you might write to corporate headquarters and ask about the life expectancy of your purchase. Ask the same questions where you have the family snapshots printed. Answers may not be forthcoming, or they may be deliberately obscure, but let enough inquiries be received and the marketing departments of Kodak, Polaroid, Agfa, Ilford, and Fuji will begin to make intracorporate politics a little more interesting.

Be selective in the purchase of photo supplies. Use and specify the most stable materials available for films, prints, and slides. Read and learn about processing for permanence. This is not the only book on that subject, and it would grow to encyclopedic length if it covered everything. If you work as a curator, historian, photographer, artist, or archivist, support conservation research in your local area and in your professional association. And do not forget: petitions, resolutions, and even lawsuits all have their part to play in making the free marketplace provide quality products.

THE LIFE OF A PHOTOGRAPH

Any photograph resembles a living organism in that it has a natural life span. The outline of this book follows that course. The photograph's interactions with its environment determine how long its life span is. The picture takes up and releases moisture, its internal chemistry speeds up in the presence of heat, and light provides energy that causes long-term changes in its appearance. It may be host to a variety of organisms that use it for shelter and food, and it suffers from fire and superior physical force in the same way that we do.

The aim of conservation is to put the print into a kind of suspended animation, to slow down the natural life cycle to the point where it seems to stop and the picture lives, we hope, forever. To do this we must, in addition to protecting it from physical harm, assure that the picture is guarded against all things that accelerate aging. If you do nothing other than keep your photos cool, dark, and dry, you will help to prolong their existence.

But obviously there is more to conserving photographs than this simple formula. At each stage we demonstrate steps that can be taken to prolong that life. The book starts with processing the negative and the print. This section includes basic instructions on how to make negatives and prints, though it is not an advanced manual on the fine points of print-making. Instead, we have a practical explanation of how processing affects print permanence.

The section on mounting and matting comprises an essential part of the approach we suggest for permanence. Here the instructions are quite detailed. We give specific instructions on how to make different kinds of attractive mats using simple tools. A competent matting job that enhances the visual appeal of the print not only protects it but makes people more likely to be concerned about preserving it.

Framing is a natural corollary to matting. Here too the choice of materials and technique is imporant. Not every secret of the framing profession is laid bare here, but you will find enough, we hope, to do a good job using your choice of frame types.

Once a print has been framed, it is ready for display and then storage — and ready to be taken from storage for display once again. In covering storage, we include critical data on types of containers and the prevailing standards for environmental conditions. Research has shown that both play vital parts in preserving a print. Display, of course, also requires careful planning, and we have included guidelines and practical suggestions that should make this less exasperating and more methodical than it often is.

Not every print has received ideal treatment. In fact, most old photographs suffer some kind of abuse or neglect. The last section discusses these old pictures — how to recognize them and what to do with them. There is a detailed discussion of what was found with three old pictures when we looked at them closely; the problems we encountered will be the same that face anyone with old prints.

CURRENT STATUS OF PHOTOGRAPHIC CONSERVATION

Over the centuries, professionals in the art world have developed a canon of acceptable practices for cleaning, restoring, displaying, and storing oil paintings. More recently, during this century, great advances have been made in the related field of paper conservation. This encompasses the care of such items as books, fine art prints, and paper records. A good deal is now known about what should and should not be done to protect paper.

We do not yet have the same confidence about all aspects of caring for photographs. There are two reasons. One is that, compared to painting and other art media, photography is a relatively recent technology for making images. We have not accumulated the centuries of knowledge and experience that lie behind even the simplest oil painting. The other reason for uncertainty is that a photograph, compared to an etching, is a much more complicated object that embodies a great many variables whose interactions are difficult to predict. Where the etching is composed of cellulose fibers and ink, a modern color photograph has a host of components: a fiber base impregnated with whiteners and coated with resins, binders, gelatin, a slew of dyes, and more or less deactivated color couplers and processing chemicals. Each one of these complex substances has untold potential for interacting with the others and with outside agents such as light, atmospheric pollutants, buffering agents, plasticizers, and the like.

There is reason for cautious optimism, however. In the past, we had to rely on conservation practices imported from other fields of art, and the photographic conservator was in the uncomfortable position of an automobile mechanic attempting to repair a tractor trailer. But in the few brief years since we published the first edition of this book, there has emerged a new class of professional researchers into the subject of photographic permanence, as well as working photographic conservators. Professionals now hold annual conferences and symposia and institutions can hire specialists trained specifically to preserve photographs. Solid data has become easier to find, standardized tests for stability are being developed, and industrywide standards for permanence of photographic materials are being written.

The resulting increase in knowledge affects the work of people in disparate fields: museum curators, manufacturers of print washers, owners of picture framing shops, photographers, collectors and gallery owners, even people with a few snapshots to care for. To do the greatest good, results of this research must reach the largest number of people. This book summarizes and shows how to apply the current knowledge about the care of photographs to achieve the maximum possible lifetime of the image. Our aim is not to report wildly speculative and exotic theories or to suggest expensive and difficult procedures. Unless photographic conservation is a practical science with practical results, it has no justification. Practices suggested here are simple, relatively easy, and as inexpensive as feasible.

THE ESSENCE OF CONSERVATION

As we see it, the essence of conservation is conservatism: too little is better than too much. Few photographs deteriorate so rapidly that their disintegration requires emergency intervention. Rather than damage the object by ill-considered or untried treatments with uncertain outcomes, you should stabilize it by examination, cleaning, and proper storage after copying. Most important, treatment ideally is reversible; that is, you should be able to undo it without damaging the photograph.

Instances of misguided enthusiasm abound. For example, one employee at a local institution rubber-cemented old and potentially valuable photographs to garishly colored, acid-core mat board after typing identifying labels on the margins. While this person undoubtedly had the best intentions, the sulfur in the cement, the acid in the board, and the physical damage of the typing ensured rapid destruction of the prints.

For professionals who care for photographs, the temptations are more insidious.

Grant Romer, conservator at the Eastman House, often gives a talk in which he shows a ruined quarterplate daguerreotype found in the museum's files. It would be worth thousands of dollars in today's collecting market, except that half of it was "cleaned" while the other side was left untreated. Apparently the purpose was to demonstrate graphically the supposed benefits of cleaning away silver tarnish. As Romer rightly points out, neither scientific nor educational applications can justify this kind of irresponsibility.

Although caution and restraint have always been the wise choice in treating any kind of art, this is doubly true for photographs. Photographic materials are so quirky, complex, and unpredictable that restoration seldom works, and photochemists have come up with few techniques that enjoy widespread acceptance. Many techniques that at first looked promising have proved in the long run to do more damage than simple neglect. If a photographic image seems likely to continue deteriorating even when stored in optimal conditions (and there are many prints in this situation), copy the print or duplicate the negative.

Old and new converge here. The most traditional (and most ethical) standard of the professional conservator mandates that nothing be done to a piece that might damage or alter it. In the case of photographic prints, this means that you should copy all deteriorating photographs while preserving the originals in a stabilizing environment. Today, with the growing ability of microprocessors to perform digital image enhancement, this conservative approach has added validity. Technology is not yet at the stage where everyone can copy and enhance photographs at home, but with the growing use of the personal computer and video imaging, we foresee that the ability to improve image quality by copying will become widespread. Copy prints will be sharper, more detailed, and have more accurate tonality than the originals. Holography may make it possible to focus fuzzy prints, contrast will be controlled, and prints may be made on more stable substrate. All this will doubtless create new and different conservation problems. The point we wish to make, however, is that to "restore" the photograph to its original state it will no longer be necessary (or desirable) to endanger the original print with chemical or mechanical treatments.

PERSONAL PHOTOGRAPHS

Print care has a personal twist for most of us. We are moved by the universal human desire to protect our private memories and the history of our life. In contemporary society, these usually take the form of snapshots, slides, formal portraits, and photo albums. Besides simply preserving these images that are often so central to our sense of identity, we want to be able to display and enjoy them as part of life. Often this means framing and displaying them.

Most of the commonly available ways to frame and display photographs will damage or destroy them. Drugstore photo albums and do-it-yourself frames with precut mats can actively attack prints. This comes, we know, as unwelcome news to many people. However, with knowledge and a few simple skills you can do an excellent job of caring for your prints. The end result will be no more expensive than the shoddy parodies of frames and albums that are widely foisted on an unsuspecting public, and you will have the added satisfaction of a piece that looks fine enough to hang on a gallery wall. The most important things that you can bring to this endeavor — more important than any special technical skills — are patience and the willingness to select objectively your most important images for special attention.

Personal photographs have a multitude of values. Even at their most mundane they have a certain kitschy charm, and at their best they possess an electric, hyperrealistic quality matched by no other medium. These images of ourselves, our families, and our friends can buffer us against the shocks of time as we mature and age. They give us a thread that we can trace back to our earlier identities, to events that otherwise would leave no more trace in passing than radio waves. And a collection of personal photographs can do more. It can stengthen our ties with others, provide a sense of common identity, and enhance our awareness that the lives of many are bound to our own. The disorientation and anomie that we suffer in an industrialized society can be resisted by turning at least these products of industrialization into supports for our own communal continuity.

HISTORIC PHOTOGRAPHS

Photographic prints cannot simply be regarded as convenient windows into the past, ways to give us an unerring and completely factual record of how things "really" looked back then. Their historical interpretation requires subtlety and sensitivity, not only to the nuances of individual photographic styles but also to the technological and social context within which pictures were made.

Michael Lesy, for example, has devoted most of his life to studying and interpreting photographic commonplaces, the records of daily commercial and private life. His three books, *Wisconsin Death Trip; Real Life: Louisville in the Twenties;* and *Time Frames: Interpreting Snapshots,* make up a loosely linked trilogy that proposes a new historical definition of the United States from the rural post–Civil War era through the aftermath of World War II. They are assemblages of medical files, news stories, statistics, interviews, and photos culled from the piles resting in archives. Lesy is one of the rare scholars who has succeeded in transcending the nostalgic appeal of antique scenes to pull relevant new meaning from old pictures. In the context of his books, even the most stilted advertising fantasy assumes a jarring appearance.

In his introduction to *Real Life,* Lesy calls the photographs he used

> tableaus of commercial surreality. They provide evidence of a city culture whose willful manipulation of facts, confusion of people with objects, and minute-by-minute recording of fabricated events differs greatly from the matter-of-fact, human-centered, seasonally paced farm town culture that had preceded it. . . . [They portray] women who were romantically involved with vacuum cleaners, and men who talked quietly with refrigerators.

To use such photographs uncritically for historical records is to tread a path laced with punji stakes. Print and negative files do contain their own secret history, but it can be revealed only by close and cunning reading. In several places in *Real Life,* Lesy juxtaposes two images: the original is a picture taken inside a battery factory "filled with fumes, rot, and congealed lead," and next to it the finished product, retouched so that "the fumes became sunlight pouring through walls that had been turned into windows." This brings us to a fuller understanding of one purpose of photographic conservation: even the most banal records reveal astonishing information when used with a discernment that lets us see how they functioned in their original context.

The amount of data locked into even the simplest photograph is phenomenal. Consider, for example, that with only a few prints we can record the number of bricks in a large structure, the physical appearance of each member of a graduating class of hundreds of people, the arrangement and relative maturity of all the plants in a formal garden. Even the most unprepossessing picture often has such data encoded in the background. As a result, it can serve as a valuable historical record. Large accumulations of photographs such as newspaper files and studio archives often have the added benefit of providing a cumulative body of data that can be interrelated for extracting even more information. For these reasons, it is especially important that the conservator make every effort to preserve these intact.

Thoughtful organization of photographic files makes access easier, and hence makes them more valuable. Valuable collections are more likely to survive the passage of years than are moldy piles of random, anonymous prints. Editing and indexing are a major part of conserving historic photographs.

FINE ART PHOTOGRAPHS

Some prints are kept for their aesthetic merit rather than as memorabilia or historic artifacts. With the boom in collecting photographic prints, this situation has become more common. The financial value of these works of art may be quite high.* Sometimes a single print — for example, an original Steichen gum print — may be so rare and aesthetically important as to be nearly priceless.

Techniques of conservation do not substantially change for such valuable objects; they only become more critical. Prints that have, or that are suspected to have, this kind of artistic significance should always be handled with great care, and no time should be lost in identifying them. Where they may exist in the midst of a larger body of work, they should be withdrawn immediately for special attention.

The problems encountered in handling highly prized works of photographic art differ from those in the care of personal and historic pictures because of the frequent need to display the valuable work in public. Framing and display practices become an important part of conservation technique in handling fine art photographs. Collectors who have little interest in doing their own framing still need to know what constitutes an archivally safe window mat. Knowing what you really want may save money when you take prints in to have them framed; it will positively save value in years to come.

* "Valuable," naturally enough, is a relative term; certain family snapshots that might not command record prices at a Sotheby-Parke-Bernet auction can certainly be of irreplaceable value to their owners.

CONTINUITY OF CONSERVATION PRACTICES

The bias of this book is toward photography, but the line between photographs and other kinds of imagery cannot be so precisely drawn as it once was. Some artists who work in mixed media argue that it can no longer be drawn at all.

Parts of this book, like our sections on processing for permanence and on slide storage, will be irrelevant to archivists who handle manuscript collections and to collectors who hold nonphotographic prints, but there is an underlying continuity of conservation practices that extends to all the paper-based media and even beyond. The steps in making and assembling a frame are no different for a photograph than for other kinds of prints, and fungus feeds just as hungrily on a moist pastel sketch as on a color slide. So, even if your primary interest does not lie with photographic materials, we invite you to choose the parts of this work that are relevant to your specific needs.

We hope that you will find what follows useful both in saving some of the past and in helping to make the present a better time in which to live.

P A R T   O N E

Processing for Permanence

CHAPTER 1

Processing Black and White Prints for Permanence

PROCESSING STEPS

To assure the maximum permanence of a black and white print, the first — and often the most critical — step is processing to archival standards. The two accepted methods are outlined under *Fast-Fixing Process* and *Two-Bath Fixer Process*.

These procedures work equally well on all the contemporary fiber-base silver papers — Kodak Elite, Ilford Galerie, Agfa Portriga and Brovira, and others. Polyethylene-bonded papers, designated RC by Kodak for "resin coated," are not used for prints intended to keep indefinitely.*

As early as 1855, a researcher found that photographs tended to fade because of chemicals left over from the fixing process. Two types of residues cause problems. One is raw fixer ("hypo") that has not been washed out; this fades the print. The other is a group of complex silver thiosulfate compounds formed when the fixer works on the silver in the emulsion. As the fixing bath becomes exhausted, these compounds bind to the paper, where they cause yellowing and discoloration.

You can wash most residues out of the emulsion. The problems start when chemicals get in between the paper fibers where little water circulates. The problem is aggravated with the silver thiosulfate compounds because they are insoluble in water.

The *fast fixing* method described here — initially developed by Ilford photochemists — solves the problem of entrapped com-

* Certain "black and white" monochrome prints are made by commercial minilabs on color paper from chromogenic films like Ilford XP1. These are — technically speaking — color prints, and should be handled and stored as such. When this type of film is printed on color paper, processing is done with color chemistry, not the procedures followed here. These chromogenic negatives can also be printed on conventional silver paper.

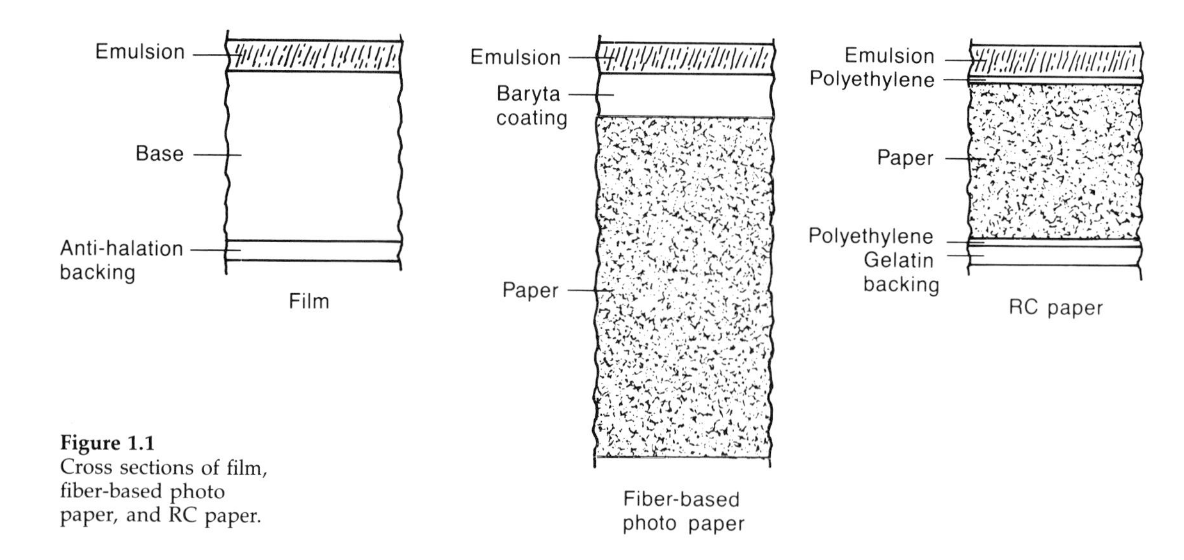

Figure 1.1
Cross sections of film, fiber-based photo paper, and RC paper.

pounds by prescribing short fixing with a strong bath and continuous agitation. Fixer cannot get into the paper fibers, the strong bath prevents precipitation of the thiosulfates, and you save time.

This technique does not correspond with the more leisurely method of *two-bath fixing* described in all literature before 1979. Because this method is widely accepted, demands less critical timing, and also makes archival grade prints, we have included it here.

We assume that you already have basic darkroom skills or that you intend to specify procedures for a professional photographic printer. A number of excellent manuals offer detailed instructions on basic printing; we especially recommend the one by David Vestal [1].

An archivally processed print is not necessarily a fine print. Making a print that does not slowly self-destruct is a craft; making a great photograph is an art. Here, we are concerned with the craft.

FAST-FIXING PROCESS

Step	*Time*	*Remarks*
*Developer**	Varies	Time varies with developing agent
Indicator stop bath	30 sec	Agitate in fresh solution
Ammonium thiosulfate fixer	30 sec	Fresh, film-strength fixer without hardener; constant agitation
Wash	5 min	Rapidly running fresh water, 68°–75°F (20°–24°C)
Washing aid	5 min	Immerse with agitation
Wash	20 min	Running water

* For explanation of words in italics, see "Terminology."

(Processing can be halted at this stage to build a backlog of prints for toning. Dry prints on fiberglass screens or store wet in a tray of water. Use cold water to prevent swelling of the emulsion, and do not leave immersed more than 24 hours. Segregate dry prints from completed ones, and wash the screens after drying to prevent transfer of residues.)

Toner	Per instructions	Gold protective toner GP-1 or selenium*
Wash	10 min	Minimum
Dry	Varies	Fiberglass mesh screens in dust-free environment

TWO-BATH FIXING PROCEDURE

Step	***Time***	***Remarks***
Developer	Varies	Time varies with developing agent
Stop bath	30 sec	Acid stop bath
First fixing bath	3–5 min	Fresh fixer with hardener; frequent agitation
Second fixing bath	3–5 min	Fresh fixer; frequent agitation
Wash	5 min	Rapidly running fresh water, 68=−75=F (20–24=C)
Washing aid	5 min	Immerse with agitation
Wash	20 min	Running water

(Processing can be halted at this stage to build a backlog of prints for toning. See note under *Fast Fixing*.)

Toner	Per instructions	Gold protective toner GP-1 or selenium†
Wash	10 min	Minimum
Dry	Varies	Fiberglass mesh screens in dust-free environment

TERMINOLOGY

Developer

The developer has little effect on the keeping qualities of a print, barring cases of extreme under- and overdevelopment. Ingredients of a typical black and white paper developer usually consist of the following components:

* You may need to use a *hypo eliminator* (HE-1) before toning with a selenium toner to eliminate the last vestiges of thiosulfate complexes. Alternately, it is possible to move prints directly from the fixer bath into a selenium toning bath before doing any washing.

† See note about hypo eliminator under "Fast Fixing Process" page 23.

1. *Solvent*: water (distilled)
2. *Developing agents*: hydroquinone, Metol or Phenidone
3. *Accelerator*: sodium carbonate
4. *Preservative*: sodium sulfite (anhydrous)
5. *Restrainer*: potassium bromide

Occasionally other chemicals will be added for specific purposes, but most formulas contain varied amounts of these components.

The use of distilled water as a solvent not only assures that the developer will be working at the pH level it was designed for, but also prevents contamination of the print by chemicals and organic matter found in tap water even in the best of neighborhoods. Developer should also be filtered to remove any undissolved particulate matter that could deposit local concentrations on the print surface, or even scratch the emulsion. In lieu of more elaborate methods, a coffee filter or a wad of cotton in the mouth of a funnel will do.

The only times when the developer will certainly cause problems are when it is exhausted or when it has been overly replenished. In these cases, stains appear on the print very quickly, often in the fixer as soon as the lights are turned on.

We have read accounts by some photographers of working methods that call for extremely long development times for the print — up to 10 minutes or even longer. Such prolonged contact with the developer can cause the paper to entrap in its fibers some oxidizing agents from the bath.

Many people who photograph in dim light, especially photojournalists, use a concentrated potassium ferricyanide solution to bleach local areas of the print. This involves taking the print out of the fixer, wiping the target area dry, painting on the ferricyanide, and reimmersing in the fixer. Prints worked on in this fashion become irretrievably overfixed. If this local manipulation must be done and a permanent print is required, changes could be made on an enlarged duplicate negative.

The other types of tone manipulation in common practice — burning and dodging, painting local areas with hot developer, two-filter printing, and so forth — have little effect on the keeping qualities of the print.

Indicator Stop Bath

Stop baths do not contribute directly to preserving a print, but you should not neglect their use. The stop bath prevents carryover of the developer into the fixer, where it can alter the pH level and lead to early depletion. This causes miscalculation of the capacity of the fixing bath, which in turn leads to underfixing and eventual staining.

An indicator stop bath turns from yellow to reddish purple as it loses its acidity. It should be used as a matter of routine to avoid

the oversight of continuing to use an exhausted stop bath. Of course, the old workhorse formula of a solution of acetic acid works as well, but the extra safety provided by an indicator seems well worth the minor additional cost. Some experienced darkroom workers can tell the degree of exhaustion of a stop bath by feel: a print fresh from the developer feels slippery, almost soapy, but once the print is put in fresh stop bath the slippery feeling disappears instantly.

Fixer

When did "photography" begin? The camera, in the form of the camera obscura, was invented during the Renaissance; light-sensitive chemicals including silver nitrate were investigated during the eighteenth century. The major problem for inventors in the nineteenth century was to "fix" the image in some kind of stable form.

It was only because of a suggestion by Sir John F.W. Herschel about using sodium thiosulfate to dissolve the unexposed silver chloride from his paper negatives that Fox Talbot was able to produce images in 1839 that did not darken upon exposure to more light. The same concept was immediately adopted by Daguerre for his process. The theoretical purpose of a fixing bath is quite simple. Sodium thiosulfate makes unexposed silver particles in the emulsion water soluble so that they can be washed out of the emulsion. Otherwise the action of light would cause them to "print out" and turn the entire image black.

That is the theory. In practice the use of a fixing bath has also proved a big obstacle to the goal of achieving a stable print. In part, this occurs because of the extreme complexity of the reaction that takes place between the silver chloride complexes and the thiosulfate fixing bath. It is estimated that "there are probably at least three silver-thiosulfate complexes formed during fixation" [2], and the number may be even higher. The chemistry of the fixing process requires an excess of thiosulfate ions (above the ones that actually form the silver-thiosulfate complexes washed out of the print) in order that these ions can act as a conveyor for silver molecules during their conversion into water-soluble silver thiosulfates. If that sounds confusing, it is. More simply put, we can say that one of the thiosulfate compounds formed during fixation can be dissolved only by more sodium thiosulfate. Exhausted fixing baths lack this excess of thiosulfate ions, and cannot do an effective job of transporting the silver.

The realization that fresh fixer has to be applied to the print to complete fixation led to the standard procedure, used for decades, of two baths. As the first fixing bath became exhausted, it was discarded and replaced with a second bath, which in turn was replaced by a newly mixed bath. This supposed that the fresh fixing bath would supply the excess of thiosulfate ions needed to remove otherwise insoluble silver thiosulfate compounds.

The idea is not, in fact, a bad one. We still recommend it when the precise control required by fast fixing cannot be maintained. (To determine the minimum fixing time, test for residual silver with selenium toner as described below.)

The relative slowness of a conventional sodium thiosulfate fixing bath can allow the silver thiosulfate complexes to become mordanted (trapped) in the paper fibers. Hardening agents like potassium alum aggravate this problem. The use of a highly active fixing agent, ammonium thiosulfate, for a very short period of time and with continuous agitation, solves this problem. In the fast fixing procedure, the hardening bath simply gets eliminated.*

With a fixing bath of suitable strength and energy, the fast fixing process is complete in 20 seconds. Allow an additional 10 for safety's sake, and you have a total of 30 seconds in the fixer.

You probably will not be able to buy bulk quantities of ammonium thiosulfate. It is sold under various trade names as a liquid concentrate for mixing fixers, usually with a separate bottle of hardener that can be left out. Use of this liquid is preferable to raising clouds of fine dust particles when mixing dry fixers. Four widely available brands are Kodak Rapid Fixer, Ilford Ilfospeed Fixer, Ilford Hypam, and Edwal Quick-Fix. Mix them to the dilutions required for film, not paper. Use clean, nonalkaline tap water to prepare them, though again you might prefer distilled water for an extra degree of purity.

In areas of the country where the water is highly alkaline, it may be necessary to protect the acidity of the fixing bath. A simple method to test the pH of a bath is to use pH indicator sticks like the ColorpHast brand from MC/B Manufacturing Chemists. The sticks can be dipped in the solution and the color changes measured against a chart provided on the container. Some variation is allowable, but the ideal will run between a pH of 5.0 and 6.0. To compensate for overly alkaline water, add 2 to 3 ounces of 28% acetic acid to each gallon of working fixer, and test again. Use this expedient only when necessary, because extra acidity will increase the tendency of fixer residues to mordant in the paper fibers.

No matter which procedure you use, the use of absolutely fresh fixer is imperative. Do not make it a habit to save fixer from one printing session to the next. And mix fresh fixer shortly before use for best results, because fixer breaks down in the tray. Even liquid concentrate should not be stored for long times because the plastic bottles in which it is sold allow gas exchange with the atmosphere. Working-strength solutions can be stored in glass bottles for no more than a month.

Exact capacities for fixing baths cannot be determined in advance because the amount of silver removed from each print varies inversely with the amount of surface area exposed to light; this

* Finished prints will be slightly more sensitive to abrasion, but except in the tropics modern paper emulsions do not require the extra hardening given in the fixer.

surface area varies with the subject matter of each print. In general, the greater the white and highlight areas that occur in a print, the greater the demands made on the fixing bath. Thus, a print with wide white borders and a light subject will release more silver into the fixer than will a dark print with narrow borders.

Because of this uncertainty, leave a substantial safety margin over and above the manufacturer's recommendations. Each darkroom should establish its own standards for the carrying capacities of its fixing baths, based on its experience with the type of images processed and determined by testing finished prints with the selenium test for residual silver.

Edwal HypoChek will determine the silver content of the fixing bath, and indicates when this content has reached a dangerous level of saturation. A few drops of this bottled solution will turn milky in the presence of excess silver. Unhappily, the bath should already have been dumped when you get to this point, but keep some HypoChek around anyway to tell you if that whole batch you just finished should be refixed.

Fast fixing will work. It may seem hard to believe for photographers trained to believe in the virtues of the two-bath method and the dangers of underfixing a print, but try it. Skeptics can test it with a whole array of methods, such as Kodak's ST-1 or selenium toner tests, the Light Impressions print-testing pens, or even the method outlined in Appendix A3 to ANSI standard PH4.32-1974. These methods assure that the print will not be underfixed.

Also, fast fixing does prevent the mordanting of ammonium and silver thiosulfate complexes in the paper base. The photographer Ralph Steiner, for example, has published results that show a residual hypo level (measured by the silver nitrate ASA method) of less than 0.01 milligram after a 3-minute bath in Kodak Hypo Clearing Agent and 5 minutes of washing. After conventional two-bath fixing, it took a similar print 20 minutes of washing to reach a level of 0.03 milligram, the generally accepted level of safety [3]. Add to this the fact that low levels of residual complexes wash out more and more slowly as the curve of removal approaches zero, and the benefits of fast fixing become apparent.

The fast-fixing method requires increased precision of timing in the fixing bath. To allow the print to remain overlong in the fixer, or to give insufficient agitation, will defeat the whole purpose of this method. But this need for increased precision will be offset by the time saving (30 seconds as opposed to 8 to 10 minutes by the conventional method), and a darkroom routine can easily accommodate the change.

Ammonium thiosulfate costs two or three times what sodium thiosulfate does, and used as recommended has less capacity.

Wash

After a fast fix, an initial wash must rapidly remove all liquid fixer from the paper's surface before the fixer can adhere to the fibers. Drain fixer from the print by lifting one corner straight up from the tray, and hold the print clear of the bath for about 5 seconds so that most fixer runs down and off the opposite corner. Transfer to the washer and agitate immediately to dilute any liquid still on the print.

The washer should have a rapid flow of water through a small area; ideally all water should be exchanged once a minute. A method of keeping the prints separate is also required.

Archival Print Washers

Archival print washers are based on one made in the seventies by East Street Gallery. This is no longer in production, but similar models are made now by Zone VI Workshop, Gravity Works and by Kostiner. All three are moderately expensive and well made. A much cheaper alternative is the Kodak Tray Siphon, which clips onto the side of a print tray. It uses the force of the flow of incoming water to suck contaminated water out of the wash tray. The outlet tube of the siphon extends down farther than the depth of most photo trays. Set one tray upside down on the bottom of the sink, and place another on top of it right side up to provide the needed clearance. Clip the siphon to the side of the top tray, and regulate the depth of water by raising or lowering the intake port. The rate of exchange of water can be governed by both the rate of water flow at the faucet and by the water level; the lower the water level, the faster the exchange.

The archival print washers offer a big advantage over the siphon and assorted other washing tubs by keeping the prints separate. Prints lying flat in a tray or tub stick together unless they are constantly pulled apart by hand. Where the prints stick together they do not wash; it is as simple as that. Archival washers have narrow vertical compartments in which prints float separately while a constant flow of water bathes them on both sides.

You do not need to use a vertical print washer to make prints to archival standards, but for high-volume operations today, to do without one is as practical as setting newspaper type by hand. The individual who makes a few prints at a time can get by with a tray siphon or one of the tub washers that swirl the prints around in a circular flow of water, but constant attention is required to keep them separate. Separation is so important, in fact, that time when prints are left unattended in these washers should not be counted as wash time at all.

Nearly all instructions for washing prints call for a complete exchange of water in the washer every 5 minutes. For the initial wash, adjust the flow to give a complete exchange every single minute. The 5-minute exchange satisfies the requirements for later washes, once the fixer has been flushed off the print. Add four or five drops of food coloring to the wash water to test the exchange rate. (Unless you want to try an unusual toning method, do this without prints in the washer.) Time how long it takes for all the food coloring to disappear.

Calculate wash time from the moment the last print goes into the washer. You cannot flop a fixer-laden print into the wash tray,

then pull out the one below, and still expect the benefits of time already spent washing. This principle applies equally to vertical washers in which separate compartments share a common water flow; despite manufacturers' claims to the contrary, we have seen carryover occur from one compartment to the next. The given wash times should be considered minimum times; no harm is done to paper-base prints by overwashing, unless you go to extremes and leave them in a day or more.

A word about double-weight versus single-weight photographic papers: usually, shorter wash times are prescribed for single-weight papers because they absorb less fixer. In fast fixing, since the paper should have little fixer in it anyway, you can disregard this difference.

Time and temperature in the wash, as in other steps in the photographic process, should be monitored to assure uniform results. The wash water should be checked frequently to keep it between 68° and 75°F. A washing aid allows cold water to be used, but even so wash times should be extended. Below 65°F the ability of water — by itself — to remove fixer residues drops so precipitously that even very long wash times do little good. High temperatures, on the other hand, can seriously stress an unhardened emulsion. Clip a thermometer inside the washer, facing out through the clear sides so that you can read it at a glance.

Washers themselves need regular cleaning. If one is left standing filled with water for long periods of time, as it will be in most darkrooms, algae and other kinds of organic matter accumulate. Remove these by adding a capful of household bleach to the water; stir it in thoroughly and let it stand for an hour. Drain and wipe down the washer with a clean sponge, and let it dry after rinsing. Remove chemical residues with a bath of 1% sodium bisulfite or Kodak Hypo Clearing Agent; allow the tank to soak for an hour, and follow with a thorough rinsing.

Washing Aids

Since World War II, different washing aids have been sold that speed up the removal of fixer residues from prints and films, even in cold water. All have a common origin in the discovery by U.S. Navy photographers that when prints were washed in cold sea water — often all that was available to them — the removal of fixer took place more rapidly [4]. Because sea water includes a veritable complex of chemicals, not all of which help the print, research was begun to find out which chemicals produced the desired effect. Certain salts were found to be responsible.

Washing aids, as we will explain later, are not hypo eliminators. Rather than chemically convert the fixer residues to a harmless substance, they speed up the washing process. Washing aids are sold under a variety of names, including Hustler, Orbit Bath, Permawash, and others.

Extensive tests at an independent laboratory have not compared the relative efficiency of the various aids. Kodak Hypo Clearing Agent, sold in powder form, has been most widely used and

tested, so that it can be said to be the industry standard. Mixed according to the directions on the package, a gallon of working solution will treat up to 200 8 × 10 prints. A much lower capacity will result if the prints have not been washed for 5 minutes prior to use of the clearing agent.

We view certain claims for the efficacy of washing aids with skepticism, especially when the instructions say that an undefined "archival" state of permanence can be reached after only a few minutes of washing. If you do not test for fixer residues after using this kind of product, you may be trusting in snake oil.

A cheap alternative to Hypo Clearing Agent is a 1% solution of sodium sulfite, made simply as follows:

Sodium sulfite, anhydrous	10 grams
Water, to make	1 liter

Use the fresh solution only once and discard. Its capacity is twenty 8 × 10 prints or the equivalent. The prints should be agitated constantly for 5 minutes, drained, and placed in the next wash.

You can buy sodium sulfite quite cheaply at any good camera store; if it is not stocked, the store will order it for you.

Hypo Eliminator

Some look askance at the use of hypo eliminator for archival processing of paper prints. Ansel Adams, for one, did not favor it, and we have heard negative comments about it from others. The reason hypo eliminator is not favored is never clearly stated. So far as we can tell, this seems to be a prejudice not backed up by any published research.

However, research has established that Kodak's HE-1 hypo eliminator [5] can damage *film* by causing microbubbles to appear between the base and the emulsion. This does not apply — again, so far as we know — to *paper*, where the gases formed can escape through the paper fibers.

Another cause for skepticism about HE-1 may be misunderstanding about a recent paper,* which reports that trace residues of thiosulfate compounds offer better protection to a photograph against industrial pollutants than does complete washing. In other words, a little is better than none. However, toning is better than either one. Further, the author of the paper specifically states that it is still desirable to remove all thiosulfate residues, if possible,

* Y. Minagawa, and M. Torigoe, "Some Factors Influencing the Discoloration of Black & White Photographic Prints in a Hydrogen Peroxide Atmosphere." Fuji Photo Film Co., Ltd.; and D. Beveridge and D.H. Cole, "Stability of Black & White Photographic Images." Ilford, Limited. Both papers were presented at the International Symposium on the Stability and Preservation of Photographic Images, Public Archives of Canada, Ottawa, Canada, August, 1982. The conclusions of the authors were that small amounts of residual thiosulfates helped to protect the print against degradation by the active chemicals common in polluted air.

because there is no way to be sure that the trace amounts at which the protective effect works can be effectively maintained.

Even after an ammonia-type fast fix, thorough washing, and the assistance of a washing aid, the print retains minute amounts of thiosulfate compounds. The rate of their removal slows drastically as their concentration goes down, until finally a point is reached where further washing, no matter how prolonged, will not remove the last residues.

Save the water. The remedy for these last recalcitrant holdouts is to blast them loose with hypo eliminator — an easily prepared solution that converts fixer residues to harmless sodium sulfate, which is readily water soluble.

The most widely used hypo eliminator is Kodak's HE-1 formula, although the American National Standards Institute does use a stronger (and more complicated) formula to prepare reference samples in testing for permanence. Because the Kodak solution requires fewer, more common chemicals, this is the one we have chosen.

HYPO ELIMINATOR*

	Avoirdupois	*Metric*
Water	16 oz	500 ml
Hydrogen peroxide, 3% solution	4 oz	125 ml
Ammonia solution	3¼ oz	100 ml
Water, to make	32 oz	1 liter

Mix the solution, adding the ingredients in the order above, just before use. Discard when finished.

Do not store hypo eliminator in any kind of closed container. Gas released by the interaction of hydrogen peroxide and ammonia will create sufficient pressure to explode a sealed glass bottle.

To use, immerse the prints with constant agitation for 6 minutes. Keep prints from sticking together. Working temperature should be approximately 68°F, though it is not crucial. Capacity of the bath is approximately fifty 8 × 10 prints or equivalent, per gallon. Follow treatment with a 10-minute water wash to remove the remaining sulfates.

Toner

After the final treatment in hypo eliminator, the print will be free of the chemicals that can cause it to self-destruct. However, the metallic silver that creates the image remains susceptible to un-

* A suitable grade of hydrogen peroxide can be purchased at any drugstore. The ammonia solution is made by adding 1 part concentrated (28%) ammonia, USP grade, to 9 parts water. USP grade refers to common household-grade nondetergent ammonia.

wanted changes from outside agents. In particular, sulfur in the form of sulfur dioxide commonly occurs in the environment of most urban centers, and it combines spontaneously with silver. The practical effect is the same as that of leaving the print only partly washed.

The final stage in processing to archival standards should be use of a gold or selenium toner. This protective solution plates each silver particle with an ultrathin layer of metal that is much more chemically stable than silver.

"Gold toner" and "gold protective solution," as used here, mean the same thing, although in fact the toner formula given below does very little to change the color of the print except to alter it slightly toward a cool blue. When albumen paper was in widespread use, gold solutions were used more as true toners — to convert the unsightly orange color of the untreated albumen print to a more acceptable purple-brown. There are still formulas for gold toners that can produce drastic shifts in color, but the chief function of a gold toner in contemporary applications is to armorplate the picture against chemical attack.

Despite the rising price of gold, the cost per print is relatively modest because so little is used for each print. Gold protective solutions are available ready to use from two commercial sources, Light Impressions Corporation and Berg Color-Tone, Inc.

If you wish to make your own, the formula is:

GOLD PROTECTIVE SOLUTION*

	Avoirdupois	*Metric*
Water	24 oz	750 ml
Gold chloride (1% stock solution)	2½ drams	10 ml
Sodium thiocyanate	145 grains	10 g
Water, to make	32 oz	1 liter

Add the stock solution of gold chloride to the water. Mix the sodium thiocyanate separately with 4 ounces (or 125 ml) water, and stir vigorously while mixing with the gold chloride. Rubber gloves and excellent ventilation are recommended while mixing these chemicals.

Immerse the prints in the gold protective solution for approximately 10 minutes at 68°F. Comparison with an untoned sample will show a slight shift in color toward bluish black. Any trace of hypo will destroy the protective solution, so prints must be treated with a hypo eliminator before the gold protective solution.

Gold protective solutions start to deteriorate shortly after being

* Gold chloride can be purchased commercially either as a 1% stock solution (technically known as chlorauric acid) or by the gram in its dry state. Dry gold chloride is a shapeless orange mass that absorbs moisture from the air rapidly. To mix a 1% solution, add 1 dry gram to 100 ml distilled water; store in a dark-brown sealed glass jar.

mixed, so use them immediately. Capacity of the bath is thirty 8 × 10 prints or equivalent per gallon. Follow with a 10-minute wash.

Selenium Toner

A selenium toner can be used in place of the gold protective solution. Kodak Rapid Selenium Toner comes as a liquid concentrate with complete instructions that we will not repeat here. Note that selenium toner cannot be prepared by the average photographer, because selenium is extraordinarily poisonous. The toner itself should never be used without rubber gloves, or allowed to come in contact with the skin at any time.

Unlike the gold protective solution, selenium toner can work as a true toner, producing rich purplish-brown prints when diluted with water in the ratio of 1:3. To tone for added permanence with a less drastic color change, dilute to a ratio of 1:9 or greater. This will preserve the original neutral tones of the print. Shadow areas will deepen a shade or two, very subtly, and will change color slightly toward brown, but the more visible highlights stay neutral. Times vary unpredictably, so tone under a strong light with a reference print at hand until this slight color change takes place in the shadows.

A partly washed print placed in selenium toner immediately stains an ugly blotchy yellow. A paradoxical feature of the selenium toner is that the print must be either saturated with fixer or completely washed. Because the selenium toner itself contains some fixer, the best stage for using it is right after the fixer. Immerse immediately in the toner *without* rinsing. Do not selenium tone right after using the hypo eliminator and before gold toning, because the fixer from the selenium will affect the gold toner.

Some other toners also confer added stability to the print. A common one, for example, is the sulfide sepia bleach-and-redevelop toner. However, these cause such extreme print colors that most photographers find them more suitable for creative effects than for everyday use.

Drying

Prints should air dry face up on racks made of fiberglass screening material available at any hardware store for pennies a square foot. There is no preferred construction style, so you can build these to fit available space. If you use wood frames for the screening, varnish or paint them with epoxy paint to seal the frames against chemicals soaking into the wood. Wash the screens regularly to prevent chemicals carrying over from one batch of prints to the next.

A convenient fiberglass screen print dryer can be built in a few minutes with some string, a roll of uncut screen, two sturdy sticks as wide as the screen, and several thumbtacks. Wrap each end of the screen around a stick and push thumbtacks through the screen to secure it firmly. Tie a string to the ends of each stick and

suspend the entire contraption like a hammock across a room that will not be used for a day or so. You might want to spread some newspapers on the floor beneath to catch drips. It does not look very elegant, but it works well, is safe and cheap, and will dry an amazingly large number of prints.

A few words of caution about drying prints. Two kinds of common print dryers cannot be used to make archival quality prints: fabric belt drum dryers and photo blotters. The belts of drum dryers retain chemicals that transfer to the wet print no matter how well the belt is washed. It is a curious fact that many photographers will spend a great deal of effort washing and clearing a print, then lay it down on one of these filthy monsters to undo all their work. Photo blotters, a venerable institution in photographic practice, must be used fresh each time to be safe. The expense of this method is enormous, and we suspect the temptation to reuse blotters is irresistible.

While we are being negative, let us talk about "print-flattening agents." If you use these, you can forget about bothering to air condition or climate control the environment your prints get stored in, because such measures will no longer do any good. Print-flattening agents work by depositing hygroscopic chemicals in the paper to pull moisture out of the air. This causes what might be called the "dishrag phenomenon," in which the prints lie limp because they are soaking themselves with atmospheric humidity. Needless to say, molds and fungi just love print-flattening agents. We do not. If your prints are curled up at the edges after drying, interleave them and put them under weights for a day or so. If you are in a hurry, put them in a dry mount press for a minute or so between two sheets of preheated archival board. Just make sure that there are no stray particles of grit also trapped in the press to damage the surface of the print.

TESTING

Many factors affecting print life operate on the molecular level. A print processed to archival standards looks, feels, and smells no different from a photograph made to less demanding standards. You cannot tell if the print will last without performing some tests. These tests can do one of two things. Either they indicate the presence of certain chemicals that shorten print life (see Table 1.1) or they artificially age the print to simulate the passage of time.

Artificial aging tests, which include incubation under controlled combinations of humidity, light, and temperature, belong more properly to the realm of industrial research laboratories than to the working darkroom. In general, artificial aging tests determine the inherent potential life span of photographic material processed to the manufacturer's specifications.

Tests for the presence or absence of specific chemicals can play a more significant role in regulating darkroom procedure. Since we know, for example, that residual fixer compounds will stain or

Table 1.1
FIXATION PROBLEMS AFFECTING PRINT STABILITY

Problem	*Symptom(s)*	*Explanation*	*Cause and Treatment*
Residual ammonium thiosulfate or sodium thiosulfate	Highlight fading; in extreme cases overall fading and staining to yellow-brown.	Fixer attacks silver image, changes it to silver sulfide, which in turn can convert to silver sulfate (cannot be reversed).	Inadequate washing fails to remove fixer. Rewash print.
Residual silver thiosulfate in emulsion	Yellow-brown stain over entire image, appearing stronger in highlights because masked by shadows. Stain is dichroic (may appear gray, metallic silver, or shades of copper), especially in darker areas of print.	Residual silver thiosulfate compounds are unstable and decompose to form silver sulfide.	Exhausted fixer or inadequate agitation during fixation. Refix and rewash print.
Residual silver halide in emulsion	Localized yellow stain in shadow areas of image.	Silver halides have not been made water soluble by action of fixer, and ambient light develops them in a process like that used in printing out papers. This is the least troublesome of fixing problems.	Inadequate fixing. Refix and rewash print.
Chemical contamination of local areas on finished print	Local yellow-brown stains.	Same process as residual fixer (see above) except usually more active because of greater concentration.	Splashes of chemistry during processing, especially fixer. Rewash.
Ammonium or sodium thiosulfate present on emulsion before development	White streaks, spots, or fingerprints.	The fixer present on the emulsion causes the silver halide to wash away before it can be developed.	Careless handling of fixer during darkroom procedures; only solution is to reprint.

fade the print, and since we can control their retention, it is useful to test for them and to adjust working methods accordingly. If we assume that fresh chemicals are used and the same procedure followed each time, not every print or even every batch of prints need be tested.

Our experience in a large number of darkrooms indicates that the latter is a pretty large assumption. Only constant attention to details of cleanliness, mixing schedules, and other particulars can assure you of clean prints without testing. And in cases where prints have to be ordered from an outside processor working to

your specifications, we suggest ordering an extra print with each batch, just for testing.

The following tests should be used when first starting to process photographs to archival standards, to make sure that the agitation times and other processes being used give the results desired. After that, occasional prints can be tested at random on a regular schedule set up according to the volume being made. Fresh tests should be made when any change in procedure is initiated. "Any change" means just what it says: for example, changing paper type, or brand of fixer, or the washer being used.

Residual Silver

Residual silver comes from incomplete fixing. The most convenient test for its presence simply requires a stock solution of Kodak Rapid Selenium Toner diluted 9:1 with water. On the blank margin of a print or an unexposed blank scrap of photo paper processed in tandem with the rest of the prints, apply a drop of solution. The paper should be moist but without liquid water on the surface. After 2 minutes blot off the drop of toner. Any coloration other than a slight cream tint indicates the presence of silver and the need for more fixing. Incomplete fixing can be caused either by fixer that is too old and exhausted, or by too short a time in the fixer. For reference, make a sample that you know is underfixed and stain it.

Residual Fixer Compounds

These compounds present a more complicated problem. As mentioned earlier, simple ammonium or sodium thiosulfate are not the only substances that can be retained by the print after fixing. The complex thiosulfate compounds produced as part of fixing, which are not water soluble, can also remain. You can test for these thiosulfate compounds by using a solution of silver nitrate and acetic acid, both of which you can purchase through camera stores, though the silver nitrate will probably have to be ordered.

For several decades Kodak has promoted its version of this test, HT-2. After treating a part of the test print with HT-2 solution, a comparison is made between the stain produced and some color samples. The lighter the stain, the less thiosulfate. With a transmission densitometer and calibration curves, you can even use the test to determine the specific quantities of thiosulfate present [6].

For our purposes, however, knowing exactly how much fixer remains in the print counts for less than knowing that it is virtually all gone. The test proposed by Edwal photochemists uses a stronger solution and gives more practical results. For a copy, ask for Edwal Information Bulletin 276, "Test for Print and Film Permanence," Edwal Scientific Products Corporation, 12120 South Peoria Street, Chicago, IL 60643. This test is based on ANSI standard PH 4.8-1978.

THIOSULFATE RESIDUE TEST

Water, distilled	750 ml
Acetic acid, glacial	30 ml
Silver nitrate	10 g
Water, to make	1 liter*

To use, put some of the residue test solution in a thoroughly cleaned glass or plastic tray. In subdued light, immerse a dry test print halfway into the solution. Keep there for 4 minutes, agitating gently. Any stain that appears will show approximately the maximum yellowing caused by residual thiosulfate after long storage. The part of the print that was not submerged will serve as a comparison. It is important to perform this in subdued light, and preferably to keep the print dark for a few minutes before the test, because ultraviolet light will influence the test. Discard the used portion of the testing solution after completing the test.

A word about the size of the test print. Edwal recommends a 4 × 5 sheet processed in tandem with other prints. Often prints get washed better at the edges than in the center, and small prints wash better than large ones. We suggest processing a test print the same size as the others being done, and then cutting it in half for the test. This ensures that the center will be tested as well as the edges.

Light Impressions Print-Testing Pens

The same tests can be effectively performed with less trouble about mixing test solutions by using Light Impressions Print-Testing Pens. This kit consists of four felt-tip pens. Used in the prescribed series, they indicate paper acidity, the presence of undeveloped silver, unremoved washing aid, and thiosulfate compounds. Complete instructions on their use come with the pens. Their one drawback is that they test only a very small spot, so you should sample a number of locations.

RESIN-COATED PAPERS

The "plastic" black and white papers of recent years have some advantages over conventional fiber-base photographic papers. Long-term durability is not one of these advantages. Prints made for long-term use or for sale to collectors of photographic art should not be made on these papers.

Papers like Kodak's RC (resin-coated) types have a thin polyethylene coating bonded to a paper base on both sides to prevent moisture from penetrating its fibers. Because of interactions between the polyethylene, the fluorescent brightener (titanium dioxide, or TiO_2), and associated substances like antioxidants and

* Store in a clean brown glass bottle with a bakelite or styrene (not metal) cap.

energy-quenchers, the plastic papers have an unfortunate tendency to discolor and sometimes even to shed their emulsions. Kodak has indicated that it knows about these problems, and it has promised that it will not stop making fiber-base until they are solved [7].

Kodak makes this argument for the continued use of RC papers: since they require less washing and fixing, these papers are more convenient to use and so the marketplace demands them. Further, Kodak says that its tests show that if prints on RC paper are kept in the dark with the humidity levels strictly controlled, these prints will last just as long as will those on fiber-base papers. This is fine, except that the reason for making most prints is to view them.

In the simplest possible terms, what happens to plastic prints on display is this: during the daytime hours, ambient light and the associated increase in temperature drive absorbed humidity out of the emulsion into the atmosphere. At night, with the drop in temperature and the decrease of energy available from surrounding light sources, moisture gets reabsorbed. Because the plastic base remains, relatively speaking, free from changes in humidity content, it stays the same size. Not so for the emulsion. As moisture gets absorbed and released, the emulsion stretches and contracts, flexing on the surface of the paper, eventually pulling it off the base.

Resin-coated papers do have some important uses. Use them to make such items as utility prints, copy prints, filing and cataloguing samples, reproductions, and copy sheets. Plastic prints can be effectively used in any application where rough handling, quick processing, and a fairly short life are expected.

Despite their inherent limitations, the growing popularity of RC-type prints may actually have some benefits for institutions that collect large numbers of photographs from diverse sources. The amount of work involved in making a print to conservation standards is fairly great. Needless to say, many prints do not get this attention, especially when rapid access to the image is required for applications such as newspaper use to illustrate fast-breaking stories. Because the resin-coated prints need less washing to remove fixer residues, collections of those RC prints produced for rapid access probably will fade more slowly than those made on fiber-base papers.

Processing Resin-Coated Prints

For tray processing, the best course is to follow *exactly* the manufacturer's recommendations for the particular paper, whether it comes from Kodak, Ilford, or Agfa. These instructions all call for short fixing and washing times. The reason for this is to keep wet time to a minimum. Prolonged exposure to moisture allows water and chemicals to seep under the plastic coating and become embedded in the paper underneath. These chemicals will not wash out, and the trapped moisture will eventually cause the print to roll itself up into a tight little tube that cannot be flattened.

Make certain not to overwash. Check the temperature to maintain it near 68°F to assure maximum efficiency. If sample prints test positive for residual fixer, treat with a washing aid and follow with a 2-minute wash.

The weakest point of resin-coated prints is the edge, because the paper is made in large rolls and then cut to size. This exposes fibers along each side, where water and chemicals can enter. To compensate for this, we suggest making prints with wide blank borders and then trimming approximately ¼ inch per side.

Automated processing of resin-coated papers is gaining widely in popularity, and some processors now on the market deliver a dry print in under 5 minutes. Automation of the processing will ensure uniformity of quality if the freshness of the chemicals is monitored regularly. Most observers agree that the widespread adoption of these processors has already substantially changed the way in which most black and white prints are made.

STABILIZATION PRINTS

Stabilization prints were first created during World War II to supply the need for rapid access to the results of aerial reconnaissance, where the quality of permanence was not essential. They are made on a conventional fiber-base paper with an emulsion that has a developer incorporated into it. After exposure, the paper is fed into a machine processor that has two baths and a set of rollers to transport the paper. The first bath activates the developer and within seconds an image appears. The paper is then transported across a stabilizing bath, which converts the remaining silver halides into silver salts that are not sensitive to light. These silver salts remain in the paper, and will stain the image sometime within about six months.

The damp prints that come out of the machine after about 10 seconds can be used immediately. Any time until they begin to stain, they can be fixed and washed like conventional prints.

In their condition as they come out of the machine, stabilization prints represent an active menace to other photographs. The stabilizer chemical, which resembles strong fixer, can contaminate other pictures right through paper envelopes or any other porous material. Even residues on such surfaces as drying racks can prove dangerous. Absolute cleanliness and isolation are essential when working with these materials.

Stabilization prints can be identified by a strong hypo smell while still damp; by the sticky, damp feel that they retain even after several months; by Kodak's ST-1 test; or even by the test for residual silver, which will show an extremely marked stain.

Stabilization processors cost considerably less than automatic processors for resin-coated paper, which run up to many thousands of dollars. Spiratone makes one processor that sells for only about $200, so it seems that this price differential will assure that stabilization processing will be around for some time to come.

Stabilization prints can be made safe. They have to go through

the same fixing and washing steps as any conventional fiber-base paper, though, and the sooner this is done after development the better the results. Staining and fading can start to occur unpredictably soon after machine processing, depending in part on the age of the chemicals used. Follow the instructions given for fast fixing or two-bath fixing starting with the stop bath step. You may find upon testing, especially with stabilization prints that have not been treated immediately, that longer wash times will be needed because the fixer residues have mordanted to the fibers in the paper base.

For archival processing, stabilization prints have one advantage over conventional materials. A number of them can be accumulated in the darkroom and then fixed and washed at one time. This makes possible more accurate control of these processes. Make sure, however, that if this is done, you wash thoroughly any surface where the prints have lain in order to remove any chemical residues. You might find it convenient to set aside a photo tray to hold damp prints just out of the processor until they can go into the fixing bath.

REFERENCES

1. David Vestal. *The Craft of Photography*. New York: Harper & Row, 1975.
2. James M. Reilly. *The Albumen and Salted Paper Book: The History and Practice of Photographic Printing, 1840–1895*. Rochester, NY: Light Impressions, 1980, p. 84. One of the best general expositions of the theory of fixation published to date.
3. Ralph Steiner. "Comparing Fixing Methods." *PhotographiConservation* 2(1) (March 1980):3. Graphic Arts Research Center, Rochester Institute of Technology.
4. "Hypo and Silver Elimination in Salt Bath." *Photographic Engineering* 7 (3–4) (1956).
5. *Processing Chemicals and Formulas for Black-and-White Photography*. Kodak Professional Data Book J-1, 6th ed. Rochester, NY: Kodak, 1963, p. 51. This book is packed with formulas, and is an invaluable reference source, but it is a bit short on explaining how things work.
6. J.I. Crabtree, G.T. Eaton, and L.E. Muehler. "The Quantitative Determination of Hypo in Photographic Prints with Silver Nitrate." *Journal of the Franklin Institute* 235 (April 1943):351–360.
7. Eastman Kodak. *Studio Light*. No. 1. Rochester, NY: Kodak, 1976.

C H A P T E R 2

Processing Film for Permanence

Perhaps no art, science, or craft has evolved by such an extraordinary combination of pure science, pure witchcraft, and wishful thinking as that which constitutes the popularly accepted procedures in photography. In the development processes above all others, weird mumbo jumbo persists and flourishes.

— Ansel Adams [1]

FILM PERMANENCE: PREPARING TO PROCESS

Contemporary Films

Unlike many of the materials used by early photographers, contemporary black and white negative films are both easy to process and relatively stable in storage. The triacetate base used for modern film is both fire-resistant and free from the tendency to spontaneously disintegrate.* When kept at recommended levels of heat and humidity, it is also dimensionally stable.

Film needs relatively less elaborate washing and postprocessing care than do paper prints in order to achieve the same degree of stability. This is because the film base does not absorb and retain chemicals from the processing baths. Washing needs to be directed primarily at removing chemicals from the thin layer of the emulsion. Care should still be taken during processing of film to achieve the optimum final result: a clean, well-washed negative that is free of fixer residues. Such residues have the same deleterious effects on film as they do on paper. The ease of processing film to archival standards means that there is no excuse for an unstable negative.

* Triacetate film is also called "safety film" because, unlike the older nitrate films, it neither burns easily nor does it spontaneously ignite.

PROCESSING FILM: BASIC PROCEDURE

If you are a photographic novice, you can still develop your own film without a darkroom. Learning to do it will help you understand the conservation of negatives.

- ☐ Go to the camera store and buy a two-reel stainless-steel developing tank, a couple of 35mm film reels, a photo thermometer, some film hanging clips, D-76 developer mix (it comes in powder form), a bottle of stop bath concentrate, and an envelope of powder for a half-gallon of fixer. The whole outfit should cost you around $45.
- ☐ Mix the chemicals according to directions on the packages. In a dark closet load film onto the reels and put them into the developing tank. Believe it or not, this is the hardest part. You might want an experienced friend to help you do it, or practice loading a length of film onto a reel in the light before you "go dark," as the jargon has it.
- ☐ Once the cap is on, solutions can be poured in and out through a light trap on the top.
- ☐ Take the tank to a work surface (*not* the kitchen sink) where you have the chemicals set up. Measure their temperatures with the thermometer, and consult the time and temperature charts packaged with the film. Use those times for development.
- ☐ Pour in the developer. Invert the tank twice every 5 seconds for the first 30 seconds of development, then twice at every 30-second interval afterward. Pour out the developer through the light trap in the lid, pour in the stop bath, and let it sit for about a minute.
- ☐ Pour out the stop bath and replace with fixer. Agitate regularly during the minimum fixing time. Pour out the fixer and take off the top.
- ☐ Adjust the temperature of your running water so that it is within 5 degrees of that of the processing solutions. Fill and empty the tank once every 5 minutes for 1 hour.
- ☐ At this point you have wet negatives that are completely processed.
- ☐ Refer to the discussion of washing and drying for further procedures. Even if you are not regularly a photographer, it helps to have practical experience so you know what we are talking about.

Fresh Chemicals

Your primary rule should be never to push any bath beyond its rated capacity. Any chemical solution used beyond its capacity not only does not do its job properly, but it also actively harms the film. Developer that is replenished too much or used for too long, even with increased development times, deposits silver bromide back on the film and wipes out shadow details. An exhausted acid stop bath allows developer to carry over into the fixer and cause fixing activity to drop below intended levels. (In the chapter on prints, we discussed the dangers of exhausted fixer in detail; the same principles apply to film.) Hypo clearing solution obviously has only one purpose, to preserve the negatives; when it gets exhausted, it cannot do that. The photographer who tries to save pennies by using developer until it turns to silver sludge, or who gets just a few more rolls of film through an old fixer bath, pays in the long run for those few cents.

Temperature Control

Getting all the different chemicals to the same temperature at the same time poses a bit of a challenge. Your choices go from using everything at room temperature, whatever it happens to be, to using elaborate electronic monitoring devices to automatically mix incoming water of different temperatures. The simplest method, for those with a bit of patience, is to use a high-volume bath of standing water. After mixing the chemicals, let them all stand in a large tub of water at the desired temperature. It will take about an hour for them to reach equilibrium. Hot or cold water may be needed to maintain the bath temperature, so check it from time to time. Any kind of container that accommodates all the chemicals will serve the purpose; a large dishwashing basin works quite well.

Maintaining a constant temperature for the wash water is a little more difficult. If you are using the dump-and-refill method of washing, a large quantity of water can be drawn into the container, where it is possible to adjust the temperature in advance by adding more or less hot and cold water. Needless to say, the reservoir should be thoroughly clean. When running water is required, the cheapest solution is to use a thermometer well of the kind made by Wat-Air. It holds a dial-type thermometer in the water line so that the hot and cold faucets can be adjusted as necessary. Keep a wary eye on the dial while washing, because other appliances can make demands on the plumbing system that cause a sudden drop or rise in temperature.

DEVELOPERS

Conventional Developers

The function of a developer is to reduce exposed silver halide grains to metallic silver. The pattern of the exposed grains creates a continuous-tone stencil on the transparent film base through which more or less light is allowed to pass onto the surface of the print paper. The less light allowed to pass, the lighter will be that area of the image.

There is no magic developer formula that will do all things for all photographers, though you might be tempted to think so because of the hundreds of formulas that have been published. But neither do you need to master a majority of these formulas; one or two will serve.

Most conventional developers use a combination of *p*-methylaminophenol (also known by Metol, Pictol, Rhodol, Elon, and other trade names) and hydroquinone with additives. Developers in this class have virtually no distinguishable effect on the keeping characteristics of film processed in them.

The components of conventional modern developers usually include chemicals that fill four roles.

- ☐ The *developing agents* actually reduce the silver halides; they include Metol (or Elon, and so on), hydroquinone, Amidol, pyrogallol, and glycin (the last two increasingly little used).
- ☐ The *preservative*, usually sodium sulfite, inhibits oxidation of the developer.
- ☐ The *alkaline accelerator* creates the high pH needed by most developing agents to function. It can be sodium carbonate, sodium hydroxide, borax, or sodium metaborate (or Kodalk).
- ☐ The *restrainer*, usually potassium bromide, prevents high levels of fogging density in the negative. It also cuts back the effective film speed and brings out the contrasts in the negative.

The formulas for developers mixed along these lines often include instructions for a replenisher. Usually the replenisher formula is quite similar to that for the stock solution of the developer; the difference is usually a reduction or omission of the restrainer, because the silver bromides created during development pass from the film into the developer and take over this function. Adjustment of the accelerator quantities is also done to maintain the desired pH.

Replenisher is generally added to the used developer in specified quantities based on the amount of film processed; for example, an ounce of replenisher might be added for every 36-exposure roll of 35mm film. When a replenisher is used, take care to record the amount already added, and replenish every time film is processed. Otherwise, consistent results will be impossible to achieve.

Consistency will also be affected by storage of the developer. Exposure to either light or oxygen degrades the developing agent after it has been dissolved in water. This is the reason that developers should be kept in full, tightly stoppered, dark-brown glass bottles or in covered, opaque developing tanks. Plastic cartons like those used for milk are clear to light and let oxygen pass through the container walls.

The alternative to replenishment is to discard the developer after use. One-shot development is not nearly so wasteful as it seems at first. Most one-shot developers like Kodak D-76 are diluted 1:1 with water and development time is extended. By simply adding water, you get exactly the same number of rolls of film processed per gallon, without having to mix the replenisher; the only penalty paid is a little longer developing time, which is probably no more time than you would spend adding the replenisher anyway.

One-shot development assures consistency from roll to roll (at least while using the gallon that you have prepared) without the worry of how the replenisher is interacting with the stock solution. You do not have to deal with potential problems caused by silver bromide buildup in the developer. It is the method favored for most applications.

Two-Bath Developers

Two-bath developers give the photographer excellent contrast control that is as nearly uniform from film to film as is possible. They do this because the degree of development is not nearly so dependent on the vagaries of agitation and temperature fluctuations as with conventional developers.

Most two-bath developers work on similar principles. First, the film is immersed in a bath containing one of the standard developing agents like Metol, until the film is saturated. Because these agents work only very weakly by themselves, time is not a very important factor here. The film is then transferred to a second solution containing an accelerator such as hydroquinone. The actuator activates the developing agent absorbed in the film, and development proceeds until this developer is exhausted. At this point, development is effectively finished.

Two-bath formulas generally call for no components that would not be found in ordinary developers; the primary difference is that these components are divided into two solutions instead of one, and have different strengths. There is not any reason to believe that they have any different effects on the keeping properties of the film from conventional developer.

Monobath Developers

Monobath developers combine developing and fixing into one operation that usually lasts 4 to 6 minutes. They were originally developed to speed up processing for news photographers. The complex reactions set up between silver bromides and thiosulfates in the developer would require an extended technical treatise to explain, but in brief we can say that they are complex enough to make it virtually impossible to remove all residues from the film by washing. The theory is, at first thought, a nice one, but it is risky and obsolete. If speed is that important, use one of the instant-print processes.

STOP BATH

Mixing a stop bath to its requisite strength is often overlooked in processing, but it should not be. Too weak a dilution of the acetic acid not only fails to stop the action of the developer immediately, but it can also cause a shift in the pH of unbuffered fixers. On the other hand, when the stop bath is too strong it can cause small pinholes in the emulsion. These print black.

Some of the older technical literature suggests dispensing with the stop bath altogether and replacing it with a quick water rinse. Doing this can lead to complications that may be difficult to track down. For example, one of us tried this while using the Kodak formula D-23 for film developer. Opaque deposits kept appearing on the negatives despite the most careful attempts to get rid of them by using distilled water, filtering, and other techniques. Not until we came across a footnote in an obscure formulary did we realize that the absence of a stop bath caused calcium sludging as the film was put into the fixer.

Use a stop bath.

FIXING FILM

The rule of thumb to determine fixing time is that it equals twice the time needed to "clear" the film. This time varies depending upon the type of fixer and the kind of film. Generally, the faster the film, the slower it clears, because the amount of silver halide to be dissolved by the fixer is greater. To determine clearing time, use an open tank. Give a piece of wet, undeveloped film the same agitation in fixer as it gets during normal processing. Observe the film closely during this procedure, which can be carried out in normal room light. It will turn nearly white, then milky, and then transparent. Make a note of the amount of elapsed time it takes for the film to turn completely transparent, and then double that to get your standard fixing time. This time applies only to that kind of film in that type of fixer.

Using a liquid fixer instead of the powdered form tends to reduce spills and the incidence of hypo dust. Ektaflo Fixer is a diluted ammonium thiosulfate fixer that comes at working strength in one-gallon cubitainers; it is merely drained from them into the tray.

The use of an ammonium thiosulfate "rapid" fixing bath speeds washing of film. "The residual hypo content is reduced to zero in 50–65% of the time required for eliminating sodium thiosulfate," according to one researcher [2].

The pH of the gelatin in photographic emulsions averages around 4.9. The more acidic the fixing bath — that is, the lower the pH — the more acidic the gelatin becomes. As the pH of the gelatin drops, it binds hypo residues more firmly. Consequently the use of an acid-hardening fixing bath necessitates an increase in wash time [2, p. 88].

Fixing baths for film and for paper should be stored separately and not interchanged. Because it is so much faster than paper, film has a much greater density of silver than paper and depletes the fixer more rapidly. In addition, most films contain an antihalation layer of dye that is removed during fixing; we have not seen any research on the possible effects on paper, but there is no reason to take any chances of staining.

Two-Bath Fixation of Film

When a fixing bath gets used repeatedly until it nears exhaustion, as is frequently the case, the two-bath method makes thorough washing easier. When fresh fixer is used for each new bath of negatives, you can dispense with a second bath.

Unexposed silver in the film emulsion is dissolved by the sodium (or ammonium) thiosulfate in the fixer and is converted into silver thiosulfate compounds in solution in the fixing bath. As the fixing bath is used on succeeding batches of film, the amount of silver thiosulfate compounds in solution increases. When the fixer gets saturated with them, these compounds start to redeposit back onto the film. Because they are insoluble in plain water, no amount of washing removes them.

Obviously this problem does not occur if fresh fixer is used each time, because the fixer has the capacity to hold in solution a certain amount of the thiosulfate compounds. The purpose of the second bath is not really to "fix" the image, in the sense of taking out more unexposed silver. Rather, it redissolves any of the residual thiosulfate compounds and removes them from the emulsion.

To use the two-bath fixing technique, negatives are first fixed in one bath for twice the clearing time minus 1 minute. Then they are transferred to a fresh second solution for 1 minute and fixed there with constant agitation. Washing follows. Keep the two fixing solutions in separate containers marked A and B (or 1 and 2, or whatever designation you like). When the first nears exhaustion, replace it with the second solution and discard the first. Mix new fixer for the second bath. Replace both solutions after three such cycles and start fresh.

Two-bath fixing is still recommended for films, rather than using the short fixing method given for prints. Because the base material of contemporary black and white films is nonpermeable, there does not exist the same danger of fixer residues mordanting in the film.

Care needs to be taken to avoid overfixing film. For one thing, the subbing (or substrate) layer that binds the gelatin emulsion to the triacetate base can retain the by-products of overfixing in the same way that paper does, although how serious a problem this may be awaits further research. More important, overfixing can bleach out parts of the negative, particularly the shadow areas of least density. This results in loss of detail. Coarsening of the grain structure can also result; this is likely to be a more severe problem with small-format negatives.

Many photographers still consider agitation during fixing as an afterthought, something done during a break in cleaning the darkroom. This should not be so. Regular agitation is the only method of supplying fresh fixer to the film surface. Film lying in stagnant solution soon exhausts all fixer in the immediate vicinity. From that point on, it might as well be fixed in the stuff you discarded yesterday. Agitation pumps a continuous supply of fresh fixer across the film surface.

When fixing film in reel-type tanks, agitate continuously for the first minute, and at 30-second intervals thereafter until fixing is complete. Sheet film processed in trays should be agitated constantly; when using hangers and tanks, agitate as for film on reels.

Fixing Bath Capacity

It is difficult to predict exactly the point when a fixing bath will reach exhaustion. Different negatives have varied ratios of highlight to shadow areas, so that some use fixer at a more rapid rate. Developers of different alkalinity and the strength of the stop bath also play a part, and different working habits can have an effect. The published capacities represent an educated estimate based upon the manufacturer's idea of average working conditions, with a safety margin incorporated. These estimates compromise between what is considered "safe" and the consumer's understandable desire to get the most fixing per dollar spent.

The safest method to follow when reusing fixing baths is to establish a capacity that is known to be safe under current operating conditions, to record the amount of film that is run through the fixing bath, and then never to exceed it. To test the rate of exhaustion of a fixing bath, the same film clearing test can be used as was described earlier. With sodium thiosulfate fixers, discard when the test strip clearing time doubles; and for rapid fixers discard when it gets to be four times the original time.

Rapid fixers retain their activity beyond the point at which they have become dangerously saturated with conversion products. Unlike conventional sodium thiosulfate fixers, a rapid fixer will clear film even after it contains enough insoluble residues to deposit them back into the emulsion. For this reason, clearing activity alone should not be considered proof of a fixing bath's safety, and the known capacities should always be respected. For safety's sake, it would be wise to reduce your regular usage by at least 25% below any published capacity to allow for unforeseen fluctuations caused by such factors as human error or improper mixing.

Discard immediately any fixing bath that changes color or turns cloudy.

WASHING AND TONING

Hypo eliminator (HE-1) does not need to be used with film for maximum permanence. Ansel Adams reports in *The Negative* [1, p. 15] that it has caused small blisters on his negatives.

Because the acid-alkaline balance of the emulsion affects the washing rate of film, raising the wash water's pH by adding a 0.03% solution of ammonia has been used to cut washing time. However, Kodak Hypo Clearing Agent works more effectively than the ammonia rinse and has the additional benefit of making it possible to wash effectively in water as cold as 40°F [2, pp. 92–93]. It is possible with Hypo Clearing Agent to effect a complete removal of fixer residues from film.

Wash water temperature should be within ±5°F of other processing baths for large-format negatives, and nearly the same temperature or just slightly cooler for small-format negatives. If greater variation cannot be prevented, it is better to err on the cool side because hot water softens the emulsion, causes frilling, and creates greater sensitivity to dust particles. Keep in mind that lower temperatures increase washing time unless KHCA or a similar treatment like Permawash is used.

Film Washers

The simplest film washer for roll films is the film-developing tank. A perfectly satisfactory method of using this tank is to dump and refill it once every 5 minutes during the washing period. If this is too laborious, and it probably will be if much film is being processed, a short section of stiff pipe can be attached to hose running from the mixing faucets. Use a radiator hose clamp of the kind sold in hardware stores to hold the pipe in the hose, and then stick it down through the center of the reels. When the water is turned on, the excess overflows the sides of the tank. The only problem with this kind of washer is that bubbles may form on the film and prevent washing, so give it a bang once in a while to shake loose the bubbles.

The most convenient film washer for roll film is the tubular design made by Wat-Air, Zone VI Workshop, and Kostimer. With this device you hook up the intake hose to a faucet, drop the film reels into the tank, and turn on the water. It has a capillary tube along the side and a mixing block at the water inlet, so that aerated water constantly bathes the film. This type of film washer also needs to be tapped occasionally to shake bubbles loose.

Washing sheet film is a bit more of a problem. If you are tray processing the film, one method that is really safe is to use two trays, transferring the film from one to another and dumping each time. A tray siphon is likely to cause the films to slide across one another, and the corners will scratch the emulsions. If you are processing on hangers, you can make your own washing tank with a section of pipe stuck into the bottom side of a tank. Drill some holes along the length of the pipe and stick it through the side of the tank. Seal around the hole with threaded washers and some silicone cement and attach the pipe to an intake hose. Excess water can overflow the top. Agitate regularly to prevent bubbles on the film. Another safe method is to wash in one of the washers designed for archival washing of paper.

Testing Film for Thiosulfate Residues

Film fixed in fresh hypo, treated with a clearing agent, and washed thoroughly should have virtually no fixer residues left. If any doubt exists as to the efficiency of the washing methods employed, or when a standardized processing system is being established, it is handy to have a convenient test to check for fixer residues.

In recent years very precise tests have been published that measure the retention of extremely minute amounts of thiosulfate compounds. For most practical purposes, however, a silver nitrate test is still adequate. The version given here is Kodak Formula HT-2:

Distilled water	24 oz
28% acetic acid	4 oz
Silver nitrate	¼ oz
Water, to make	32 oz

Store in a dark-brown bottle, tightly stoppered, away from strong light.

To make a 28% acetic acid solution, add 3 parts glacial acetic acid, available in camera stores for stop baths, to 8 parts water. To avoid spattering and possible burns, *always* add acid to water rather than water to acid.

To use this solution, put a large drop on a clear section of the processed film. Anything other than the faintest ivory-colored stain indicates the presence of excessive fixer residue. Comparison with a Kodak Hypo Estimator [3] gives an indication of the relative degree of retention. For those who do not wish to mix their own, Kodak sells a Hypo Test Kit with premixed solution and an eyedropper.

The archival standard for microfilms and other film records calls for residual thiosulfates not to exceed 0.7 microgram per square centimeter. When tests to this degree of exactitude are required, either the methylene blue or the silver densitometric methods can be employed, as outlined in the ANSI standard PH4.8. The methylene blue test is more precise, but the silver densitometric test can be used for more routine testing. Copies of the standard with testing instructions can be obtained either from ANSI or the National Micrographics Association, 8719 Colesville Road, Silver Spring, MD 20910.

Toning Negatives

Toning negatives will not, of course, improve or change prints made from them, but it can help preserve the image. Toners for films are not chosen for the color changes they produce. For a long time, a favorite toner for negatives has been the Kodak formula GP-1. This solution is easy to use, does not change a negative's density, and does not affect grain size or resolution. Among the few current distributors of protective gold toner are Light Impressions and Berg Color-Tone, Inc., P.O. Box 16, East Amherst, NY 14051. The alternative is making up your own chemicals, which involves purchasing raw materials from a chemical supply house, and a rather elaborate preparation procedure.

GOLD PROTECTIVE SOLUTION

Distilled water	24 oz	750 ml
Gold chloride (1% stock solution)	2½ drams	10 ml
Sodium thiocyanate	145 grains	10 g
Distilled water, to make	32 oz	1 liter

Gold chloride is sold by chemical supply houses in hermetically sealed glass tubes that you break with a little file when ready to use. For accuracy of measurement, drop the entire tube into the water so that all the gold chloride washes off, and then decant into another container. Make sure to use distilled water, because gold chloride is very sensitive to contamination.

A 1% stock solution of gold chloride is made by dissolving 1 g gold chloride in 100 ml water. Add the stock solution to the volume of water indicated. Separately, dissolve the sodium thiocyanate in 4 ounces (125 ml) water. Add the thiocyanate solution slowly to the gold chloride while stirring rapidly.

Immerse film in the toner for 10 minutes. The working capacity is approximately 7 to 8 rolls of 35mm film or an equal number of 8 × 10 sheets of film per quart. Use immediately after mixing the two solutions. Wash for 10 minutes following treatment.

Selenium toner, available as liquid concentrate from Kodak distributors, has the multiple advantages over gold toner of being less expensive and easier to prepare and of providing at least as good protection. Selenium toner used to protect negatives can be quite dilute. Mix 1 part liquid concentrate to 20 parts water. The film can be treated either immediately after fixing, while still saturated with hypo, or after it has been completely washed. It is more convenient to treat the film right after fixing, because then it is not necessary to add any more washing time after the toning. Transfer immediately to the selenium bath after fixing and without any rinse; then agitate regularly for about 6 minutes or until a subtle color change appears, whichever is first. Do not use the same toner bath for treating prints as for film; in fact, at this dilution it will be economical enough to discard the toner after each use.

DRYING

A chronic problem in drying is the presence of water spots. These occur when most of the emulsion side of the film dries completely, except for a few small areas that retain droplets. If these drops are allowed to dry in place they leave rings that cannot be removed, and the rings print as light areas.

Kodak Photo-Flo and Edwal LFN are nonionic wetting agents. They fall into the same category of chemicals as household detergents. Both are surfactants, which means that they reduce the surface tension of water so that the droplets cannot form. Instead of beading up on the surface of the film, water just slides right off in a flat sheet. Evaporation also proceeds much faster, because of the greater surface presented to the atmosphere.

Conversations with numerous photographers have convinced us that nearly everybody at some time has experienced problems with wetting agents. The problem often shows up as a thin, greasy film on the negatives. The cause may be human error in not following the manufacturer's instructions, or it may be that these instructions call for too strong a solution. Either way, if you experience this problem, here are some options to consider. One is to mix the Photo-Flo or LFN at a 50% weaker dilution. Another is to use it as directed, but then follow with a brief rinse of clean water. The emulsion of the film absorbs enough wetting agent to break down water droplets without leaving a surface film behind.

An excellent final rinse can also be made up as follows. Mix a working solution of 50 ml 91% isopropyl alcohol, 1600 ml distilled water, and 4 ml Kodak Photo-Flo. This makes more than half a gallon of working solution; excess can be stored for later use. Immerse the film for about 30 seconds and remove for drying. Discard the solution after using once. This dilution is only half as strong as Kodak's recommended strength, but as the instructions note, "scum can form on the film if the Photo-Flo concentration is excessive." The weaker solution seems to work as effectively, and provides an extra safety margin.

When drying, try keeping humidity levels at 70% relative humidity or even higher to prevent water spots; this also holds down static electricity that causes dust to attach itself to film. Temperature should be around 85°F, and no warmer. Heat can safely be provided by using a 100W light bulb in a closed space like a closet or cabinet. Do not use forced hot air for drying film.

Rapid drying of the film before fixing and washing have been thoroughly completed is sometimes necessary when a print has to be made right away. In these extreme cases, the recommended procedure is to use a rapid fixer to clear the film. Then wash the film briefly and soak in a 1:9 solution of ethyl alcohol diluted with water. Dry the film at a temperature not exceeding 80°F. After the necessary print has been made, fixing and washing are continued in normal fashion.

Two cautions to observe when using the alcohol drying method: (1) use *ethyl* alcohol, not methyl alcohol, which acts as a solvent on the film base; and (2) do not use hot air to speed the film drying. When drying temperature gets over 80°, the film can turn an opaque, pearly color. This opalescence can be removed by rewetting the film and drying slowly, but it nullifies your efforts to get a fast print.

CLEANING

Wash trays and tanks thoroughly with hot water after each use. With heavy use they may show a tendency to build up dark-colored deposits; these should be removed with a tray-cleaning solution. The reasons for doing this are not aesthetic. These deposits often consist of silver sulfide; besides being harmful in themselves, their spongy texture causes them to retain part of the processing solutions so that effective control and complete washing are made more difficult.

Film and print washers are left full of water in many darkrooms. Algae and other organisms can grow in the water. They cause a slimy feel on surfaces of the vessels. A capful or two of household bleach can be added to the standing water and left for an hour. After that, the inside of the vessel should be washed thoroughly and allowed to dry.

Solution A		
Water	1 gallon	1 liter
Potassium permanganate	¼ oz	2 g
Sulfuric acid (concentrated)	½ oz	4 ml
Solution B		
Water	1 gallon	1 liter
Sodium bisulfite	4 oz	30 g
Sodium sulfite, desiccated	4 oz	30 g*

An acid stop bath can be substituted for solution B, but additional washing will be needed to get rid of fixer residues. This treatment removes silver stains, silver sulfide, and many dyes. Pour a small quantity of solution A into the vessel and allow it to stand for a few minutes; wash thoroughly and replace it with solution B. Agitate and then wash thoroughly. To remove buildups of calcium scale, an acid stop bath can be allowed to stand in the affected container overnight; the remaining scale should then wash off easily.

It is always better to prevent the creation of stains by complete washing with soap, water, and a clean cloth after processing. Use

* Store in a dark stoppered glass bottle away from light.

these cleaning solutions only as necessary, rather than as a routine housecleaning procedure.

Handling Negatives in the Darkroom

Careless handling of negatives while printing can undo much of the good of meticulous processing. Negatives should be handled by the edges only. If the storage enclosure cannot be unwrapped to lift out the negative, shake or tap the negative partway out before grasping it by the edges. When it is necessary to reach inside the enclosure to pull it out, don a lintless cotton glove to avoid putting fingerprints on the film.

When fingerprints or other grease marks do get on the film during printing, clean them off before returning the negative to storage by lightly swabbing with a Q-tip dipped in Kodak Film Cleaner. Body oils can feed fungus growth and chemically interact with the emulsion.

Prevent scratches on the film by *fully* opening the enlarger's negative carrier before moving or removing the negative. Before doing any printing with a new enlarger, it is a good idea to check the surfaces of the negative carrier for small burrs left over from machining. When you find these, remove them with a few strokes with an emery cloth.

Film that is in otherwise good condition but that suffers from some minor surface scratches can be treated with film lacquer to keep the scratches from showing in the print. Of course, this will not work if the emulsion has been scratched through to the film base and part of the image is removed. Anytime that a film is lacquered, note the treatment on the negative enclosure.

Work with negatives in a dust-free environment. Take any dust off the film with a clean, dry brush, with compressed air, or with Dust-Off; never smear it off with the fingertips. Make sure the negative is free from dust particles before returning it to storage. This is especially important because the dust found in darkrooms often contains the most pernicious components, such as very fine pieces of dry fixer that need only the slightest amount of moisture to start reacting with the image. And believe us, they are not likely to interact in any way that will improve the picture. Darkrooms should always be vacuumed rather than swept, because sweeping tends only to rearrange the dust so that it penetrates more inaccessible places.

PROCESSING INSTANT FILM

Polaroid Positive/Negative Film

Polaroid's Type 665 Positive/Negative Film gives you a positive print and a large 3¼ × 4¼ negative at the same time, with all the advantages of instant photography. Adapter backs make it possible to use this material in conventional 4 × 5 cameras. Though the film has obvious applications in the field, it can also be used to make quick negatives for copying prints. Use a standard 4 × 5 camera with adapter back, if you have one. If not, it may be more economical to buy Polaroid's MP-4 copy camera, which comes complete with a vertical pillar for the camera, copy board, and lights. The MP-4 is a top-notch unit, but the initial investment is such that it is not for everybody.

When using P/N film, you start processing in the same fashion as for other Polaroid black and white films — by pulling the film out of the holder in a single smooth motion. After waiting the prescribed 30 seconds (more or less according to time and temperature tables packed with the film), the film packet is peeled apart.

Take out the print first, making sure not to smear any of the goo from the packet onto it. Coat it with the little squeegee saturated with pink glop that Polaroid packs with the film. Much experience with these little squeegees has convinced us that they are too heavily loaded with the protective coating, so we suggest squeezing out some of the excess before starting, to avoid too heavy a buildup on the print surface. On the other hand, do not fail to cover the entire surface of the print thoroughly; otherwise, local fading will take place. Later, if you see that part was missed, recoat the print. Set aside to dry in a dust-free drawer. This print can be considered of archival quality.

The next step, before too much time has elapsed, is to clear the negative. Polaroid sells a portable processing bucket with film holders built in. Mix a sodium sulfite solution (Polaroid also provides the chemical and a little measuring spoon) according to the directions before starting to shoot, and you can use the bucket as a holding tank for the negatives for up to three days. The sodium sulfite removes a film of processing gel from the negative and makes the image visible. Do not let the negative dry out before putting it in the bucket, because then the film becomes difficult to remove. All further processing can be done in room light.

When the bucket is filled with negatives, lift out the film holder unit and dump the solution. Make sure that there is a drain strainer to catch the remnants of black film, which should be discarded separately to avoid clogging the pipes. The bucket can be used as a tank to finish processing the negatives. Rinse the negatives thoroughly for about 1 minute in running water. All washes and solutions should, of course, be kept at a uniform temperature around 68°F.

Fix the negatives for 2 minutes in an ammonium thiosulfate fixer with hardener (e.g., Kodak Rapid Fixer or Edwal Quick Fix). This removes any residual silver, though there should already be little left, and hardens the negative against scratching. Pour out the fixer and wash for another minute. Drain the water and replace with Kodak Hypo Clearing Agent. Agitate the film in KHCA solution for 2 minutes, discard the solution, and wash for at least 20 minutes. This can be done by lifting the film holder out of the bucket every 5 minutes and replacing the water. Agitate periodically between dumps.

Finally, treat the negative with the distilled water-Photo-

Flo-isopropyl alcohol solution used for conventional films, and hang to dry in a dust-free area. The result will be an archival quality, large-format negative packed with crisp detail.

Type 665 film costs appreciably more than conventional negative film, but the advantages of rapid access and the fact that an elaborate darkroom is not needed to produce uniformly good negatives make the cost well worth it for many applications. The prints can be used for reference in a card index when it is necessary to provide a key to the negatives, and this will save much handling of the originals. If you or your institution do not have a large-format enlarger, prints can usually be made at fairly low cost by a custom laboratory or commercial photographer.

The one shortcoming of this film, which Polaroid admits to, is that the optimum exposure differs for the negatives and for prints; the negatives need more exposure. This problem is still being worked on, but in the meantime do not build up a large file of negatives, depending upon the prints to serve as the only method of exposure control. Make prints from the first negatives until you have established a reliable standard by which to gauge the Polaroid print against the probable quality of prints that can be made from the negative.

REFERENCES

1. Ansel Adams. *The Negative.* Hastings-on-Hudson, NY: Morgan and Morgan, 1968. Basic Photo Series no. 2.
2. George T. Eaton. "Preservation Deterioration, Restoration of Photographic Images." *The Library Quarterly* 40 (1) (January 1970).
3. *Kodak Hypo Estimator*. Kodak Professional Data Book J-11. Rochester, NY: Kodak, 1979.

PART TWO

Mounting and Mats

CHAPTER 3

Flattening Prints

Frustration with a curled or uneven image plane drives some people to desperate stratagems like taping down all four sides of the print or gluing it tightly to a backing board. Some procedures such as making copies require that a print be as flat as possible to reproduce the image faithfully. Filing, storage, and matting can be difficult, too, with pictures that won't lie flat.

Waves and ripples in a sheet of paper arise naturally from the paper-making process. Paper is born in water in the papermaker's vat, and through water it interacts with the environment. Papermakers create fine handmade papers by dipping a screen into a vat of water and cellulose fibers, lifting and draining the mold, and couching it onto a large sheet of felt. They build up a large pile of interleaved sheets and squeeze out the remaining water with a counterweighted lever or with weights. The paper dries under pressure. Its changes in shape and size, its absorption of the photosensitive materials in a gum or platinum print, the way it takes inks and emulsions — in fact, all the chemical and physical events that occur on the surface of paper — arise from this marvelous ability to interact with water.

MODERATE MEASURES

It is always best to use a technique that places the least possible strain on the print. This principle applies especially to old and valuable photographs, and obviously, the photographer or printer who has made the print works under fewer constraints than does the archivist or collector who has responsibility for the work of another.

For prints in a delicate condition that have problem waves, humidifying will relax the paper fibers. You can easily construct a simple humidifying chamber.

Start with a tray that is quite a bit bigger than the print. Fill the tray with water and stretch clean fiberglass screening across the top. The screen can be weighted down at the edges where it overlaps the tray, or it can be stretched around some kind of frame (like a screen printing frame) that can be placed into the tray so that it holds the print above the surface of the water. Place the print on the screen and cover with clear plastic sheeting, and weight the edges of the sheeting. The heat from a low-intensity light bulb placed close to the tray will speed the humidifying process. Check the print periodically by touching it in several places to make sure that it does not become too damp. If moisture droplets appear on the plastic sheeting, take apart the entire assembly immediately to avoid damage. This process may flatten the print satisfactorily, or you may decide that additional weighting is required. In that case, put the print between several layers of clean blotters that are larger than the print and place under a large flat weight for 24 hours.

Some old photographs that were rolled for storage may resist being unrolled. For these, moderate humidifying may also prove beneficial. This procedure takes longer — one or more days — because moisture has to penetrate between the layers of the rolled print. Fill a plastic wastebasket with about 2 inches of water and balance the prints vertically on blocks set in the water. (To avoid having them accidentally slide into the water, pass a cord through the center of the roll and tape both ends onto the outside of the wastebasket.) Cover the top of the basket with transparent plastic sheeting and let stand. Check the prints periodically to see how much they have relaxed. When they are limp enough, unroll and let dry under blotters and a flat weight.

THE COLD AND WET METHOD

The use of directly applied moisture and weights can be done at the discretion of the original printmaker or by a conservator. When you have a photograph in good condition with a strong paper base, this cold-pressing technique gives you an excellent way to flatten the print. You will need several sheets of acid-free blotting paper, a thick plate of metal moisture-sealed wood, a bowl of clean water (distilled if possible), a clean sponge with any sizing rinsed out, and weights. Large, heavy books will do for the weights (see Figures 3.1–3.7).

Brush the workspace clean and cover it with paper to soak up any water spills. Place the bottom blotter somewhere where it can remain undisturbed for a day or more. Now take the first print and lightly sponge a small amount of water onto the back. The paper responds almost immediately to the water by relaxing throughout. After it becomes limp, lay it face down on the first sheet of blotter, put another sheet of blotter on top of it, and lay the metal plate on that.

Repeat this procedure for as many prints as need flattening, until a stack of interleaved blotters and prints builds up. Place weights on top of the plate to distribute the pressure evenly over the stack. For complete drying, exchange the blotters with fresh ones several times during drying.

An alternative method of wetting with which we experimented was to use a water sprayer like the ones for misting houseplants. It worked, but it also tended to spread water indiscriminately around the work area.

Naturally, there are some precautions to observe when cold-pressing prints. Before applying moisture to any material with which you are not intimately familiar, test for water staining. Put a small drop of warm water in one corner of the back, allow it to sit for about 5 minutes, and blot it off. If any stain appears, do not put water on the print. Try pressing it dry under weights and blotter paper.

Ferrotyped photographs have a glossy sheen that comes from drying in contact with a sheet of polished metal. Stray drops of moisture on the front will eliminate the sheen in places and will give the print an ugly, mottled appearance. Some of these photographs can be flattened with a dry mount press if they are not so curled that flattening will crack the emulsion.

Cold-pressing will not work on some materials. Photographs on resin-coated papers do not absorb much water; if they have a tight curl, it probably comes from overlong immersion during processing, usually in the rinse water. These prints will probably have to be flattened by dry mounting onto a sheet of thick backing board.

Do not cold-press any material that might transfer something to the blotter sheets. This means avoiding applied media like charcoal, pastels, tempera, and crayon. The loss of detail and pigment more than offsets any gain from image flatness.

Use discretion when working with paper materials that have raised or indented surfaces, such as embossed prints or etchings with plate marks. A happy feature of the cold-pressing method is that it is gentle enough to use with some embossed materials without completely flattening them.

THE HOT AND DRY METHOD

A dry mount press gives you a powerful tool for flattening large numbers of prints, though it would hardly make sense to buy one for this purpose only.

Some frame shops rent time on their dry mount presses, and even if they do not make a practice of it, sometimes you can strike a deal. Do not be afraid to ask. Check out local schools or universities that offer courses in photography. And many amateur photographers own small presses they let their friends use. Before using a borrowed or rented press, check the platen (the metal plate with heating elements) to make certain that it is clean and free of deposits.

Once you have access to a press you will need three or four sheets of acid-free blotting paper or 4-ply rag or conservation board, all larger than the prints; you will also require release paper or Seal cover sheets from photo stores, and a cooling weight. The cooling weight can be either one made commercially, like the one

Figure 3.1
Metal plate, water and sponge, prints, and acid-free blotters for flattening prints. Notice the extreme curl of the prints.

Figure 3.2
Squeeze excess water out of the sponge before wetting back of print so that it doesn't splash around the work area. Only a little moisture is needed to flatten print.

Figure 3.3
Apply moisture only to the back of the print.

Figure 3.4
Put prints individually between blotters. A series of prints and blotters can be stacked.

sold by Seal, or a large plate of metal with weights. A sheet of wood will not work as well, because a function of the cooling weight is to absorb heat rapidly. Allow the press to warm up to 200°F but no warmer. To check whether the thermostat is accurate, use Seal temperature indicator strips. These cunning little inventions are strips of paper with bands at each end that melt at 200° and 210°F, respectively; if one melts and the other does not, the temperature is in the correct range.

Predry the blotter or board by heating it in the press and opening and closing the press a few times. This drives the moisture out. But is the blotter not already dry? Take our word that it will not be completely dry. Even if you cannot perceive it, paper contains moisture absorbed from the atmosphere.

Make a sandwich of sheets in the following order: bottom — dried blotter or board; middle — the print face-up; and top — cover sheet or release paper. For flattening prints (but not dry

Figure 3.5
Set the plate on top of the blotters.

Figure 3.6
Allow to dry for a day with weights on top of the metal plate.

Figure 3.7
The prints turn out flat after cold-pressing.

mounting) you can use a good grade of smooth-textured acid-free paper in place of the release paper.

Before the sandwich goes into the press there is one point we cannot emphasize enough: be clean. Any speck of dust will become a permanent part of the image if it gets onto the print before the press closes. The same holds true for folds and wrinkles, all of which must be smoothed out at this point.

Put the sandwich into the press and close it for 10 seconds. Open and repeat several times. The circulating air vents moisture from the print and prepares it for final flattening.

Close the press tight for 30 to 45 seconds. Time this closing. Then open the press, pull out the print, and stick it immediately under the cooling weight. Have a couple of sheets of cool blotter or board under the weight ahead of time, so that the print can be slipped between them.

The print has to cool under pressure to get the full benefit of hot pressing. If not kept flat while it cools, it returns to its original shape or something approximating it.

Using these procedures, a large number of prints can be flattened very quickly. Setting up an assembly line makes the work go faster. For example, while the press flattens one photograph, the previous one can be taken out from under the weight and added to the pile of already flat prints. Using this method, one of us has flattened several hundred badly curled prints in the course of an evening.

You will want to remember the following precautions.

☐ *Kraft paper*. Do not substitute it for the blotters or board. Heat and water vapor transfer residual acidity from this paper right into the print, and the coarse texture can emboss the print surface.

☐ *High temperatures*. A press with a faulty thermostat is a common occurrence. If the press is too hot it can actually scorch the print. Be sure to verify the accuracy of the thermostat at least once.

☐ *Heat-sensitive media*. Do not try to flatten prints that incorporate these media: Van Dyke brown, Cibachrome, resin-coated photographic paper, pastel, crayon, silkscreen, oil or acrylic paint, wax rubbings, or work done on parchment. Avoid flattening prints with raised or embossed surfaces, since the press will flatten the raised surfaces as well as the paper.

☐ *Other media*. Also, use discretion with other media. If uncertain about an object's composition or its ability to withstand heat and pressure, defer flattening it until further research is done.

CHAPTER 4

Techniques for Conservation Mounting

Mounting means ways of fastening a print to the backing board of a mat. Let us assume for the sake of discussion that we have at hand a conservation-quality mat already cut and assembled, and a flat print, processed to archival standards, ready to be mounted in the mat. At this point we can still undo all our careful work by improperly mounting the print. However, with a few pennies' worth of material and a pair of scissors we can do the job right. More elaborate techniques can also be used; they will be discussed later.

First, let us go over the factors that distinguish conservation-grade mounting from unsafe mounting. For practical applications we can summarize the principles briefly.

- ☐ *Acid-free adhesives.* An adhesive, paste, or dry mount tissue that touches the print needs to meet the same requirement for pH neutrality as do board and other materials. Many adhesives, like rubber cement and pressure-sensitive tapes, contain sulfur compounds that will form sulfuric acid. Always choose an adhesive of known chemical composition that has been tested for archival mounting and suitability.
- ☐ *Reversibility*. Sometime, somewhere, somebody will have to take the print out of the mat. Take this as an article of faith. To this end, use mounting methods that are reversible. In the case of hinge pastes, use the kind that redissolve in water. Do not put any tape on the front of the print where it will leave marks after removal.

☐ *Adequate support*. Adequate, but just barely so — that is the key. Except for dry mounting, try to use mounting methods that are not any stronger than the print paper. The mount should tear or give before the paper does. This confines the damage to the mount, rather than involving the print. On the other hand, of course, we want mounting that will hold the print in place so that the print does not fall down in the frame or slide out of the mat.

☐ *Inspection*. The print should be mounted so that someone can easily open the mat and check it for such damage as insect holes, fading, and fungus. Make it possible to lift the print away from the backing board without its having to be remounted so that the back can be checked. In the case of prints dry mounted to a backing board, the board becomes a part of the print for practical purposes. To eliminate the need to turn a print over, make a note someplace inside the mat when there is no information on the back of the print.

☐ *Freedom of movement*. We mentioned that fluctuations in humidity change paper size. Give the print room to expand and contract. Otherwise, stress can cause it to buckle and possibly even to tear. In practical terms this means mounting along one edge only, or using methods that do not apply adhesive directly to the print. Again, dry mounting is an exception that we will discuss later.

With these principles in mind, we will discuss adhesives that are safe to use, some practical mounting methods and their uses, and dry mounting. In one chapter, we cannot include the results of all the research done on these topics. Instead, we will give you basic tools that meet most needs. You will come across other suggestions for mounting. Many of them are excellent. For safe mounting, use such suggestions as incorporate the principles outlined here.

ADHESIVES

Here are recipes for two kinds of paste that can be safely used for hinge-mounting prints. They require some preparation but can be made up with a minimum of time and the equipment found in most kitchens. A word of warning: do not simply smear the paste onto the back of a print and plop it onto a mounting board. Follow the instructions for hinges in this chapter.

Methyl Cellulose

This paste has four features that make it ideal for conservation mounting: it is chemically neutral, it does not have to be cooked, it keeps indefinitely after mixing, and it can be redissolved in water to allow removal of hinges from the back of a print.

You can get methyl cellulose powder from a variety of sources.

Light Impressions, Process Materials, and Talas all sell it through the mail. Do not use vinyl wallpaper paste that has methyl cellulose in it, because the manufacturers add ingredients that among other things keep it from redissolving. Ingredients needed:

8 teaspoons methyl cellulose powder

16 oz distilled water: 5 oz heated, 11 oz chilled

Heat 5 ounces of distilled water to about 190°F in a stainless steel or Pyrex glass container. Stir in the methyl cellulose powder. (Be sure to stir the powder into the water, and not vice versa.) The powder will not dissolve, but it can be dispersed. Add the rest of the water, which should be chilled to 32°F, or to as near freezing as is practical. Stir until the powder dissolves completely. Let the mixture stand for 20 minutes, stir thoroughly, and pour into a clean storage jar; cap. Label "Methyl cellulose — stock solution."

You now have enough paste stock for thousands of prints. To make a working solution, thin the stock with distilled water until a workable consistency results. For convenience, make up only a little bit of working solution at a time.

A thin, even application gives the most reliable results. Methyl cellulose has what adhesives experts call a high degree of tack. That is, it is a very strong adhesive, perhaps too strong for some applications. It works fine for sturdy papers like modern double-weight photographic papers, but you might want to consider mixing it half and half with rice or wheat starch paste for thin or brittle prints, so that the hinges do not pucker the paper. Methyl cellulose absorbs moisture from humid air. Fungus will not grow on the paste itself, but it can grow on the paper, so in tropical latitudes or otherwise moist conditions fungus has to be considered.

Rice Starch Paste

Until methyl cellulose came along, rice starch paste was the single most highly recommended adhesive for print mounting. A great many conservators still favor it. You can buy rice starch powder through the mail, from art stores, or even in local health food stores. Ingredients needed:

2 heaping teaspoons of refined rice starch powder

8 oz distilled water: 1 oz cold, 7 oz boiling

In a glass, stainless steel, or porcelain saucepan (with no chips) mix 1 ounce cold distilled water with the powder and stir until there are no lumps. Pour in 7 ounces boiling water, stirring constantly. Cook for 25 to 40 minutes over low heat, stirring frequently. Stop when the paste changes from a milky to a glassy consistency. Allow to cool to room temperature before using. For a lighter consistency, thin it 1:1 with water just before using.

Store refrigerated in a capped jar that has been rinsed with boiling water to sterilize it before the paste goes in. The paste will keep for two days under refrigeration. After that, discard it. Not only does it start to smell bad, it also loses its tack.

As an alternative to cooking, you can use the rice starch powder, Zen Paste, sold by Archival Products. This has already been cooked and then freeze dried, so that you need only add water.

Thymol

Rice starch paste not only goes bad quickly, it also provides an ideal growth medium for fungus when conditions are damp. This leads us to a digression on the subject of the chemical fungicide known as thymol. Thymol crystals in fumigation chambers are widely used by paper conservators to kill fungal growths on paper. Experience has shown that thymol vapors also attack photographic emulsions, oil paints, parchment, and vellum. This has caused the American National Standards Institute to recommend against using thymol as a fungicide for photographs (IT 9.2-1988).

We endorse the prohibition on fumigating photographs with thymol; research has conclusively shown that exposure to its vapors will soften and disintegrate the emulsion. On the other hand, nearly all recipes for rice starch paste that have been published in the technical literature call for the addition of thymol. Very small amounts of thymol in the paste extend the keeping time to three weeks, and that alone is a significant advantage. Furthermore, thymol protects the paste against fungal attack at a later date. We believe that the minuscule amount of exposure created by thymol in the paste will not harm photographic prints.

To use, stir thymol crystals into methyl or denatured alcohol in a small chemical beaker to make a 20% solution by volume. For example, to 4 fluid ounces alcohol add enough crystals to increase the volume to 5 fluid ounces. Add 1 teaspoon of thymol solution to the paste after it cools. The thymol solution darkens with time, so mix only a little. And heed the warnings on the package: its vapors should not be inhaled.

Wheat Starch Paste

Wheat starch offers an alternative to rice starch. To mix, you need these materials:

12½ teaspoons wheat starch

7 oz distilled water

10 drops thymol solution (described above)

Soak the wheat starch in cold water for ½ hour; then cook in a double boiler at a slow boil for another ½ hour, stirring frequently. As the starch thickens and becomes opalescent, it will go through a very stiff stage and then become easier to stir. At the end of the 30 minutes, add the thymol solution and place the pan directly on the heat source (without the lower pan of the double boiler). Cook rapidly for 2 minutes, stirring rapidly. Store the paste in a sealed jar in a cool, dark place.

The paste must be prepared for use. Separate out a small amount of the stock and push it through a strainer with a spoon to make dilution possible. Add distilled water to bring it to a thick, creamy consistency.

Deciding Between Methyl Cellulose and the Starch Pastes

Certain features make one or the other of these pastes more useful for specific kinds of mounting.

☐ *Strength.* Rice starch is the weaker, and is preferred for use on delicate papers.

☐ *Water absorbency.* Rice starch does not draw moisture from the air. Methyl cellulose does. Hinges adhered with rice starch seldom cause puckering from absorbed moisture.

☐ *Custom.* Rice starch has been tested by centuries of experience, and methyl cellulose is a modern synthetic that so far has proved quite stable.

☐ *Keeping properties.* Methyl cellulose keeps indefinitely. Starch has to be used soon after mixing.

☐ *Tack.* Methyl cellulose has greater tack and holds large pieces more firmly and grabs more quickly.

☐ *Resistance to biological attack.* Everything from silverfish to fungus will eat starch, but we do not know of anything that likes to eat methyl cellulose.

☐ *Convenience.* Methyl cellulose wins hands down on this count.

CORNER POCKETS

Let us start mounting procedures with the simplest one of all, corner pockets. For many of us, our first contact with photography came while leafing through an old family album with black pages and little black paper corner caps. Many of the pictures had become torn, dog-eared, and stained, and sometimes, no matter how carefully we turned the pages, a heap of snapshots would fall in our laps. It all added to our confusion about who their subjects were.

While the old photo albums had their charms, the type of corner pockets we describe here claim only a distant kinship to those old-fashioned, unreliable photo corners.

Besides simplicity of use, corner pockets offer other advantages: they do not require any adhesive on the print itself, and they leave the print free to expand and contract with fluctuations in humidity. You can buy excellent acid-free paper corners and transparent polyester corners.

Because the polyester corners are clear, they are fine for use in albums or behind mats. They have a pressure-sensitive adhesive already applied, and you simply fold them along the scribes and stick them to the backing board. The adhesive is quite strong and durable. Make sure when buying polyester corners that the adhesive is designed not to come in contact with the print.

The precut paper corners have to be put on with linen tape. They are good for behind a mat, but they are not very attractive in an album where they can be seen.

If you want to make your own corners, it is a simple task. All you need are some sheets of acid-free paper like Perma-Life, a pair of scissors, a roll of nonacidic linen or paper tape, or Filmoplast P-90 acid-free pressure-sensitive tape. The last is as easy to use as Scotch tape.

Follow the diagram in Figure 4.1. Cut four squares of paper an inch long on each side. Fold each square diagonally. Lay the print on the backing board, and check the position by closing the window mat. Reposition if necessary, and put a weight on the print. Open the mat.

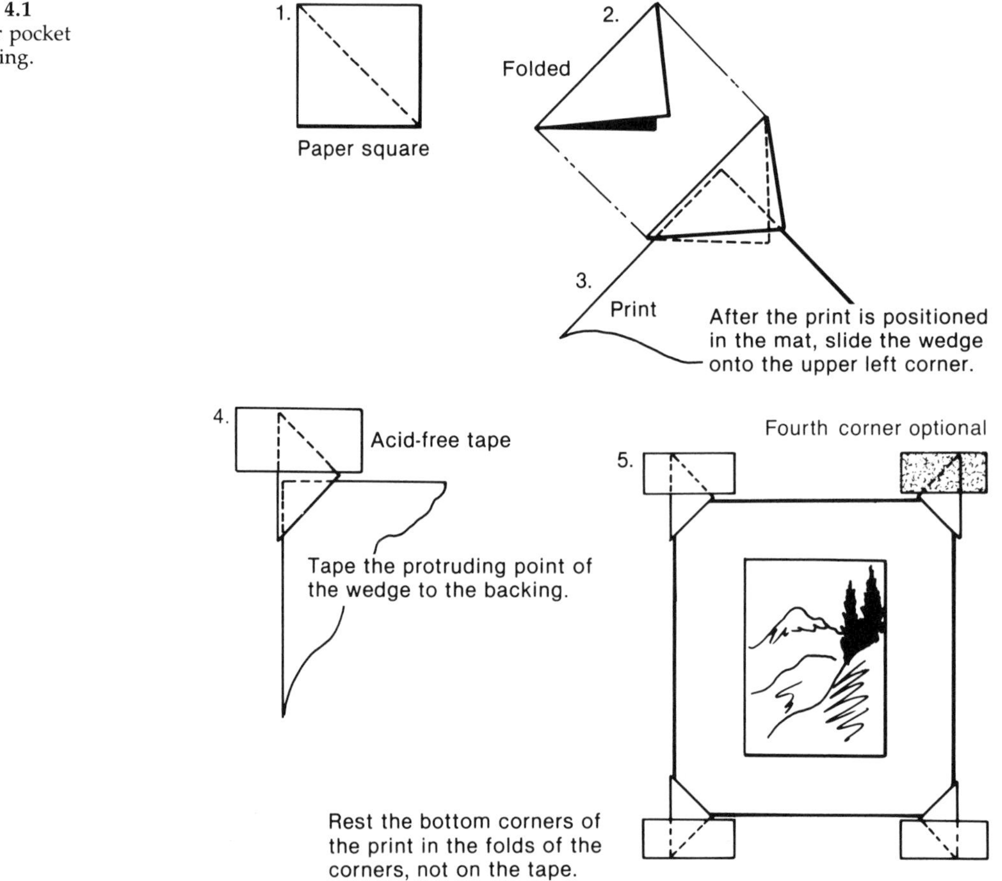

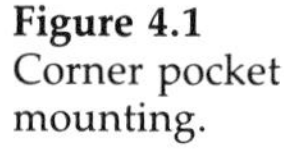
Figure 4.1 Corner pocket mounting.

Slide one of the paper wedges onto the upper left corner of the print at the corner. One half of the wedge is now behind the print, the other half in front. The tip of the triangle faces right at the top edge of the paper. The fold of the wedge is butted firmly against the side of the paper.

Lay tape across the protruding part of the wedge, parallel to the top of the print. The linen tape adheres to the backing board and the top of the wedge, forming a pocket. Leave a little room between the top of the print and the edge of the tape. Repeat the same procedure on the right side.

When it is likely that a print will do much traveling after mounting, or otherwise be exposed to fluctuations in relative humidity, one of the top corners can be left free to allow for expansion of the paper. Very large prints, or ones on brittle paper, will not stand up well to the strain of corner pockets.

On a print mounted this way, the paper of the photo corner directly touches the emulsion of the photograph, so it makes good sense to use only the best. In the section on matting, we discuss the characteristics of paper in detail; here we want to suggest that photo corner paper be acid free, lignin free, and nonbuffered.

THE PRINT POCKET

This variation of corner pockets also avoids adhesive contact with the print. A print pocket holds the picture firmly and supports brittle, delicate objects better than do corner pockets. It needs a window mat to conceal the edges of the paper, and thus it is not suitable for floating an image in the center of a window.

To make a print pocket, cut a strip of Perma-Life or similar acid-free paper 1 inch wide and about 1 inch longer than the print bottom. To get a straight, nice-looking edge, do the cutting with a ruler and razor-sharp knife rather than with scissors (see Figure 4.2).

Score the strip lengthwise down the center and fold along the score. For a scoring tool you can use a graphic arts burnisher, but a dull table knife is equally effective. Put the print in position on the mat back board. Slip the V-shaped strip snugly against its bottom.

Tape the protruding ends of the strip to the back board with acid-free linen tape next to the edge of the print. Make corner pockets for the two top corners as described earlier. To get additional support in the center, one or more pieces of tape can hold the folded paper strip to the back board. Make certain that the tape does not touch the print face. When the mat is closed on the print, it holds the pocket flat.

You can also use mounting strips, made from Eastman Kodacel plastic with a 3M adhesive on the back, to support the bottom of a print. The print is inserted into the channel of the plastic strip, which is available in 20-inch lengths and which can be cut to length with scissors.

Figure 4.2
A print pocket.

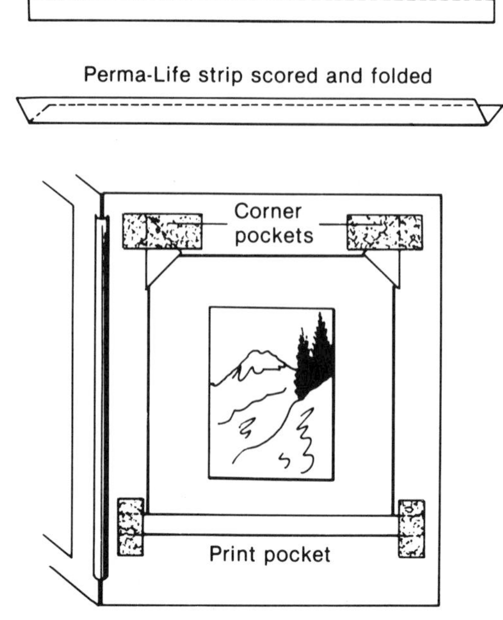

Pockets formed by the folded strip. The acid-free tape is shaded for clarity.

JAPANESE TISSUE HINGES

Hinges made from Japanese tissue paper* represent the most exquisite and graceful way yet devised to mount a print. Large prints on thick paper or gauzelike paper objects can both hang securely from these hinges on the back board of a mat. Yet, despite their strength, you will find that they appear nearly invisible upon inspection, so completely do they blend with the paper of the print.

To make tissue hinges by the simplest, most direct route, use an Insta-Hinge® kit from Archival Products. It includes three strips of tissue impregnated on both sides with wheat paste and fungicide. Each strip is already water cut to provide the requisite feathered edges, and is long enough to hang many prints. This is an excellent option, though it is more expensive than working from scratch. Correct use of the kit requires that you understand the techniques of making hinges from scratch anyway, so we suggest that you read the detailed instructions on the following pages all the way through.

Because the techniques might seem unfamiliar, we have made the instructions very detailed. Above all we want to stress preparation, because ease of working depends upon having all materials readily available.

* Japanese tissues of this kind were formerly called "rice paper" since their first importation into the West over a century ago. In origin and use they have no relation to rice, so we use the less misleading term *tissue*.

Three kinds of hinges have been developed for particular applications. They are the hanging or pendant hinge, the folded hinge, and the reinforced T hinge. A feature they all have in common is that they are made from strips taken from a large sheet of tissue by a technique of water cutting, which we will describe after first talking about tools and materials. To get the full benefit of tissue hinges, the paper must be water cut.

Tools and Materials

Have everything conveniently spaced around a clean and well-lighted work surface. Liquids should sit on a separate table or shelf so they cannot accidentally spill onto the print. Materials needed:

- ☐ Japanese tissue paper in sheets (Correct strength is important; choose according to the weight of the print. Suitable types are Mulberry, Sekishu White, Kitakata Buff, Uda Thin, and Kizukishi, in order of thickness.)
- ☐ Rice starch, wheat starch, or methyl cellulose paste (mix in advance)
- ☐ Bone knife or burnisher
- ☐ 3 sable watercolor-type brushes, 1 pointed and 2 flat
- ☐ Hard lead pencil
- ☐ Small pieces of acid-free blotter
- ☐ Small sheets of Mylar
- ☐ Razor-sharp knife or scalpel
- ☐ Tweezers
- ☐ Straightedge
- ☐ Print positioning clips, *or* small sandbag weights
- ☐ Drafting brush
- ☐ Water container

A white cotton glove like the kind sold by Kodak for slide mounting will prove useful for handling the print. In the case of fragile paper, it would also be a good idea to have a sheet of thick mat board to support the print.

You can order package selections of fine-quality tissue paper, especially for the purpose of making hinges, from most of the mail order houses listed in the section on suppliers.

Label or mark the brushes to avoid interchanging them. Mark one of the flat ones for paste, the round one for water cutting, and the other flat one for smoothing the hinge. Half- or three-quarter inch flat brushes will be the most useful widths.

An X-acto or similar style knife with replaceable blades can be picked up nearly anywhere that sells art supplies. You can sew weights beanbag style and fill them with lead shot from a sporting goods store, or use the print-positioning clips sold by Light Impressions.

Use the drafting brush regularly to sweep off the work surface.

Water Cutting

The Japanese tissue paper used for hinges has excellent tensile strength combined with extreme thinness. Due to its extremely long fibers, this paper can be water cut to produce feathered edges. The feathering causes the hinges to blend inconspicuously with the back of the print. The tapering that results reduces the chance that any visible mark will appear on the front from the added thickness of the hinge between the print and the back board. You can dispense with water cutting when mounting prints on a thick stock — for example, a salted-paper print on watercolor sheet.

Hold the tissue up to a light. You will see a grid of lines, some closely spaced and others, at right angles to them, that are widely spaced. The widely spaced lines are called *chain lines*. Very lightly mark a corner with an arrow to show the direction of the chain lines. To get maximum strength, all the strips you cut should have their lengths parallel to the chain lines.

Lay the straightedge near one edge of the tissue and parallel to the chain lines. Dip the cutting brush in water and draw it along the straightedge. This weakens the paper. Run the bone knife or burnisher along the edge to make an impression. Hold the straightedge down firmly, and pull the exposed strip up and *away* from it to the side.

Discard this strip. Never use the edge of the sheet, even if it is already deckled, because the deckled part is just as hard as a razor-cut edge.

Now look at the torn edge. You will see numerous tiny fibers along the edge, some extending out farther than others. These give the hinge its feather.

Move the straightedge about an inch to the side and repeat the process. You can cut several strips of this width for practice. Save them for reinforcement strips.

Now cut several strips for hinges. These should be between 3/8 and 1/2 inch wide. Use the same method to cut the strips down to their desired length. Hinges properly measure about 1 inch long. Reinforcements run between 1 1/2 and 2 inches long. Discard scrap.

Hanging (Pendant) Hinges

This hinge consists of a hanger strip and a reinforcement (Figure 4.4A). It shows from the front, so use it only when a window mat hides the print edges.

Clear the work area with the drafting brush. Bring out the mat and print, and position the print as it should go on the back board.

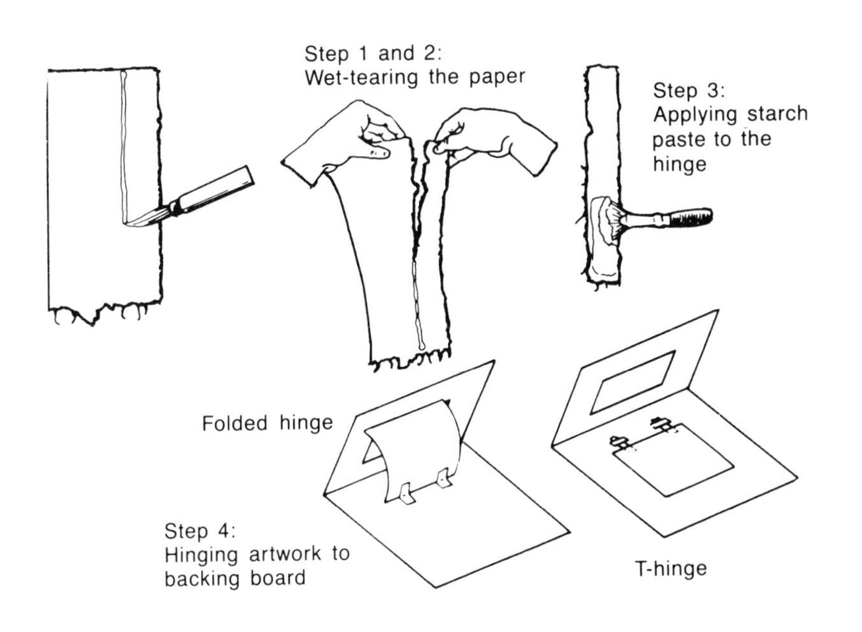

Figure 4.3
How to water-cut a hinge.

After this the print should move little or not at all. Fix it in place with print-positioning clips slid on from the side like paper clips, or use weights to hold it down.

Apply paste one-third the length of a hinge strip. Work on a small scrap of backing paper, and do not get any paste on the other side. If you dip the brush in water first, the paste flows more smoothly.

Pick up the hinge strip with tweezers. With the other hand lift one corner of the print and attach the bottom third of the strip to the back. Make sure that all the part with paste on it is behind the print.

Brush paste the length of the reinforcement strip. With tweezers, lift the protruding two-thirds of the hanger strip and brush paste on the back. Smooth it onto the backing board with the other flat brush. Pick up the reinforcement and put it on top of the protruding hanger, parallel to the top of the print but about ⅛ inch away so that the print later can be lifted for examination. Smooth down.

Slide a small square of Mylar behind the print, in back of where the hinge is adhered. This protects the backing board from moisture and ensures that the paste will not stick the print and the board together. The hinge has to dry under pressure for best adhesion, so cover with a weight or clip to put pressure on it.

Repeat the same procedure for the other hinge.

A word about the paste: better too little than too much. Apply it very thin. You get better adhesion and run less risk of smearing it.

If you mess up, do not panic. Turn to the section "Removing Japanese Tissue Hinges."

Extremely large prints require three or even more hinges along

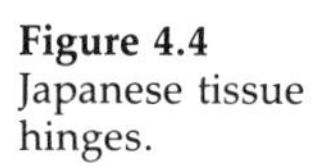

Figure 4.4
Japanese tissue hinges.

A. Hanging or pendant-style hinge, visible from the front of the print, so the edges must be concealed by overmatting. Acid-free tape can be used for the reinforcement crosspiece in place of Japanese tissue.

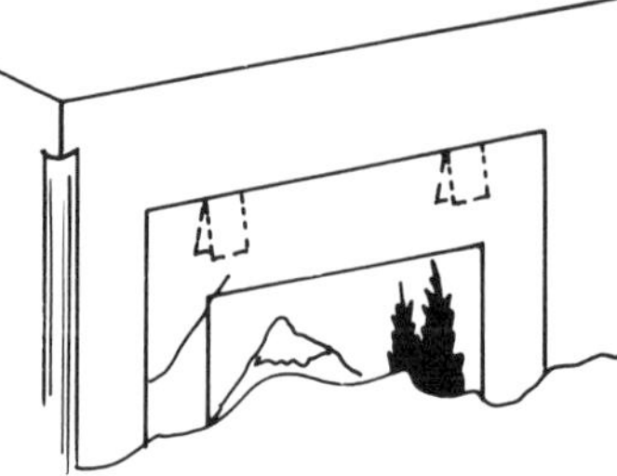

B. Folded hinge, completely concealed behind the print. Note that the fold must be flush against the top to avoid creasing edge of paper when the print is lifted to examine the back.

C. Two views of reinforced hinge. Reinforcement strip should be snug against the fold of the hinge.

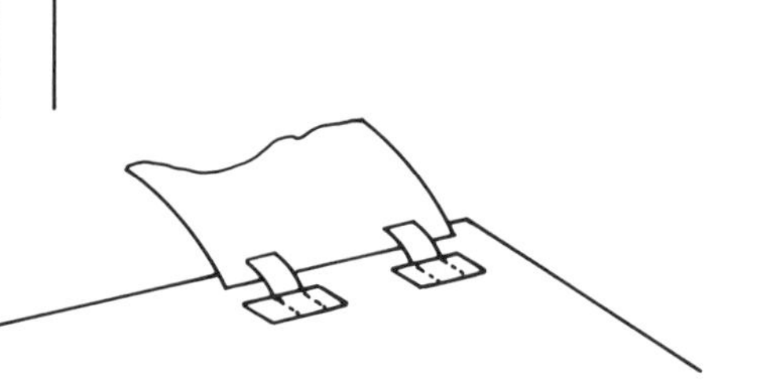

the top. In this situation, it is easier to put the first hinge in the center.

Because the reinforcement cross-strip does not go behind the print, an alternate method to speed up work is to substitute paper tape for this part; the extra thickness here will not affect the print.

Folded Hinges

These resemble the little glassine hinges that stamp collectors use (Figure 4.4B). They do not show from the front, so they are suitable for floating a print inside the window with the edges showing, or simply for mounting on a plain back board.

Position the print on the back board and clip or weight it. Fold and crease a hinge strip one-third of the way down its length. Put the dry hinge in place behind the print. The longer section goes against the back board, and the fold goes *flush* with the top of the print.

With a pencil lightly mark the location of the bottom of the longer strip on the back board. Take the hinge and brush paste onto the front and back without getting any inside the fold. Pick up the hinge with tweezers, lift the corner of the print, and position the hinge using the pencil mark as a guide. Smooth down. Slide the blotter into the fold of the hinge, and apply pressure while it dries.

Repeat the same steps for the other hinge.

As a general rule folded hinges are applied only on the top of the print. However, when a print floats inside the window it may be desirable to restrain the bottom with very small hinges so that the print cannot flop out against the glass in the frame. Keep in mind that paper secured this way might buckle because of changes in its dimensions brought about by fluctuations in relative humidity, so only use very small hinges.

Reinforced Hinges

A reinforced hinge is simply a folded hinge with a reinforcement strip added inside the fold and parallel to the crease for strength (Figure 4.4C). Linen tape cannot be substituted for tissue as a reinforcement. Its thickness will cause an impression to appear on the face of the print. Like simple folded hinges, reinforced ones cannot be seen while the print is in the mat.

To make a reinforced hinge, apply the folded hinge as before and allow it to dry. Then lift the print and apply the reinforcement strip. This does not go on the print; it is attached to the back board at right angles to the hinge. To get maximum strength, make sure that the edge of it fits snugly against the fold. In this way the hinge cannot pull away from the backing partway before the pulling force encounters the reinforcement. Let the reinforcement dry under a blotter.

Removing Japanese Tissue Hinges

It is not difficult to remove Japanese tissue hinges, but do it right. Let us assume that we are going to change the mat on a hinged print, but the same principles apply to work in progress when something goes wrong.

Open the mat and lay a sheet of paper bigger than the print next to it.

Cut the hinge as follows. For a hanging hinge, hold the print down with one hand and slice through the tissue above the edge of the print. For folded hinges, carefully slide the knife behind the print and cut up through the fold. Use the same method for reinforced hinges.

Lay the print face down on the paper. For practice start by removing the hinge remnants from the back board of the mat. Use a brush dipped in water to moisten the adhesive. Do not get water any place except on the tissue, and use just a little. Let it soak in for a few minutes.

Gently pull up a corner of the tissue. Continue pulling slowly, down and toward the center of the hinge. When all the tissue has come off, many fibers may still remain. Pluck these off with the tweezers. Now that you have a feel for it, do the same with the remnants of the hinge on the back of the print.

You may encounter an old print on which the adhesive remains after the hinge comes off. This tip may work. First, make sure that the back of the print will not water stain by putting a small drop of warm water in one corner and blotting it. If the print does stain, do not proceed further. A little adhesive should not hurt the print, but if the adhesive must come off, take the print to a paper conservator. If there is no stain, put a lightly dampened piece of blotter over the adhesive stain and touch it gently with a hot tacking iron. This makes a little steam to soften the adhesive.

Moisten the blotter with a brush to cool it, and gently peel it away from the print back. Now very lightly scrape the adhesive with a knife, taking it off a little at a time. Do not abrade the paper surface. Repeat until no adhesive remains.

CHAPTER 5

Dry Mounting

Dry mounting has become the most common method of preparing photographs for display. The practice is so widespread that it seems the natural way to prepare a print for exhibition, but this does not mean that dry mounting is the only way to do it. Although it would be unrealistic to demand that everyone immediately stop dry mounting, problems resulting from this procedure need to be considered carefully before a photographer decides to present his or her work in this fashion. Then, if you do decide to proceed, take care to do the job in ways that minimize possible problems.

TYPES OF DRY MOUNTING MATERIALS

Manufacturers of dry mount bonding materials currently offer a number of products that can be broken down into two broad categories. First are the types that require a heated press to activate the adhesive through a combination of high temperature and pressure; these are what most people think of first when dry mounting is mentioned. The older kind has two layers of adhesive, one on each side of a very thin tissue-paper core. More recently introduced types consist of a single film of adhesive only. Both types are still sold; the kind with a paper core is usually called *dry mount tissue,* and the new kind is designated *dry mount adhesive* (Figure 5.1).

Lately a new family of cold-mounting products that can be used without a press has appeared. These pressure-sensitive materials have a high-tack adhesive that works at room temperature with the application of a modest amount of pressure. All that is required to make them ready for use is to peel off a cover sheet. Some are simple adhesives with release paper, some have adhesives and cover sheets on both sides of a paper core, and others come already applied to a substrate such as mount board or wood-chip board.

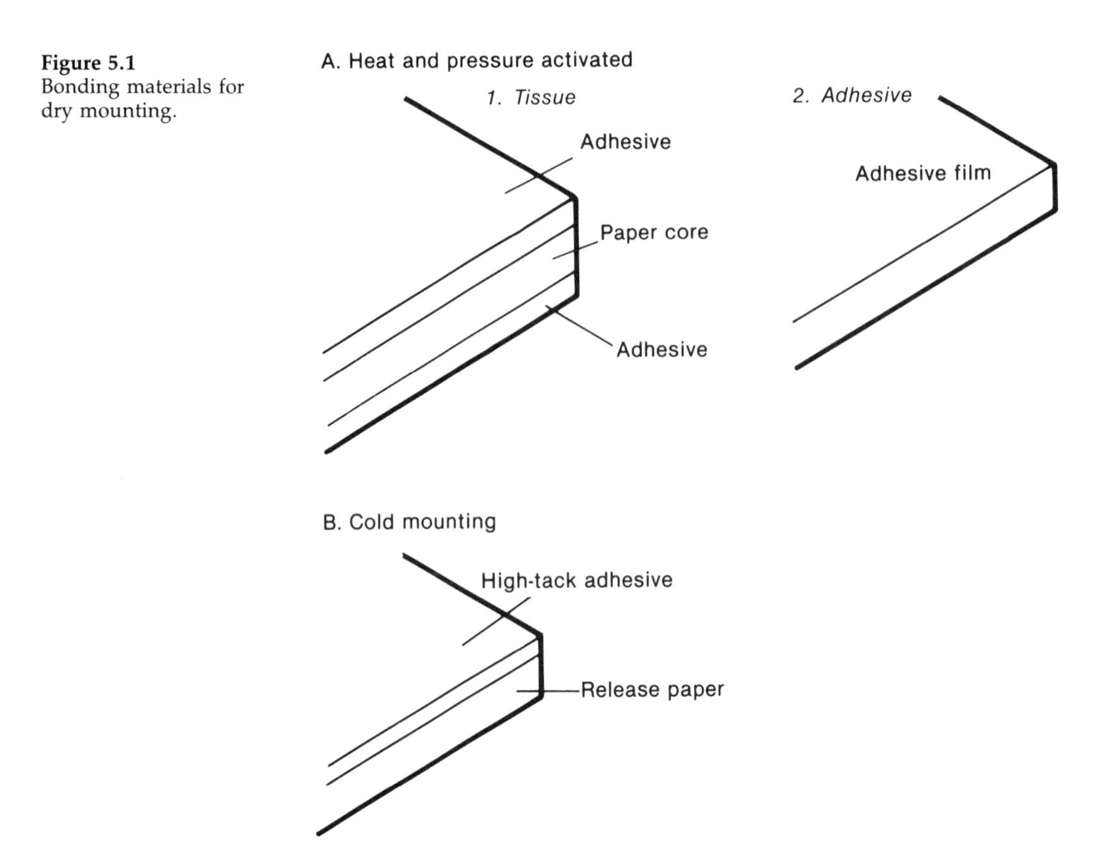

Figure 5.1
Bonding materials for dry mounting.

Most materials for cold mounting consist of low-grade combinations of adhesive and paper and do not meet archival standards. We discuss an exception later in this chapter. Another material for cold mounting comes in a can: it is an adhesive to be sprayed on the back of the print, and it too should be avoided when archival-quality mounting is desired.

PROS AND CONS

Numerous factors have combined to make dry mounting popular. An important one is neatness. The term *dry* distinguishes this kind of mounting from an older technique used on albumen prints. Because they were made on thin paper, albumen prints needed a rigid backing in order to lie flat, and the common practice was to paste them onto thick sheets of mount board [1, pp. 92–95]. Not only did the pastes have a water base, but the print itself also had to be slightly damp. So the new technique, which did away with all this mess and moisture, naturally came to be called dry mounting.

Dry mounting has the added benefit of speed. A skilled worker can dry mount a print from start to finish in a matter of minutes. Predrying the print and mount board, a necessary precaution to ensure flatness and good adhesion, removes most moisture from both and may slow down chemical deterioration, though this will be reversed if storage takes place in moist surroundings.

Dry mounting also gives an aesthetic effect preferred by many photographers. The absolute flatness that can be achieved reduces visual interference from surface reflections and gives the print greater apparent depth. This effect will be seen best on those prints with a very glossy surface, especially Cibachromes and prints that have been ferrotyped during drying.

From a conservation point of view, there are some other reasons you might choose to dry mount a print onto board. The extra support of a rigid backing does mean that the print will not be easily folded or creased. Adhering it to a flat support will also protect it against being rolled up, which is often damaging to the emulsion.

The argument has been widely advanced that dry mount tissue acts to increase a print's protection from impurities in the mount board. This may be true, though we are not aware of any studies to support the idea. It does not make much sense to permanently attach a print to poor-quality board and then to try to protect it with a thin sheet of tissue. It is better to choose quality board that will not damage the print, and then consider whether dry mounting is the route that should be taken.

Curators and knowledgeable collectors will never dry mount any work that comes to them unmounted. The professionally acceptable ways to mount collected photographs still remain paper hinges or acid-free corner pockets.*

We strongly recommend as a general rule that you never dry mount a print unless you made it yourself or unless it is a duplicate or copy of a print that already reposes in safe storage. Naturally, individual photographers may choose whether or not to dry mount their own productions, but they should keep in mind that doing so might decrease the value of their work, and in the long run this procedure may make restoration considerably more difficult.

Major objections that preservation experts have against dry mounting are (1) that adhesives can have unexpected adverse effects; and (2) that, because dry mounting is not water soluble, the print cannot be released with any guarantee of safety.

The following cautionary statement from 3M, the manufacturer of Scotch brand adhesives and tissues, deserves to be more widely disseminated.

> We know that most of our adhesive products have an indefinite age life by virtue of our accelerated aging tests and natural aging experi-

* The only exception that we can think of occurs during the restoration of albumen prints, which often requires transfer to a new mount board. For this we recommend the method developed by David Kolody and described in detail by James Reilly [1, pp. 97–100]. This wet mounting technique is a specialized application that belongs properly in the realm of restoration techniques and is not covered in this book.

> ence. However, we *do not* have a test that can accurately predict how a product will hold up after 50 to 100 years, for instance. In other words, we cannot *recommend* our products for archival applications.
>
> . . . These products are designed for general purpose use in bonding applications on items of limited value where the bond should be long aging and permanent. They *are not recommended* for use on art of significant value and considered an investment because (1) the use of full mounting techniques will reduce the value, and (2) the resulting bond may not reverse without causing physical damage to the item (italics in original) [2].

This frank statement, coming as it does from a vendor of dry mount materials, deserves to be taken very seriously.

Among conservators it is nearly universally accepted that any mounting technique that is safe for archival use must be completely reversible, and preferably soluble in plain water. It should leave no residue on the print. Further, the properties of any adhesive should be known completely, and this pertains especially to the aging properties.

Water will not touch dry mount adhesives, though these adhesives can usually be dissolved with either toluene or acetone, because they are usually made of shellac, lacquer, or a wax base. Seal markets a material claimed to be designed so that its action is reversed when the print is reinserted in a hot press. We cover this material in more detail later.

Adhesives for dry mounting have not been around long enough for anyone to be sure that they will be completely safe 100 years from now. We have seen dry-mounted photographs from the 1940s that seemed to be in perfectly good condition, but this may change in the next 60 years. Whether or not some unforeseen problems will arise, only the passage of time will tell.

Anytime that a print goes into a dry mount press, it runs a certain amount of risk no matter how carefully it gets handled. Among the things that can happen by accident are: dry mount adhesive may work its way onto the surface of the print and stick there; if the press heats up too much because of a faulty thermostat, the photograph can be scorched; and if even a minute dust speck or crumb of foreign matter slips unnoticed into the press, it will cause pits or bumps on the print's surface. In short, the heat and pressure that a print encounters in the press represent extremes of stress that a photograph will undergo at no other time.

Obviously, when you have the option of replacing a photograph by simply going back to the darkroom, these problems do not create the same dangers as when working with a substantial investment, or with a photograph that may be irreplaceable. Also keep in mind that photographs are the only kind of print that collectors will even look at when completely attached to a backing. Any other kind of art on paper, such as an etching, a drawing, a lithograph, or even a postage stamp, is considered to be effectively destroyed when mounted.

There is no disputing that some of the most eminent names in photography — people whose work brings thousands of dollars — do in fact dry mount their prints, and their work seems to suffer not a whit in the marketplace. Whether or not this will continue to be true, no one can say for sure.

Some photographic papers react badly to heat. This is especially true for the resin-coated types of paper, for Cibachrome prints, and generally for color prints. The problem gets more serious with Cibachromes, because they represent the most stable color medium we have. Too much heat causes their solid plastic substrate to warp and curl, and the stresses induced in the press by "heat, rough mount boards, and matte finish release papers cause the Cibachrome print to take on a mottled or bumpy appearance and lose a bit of its overall gloss" [3, p 27]. High temperatures can also cause color shifts in prints, although this effect cannot be predicted with certainty; for these, choose a mounting material that does not require temperatures higher than 205°F, and closely monitor possible color changes.

A good argument can be made that dry mounting makes a print more susceptible to damage. Its extra weight means that when it gets dropped, the mount board will fall harder; any damage that results becomes, in effect, damage to the print itself once it has been permanently bonded to the mount board.

Now that we have pointed out the host of problems associated with dry mounting, we want to add that doing it well protects the print better than doing it poorly. Keeping that in mind, we will discuss next the least harmful ways to dry mount.

FORMATS

The various styles of mounting fall into four categories: back mounting with photo paper, flush mounting, plain mounting, and mounting for a window mat.

Before choosing a style, consider the way that prints will be stored. When the plan is to store the prints in some kind of box, standardize the mount size. Not only does this make it easier to buy frames, but it also saves wear on the prints when they can be stacked neatly in storage.

Back Mount with Photo Paper

This is an elegant way to get a substantial, rigid print that can be hinged in the approved way, and yet that will lie perfectly flat. An unexposed sheet of photo paper of the same type and size as (or larger than) the print needs to be fixed and then archivally washed and cleaned. This procedure provides a mount board that is certain to last as long as the print itself. Because the emulsions on the front and the back absorb and discharge humidity into the air at approximately equal rates, this kind of mount overcomes the tendency to curl that comes from the difference in absorption between paper and emulsion. In effect, the two emulsions cancel each other out.

Flush Mounts

During the 1950s it became popular to trim mounted prints flush to the edge of the image. Take our advice and do not do it. It looks dated, and flush mounting leaves the corners vulnerable and the print unprotected on the edges. It also prevents overmatting, and it gives you no way to hang the print without gluing something on the back or nailing through the picture. Do not use this procedure.

Plain Mounts

On plain mounts the print is simply attached to a backing board; it usually has the borders trimmed off before mounting. This kind of mount is good for storage because it takes up less space. Try to leave at least 2 inches for protection on each side between the edges of the print and the mount. Prints stored in this way have to be interleaved to prevent scratches.

Window-Matted Mounts

The preferred way to dry mount is with a window mat. In this case the edges of the print should be left untrimmed. For greater ease in working, hinge the window to the backing board before dry mounting and close the mat to position the print. Open it and tack the tissue to the backing board. Place the open mat in the press. It is also possible to cut a window mat for a plain-mounted print — and not too difficult if you are fairly skilled. Measurements need to be quite exact, and so does the cutting; even so, about a ⅛-inch overlap into the image area has to be allowed. An alternative is to cut an oversize window and float the print inside it. This allows a penciled signature to show, if desired.

WORKSPACE, EQUIPMENT, AND SUPPLIES

Almost any sturdy surface can be used for setting up a dry mount press. Figure 5.2 illustrates an ideal workspace for speed and ease of work. An important point shown in the drawing is that the sponge pad of the press (shown in gray) should be at the same level as the surface where the piece will rest for cooling. This means that the print can be kept flat while being removed from the press, an important consideration in ensuring that it cools flat instead of warped. Additionally, accurate positioning will be easier if the tacked piece can enter the press at the same level as that of the sponge pad. If you cannot construct a special workstand, you can also get the same effect by stacking up thick sheets of plywood next to the press. It would be ideal to have a permanent site for the press, because even small presses are heavy and likely to be dropped if moved often. Frequent lifting is also likely to cause the platen to get out of adjustment so that it may not apply pressure uniformly across the entire surface.

A suitable press involves a considerable sum of money. It will pay to buy the largest one that you will conceivably have use for and that still fits within your budget. It is a common regret that a

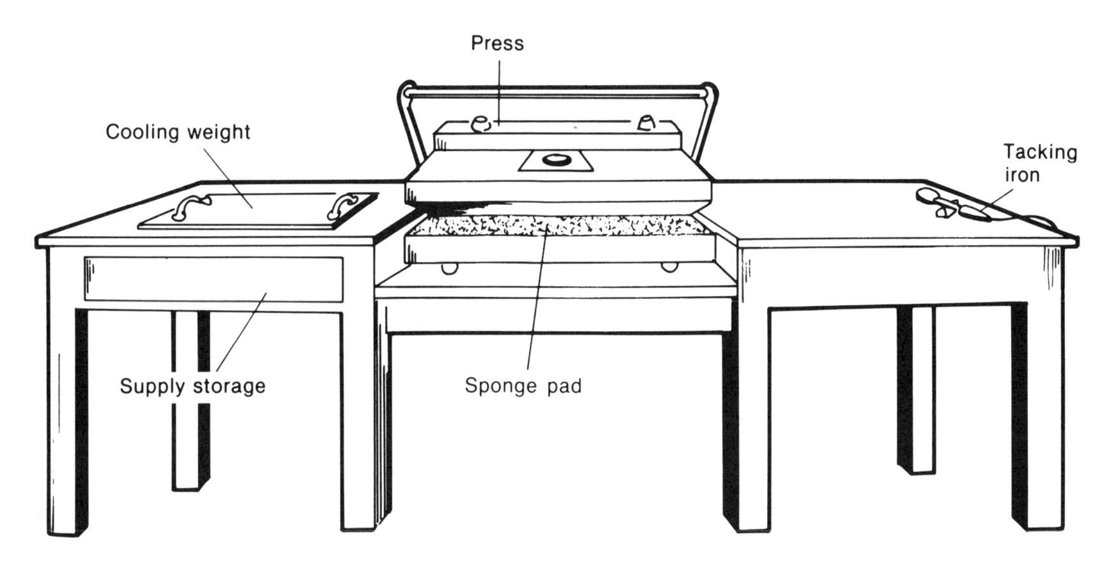

Figure 5.2
Dry mount workspace. Note that the press is recessed so that the sponge pad is level with the work table.

press is too small, and yet you may find that the times you will use a larger one do not justify the cost of a new and larger press. For some reason there does not seem to be much of a market in used presses, so do not count on readily trading up. On the other hand, if you find a used working press with the platen in good condition, it can represent a considerable bargain.

Kodak long ago dropped the manufacture of presses, and today most U.S. presses are sold by Seal and Technal. Quality is uniformly high, so shoppers can look primarily for price and necessary features. In addition to platen size, consider the following: rheostat versus thermostat temperature control; ease of pressure adjustment; and clearance required by the press handle. In general, the larger presses seem to work more smoothly than the smaller ones. (See Figure 5.3.)

There is little limit on how large a press you can get; big ones go up to 4 feet by 8 feet. The smallest used to be 8 × 10 inches, but now the smallest one that Seal makes, the Compress 110S,

Figure 5.3
The Masterpiece 500T from Seal has many of the features desirable in a large press: thermostat, pressure-adjustable platen, and easy-to-work locking handle. Note that the working area is open on three sides.

has a platen of 12 × 15 inches. Some large presses come with a vacuum attachment that eliminates the requirement for mechanical pressure by using atmospheric pressure instead. These also do away with the need for predrying. However, the initial cost and maintenance costs may run higher because of the compressor.

In considering size, keep in mind that prints can be mounted in sections. This means that the maximum size of print that a given press can handle is equal to double the depth of its platen by an indefinite width. Sectional mounting is not always a happy idea, however, because you can get marks in the middle of the print from the edge of the platen. (It does not always happen, but it can.) And predrying becomes more of a problem. Larger presses are more convenient; therefore they save time.

In addition to a press, the following tools will be handy:

- ☐ Tacking iron
- ☐ Cooling weight
- ☐ Tissue or adhesive
- ☐ Acid-free mount board
- ☐ Straightedge and hard lead pencil
- ☐ Kraft paper
- ☐ Release paper
- ☐ Seal temperature indicator strips (No. 908)
- ☐ Supplies for window matting
- ☐ Platen cleaner or solvent

Always keep the tacking iron on the little metal stand that comes with it, because otherwise you are likely to be looking at a scorch mark where it is most inconvenient. In case of failure, a tacking iron can be temporarily replaced with a household iron. However, a household iron is not good for overall dry mounting because of the impossibility of getting uniform pressure.

A cooling weight is an absolute necessity, especially for some of the new adhesives that bond when cooling; if these adhesives are not under pressure as they cool, the adhesion will not be complete. Any clean, flat, and heavy metal plate with the edges filed smooth will do. Though you can buy a nice one with handles from Seal, it does cost a considerable amount to ship.

When you buy tissue, buy it in a size at least as large as the largest prints to be mounted, and cut it down from there. Adhesives, on the other hand, can be pieced together without lines showing on the front of the print.

Lots of kraft paper is needed. If you already have a big roll for other framing uses, put it near the dry mount press. Use it to cover the work surfaces and to protect the platen and sponge pad from stray goop. Change it frequently. It is also needed for trimming adhesives like Fusion 4000.

The release sheet that Seal makes is a silicone-treated paper; almost nothing sticks to it, so it can be used for making sure that the print does not get dry mounted to the wrong thing. We explain later how it is used in the process of self-trimming.

Temperature indicator strips tell how hot the press platen really is, and they check the accuracy of the thermostat if you have one.

A special platen cleaner can be used, or, with proper ventilation, a solvent like acetone is acceptable. Make sure to turn the press off first, and let it cool before touching it.

FUNDAMENTAL TECHNIQUES FOR DRY MOUNTING

Once the workspace has been prepared and materials procured, turn on the press and allow it to warm up. Set the thermostat, if any exists on the press, for the temperature of the adhesive you want to use. Check the temperature with temperature indicator strips once the press has become hot; these are inserted in the closed press and then read according to the instructions packed with them. They are a good idea even if you have a thermostat, just as a double-check.

Before going into the requirements for working with some specific materials, there are some fundamental points that apply to all dry mounting.

Predrying

Both print and substrate should be predried before putting them into the press with the adhesive. Predrying removes moisture that can cause bubbles and poor adhesion. It also flattens out wavy photographs so that they are easier to position; also they can be measured more accurately. Remember, both print and mount board have to be predried. One exception to predrying is Cibachrome prints; if you do plan to put these in a heated press — which we do not recommend — at least do not predry them, and certainly do not try to use the press for flattening them. The other exception is nonporous mounts such as aluminum sheets.

Drying time varies. The thicker the material, the longer it takes. The minimum is about 45 seconds. With a little experience, you get so that you can feel whether the drying is complete. In a humid environment, drying takes longer.

Dry the print first so that you can work on tacking it while the mount board is in the press. Cover both sides with a carrier kraft paper, slide it into the press, close, and lock up. After 45 seconds open the press briefly and close it again. This lets water vapor escape. A last word of warning: make sure that the print is flat before closing the press, because if one of the corners gets folded over, it will be a permanent fold.

Paper materials absorb moisture from the air, so dry them just before mounting, not a day ahead of time.

Tacking and Trimming

Tacking is a way of spot-welding the print and tissue onto the mount board so that both stay in place as they go into the press. Tissue and adhesive need to be trimmed to very close tolerances before they go into the press.

Tacking and trimming are done differently for adhesives and for tissues. To tack tissue, cut a piece a bit larger than the print, lay the print face down on a clean paper, and cover it with the sheet of tissue so that the tissue edges extend beyond those of the print on all sides. Using release paper, hold the hot tacking iron against a spot about one-quarter of the way in from the edge of the print, and make sure that the tissue sticks to the back of the print. Do not touch the iron directly to tissue. With scissors or a sharp knife trim the tissue to size. Trim it about 1⁄16 inch smaller on each side than the print, because when the entire assemblage goes into the press, some adhesive oozes out around the tissue paper core.

When the tissue has been tacked on the back of the print and trimmed, position the print exactly in place on the mount board. Lift up the edge of the print away from the end where it has already been tacked, and reach underneath to press the tissue against the mount board with the tacking iron. The print is now tacked to the tissue, and the tissue to the mounting board.

There are several mistakes that seem to recur in this process, so take these precautions.

1. Tack the print in only one spot. Do not run the tacking iron along the entire edge, or crisscross the print, or make elaborate patterns. The same goes for tacking to the board. Too many tacks cause the print to pucker and bubble in the press, and sometimes crease it.
2. Make sure the tacking iron is not too hot.
3. Tack on the back of the print, not the front.

Dry mount adhesive can be tacked and trimmed in a single operation. Set the press to 210°F. Cut a sheet of adhesive 1½ to 2 inches larger than the print. Lay the print face down on a sheet of clean kraft paper and cover with the adhesive film. The kraft paper should be considerably larger than the sheet of adhesive.

Cover all three layers with release paper. This makes four layers, going from bottom to top: kraft paper, print, adhesive, release paper. Support them underneath so that they do not shift positions, put them all into the dry mount press, and close and lock the press for at least 20 seconds. This melts the adhesive, which will not adhere to the release paper although it does fuse with the back of the print.

Where the edges of the adhesive sheet extend beyond the print, they also stick to the kraft paper. Take the whole pile out of the press and turn it over so that the kraft paper is on top. Immediately pull away the kraft paper starting at one corner. Move slowly, while pressing on the center so that the print does not shift position. The adhesive will separate cleanly along the edges of the print because it has no paper core.

Discard the kraft paper immediately to avoid getting the adhesive onto anything else. Let the print cool and pull it off the release paper.

What results is a print coated over the entire back with adhesive, ready for mounting. You will not be able to lift up the adhesive to tack onto the mount board with a tacking iron, but in our experience this is not a problem. The adhesive has enough tack so that you simply need to press the print firmly onto the board and slide them into the press. In fact, handling a tacking iron around adhesive is a bit of a problem anyway; either the adhesive melts and sticks to the iron or it seems to evaporate, and in general causes no end of difficulty.

Cooling

Some dry mount materials bond as they heat, since that is when they melt into the paper fibers; others bond as they cool and solidify. The first type are more permanent, while the cool-bonding materials can sometimes be reversed using heat.

Both kinds need to be cooled flat and under pressure. This ensures that heat-bonding materials will not warp during cooling. It is also essential for cool-bonding materials because it allows the adhesive to maintain contact between print and mount until it has a chance to grip them both firmly.

A metal plate is ideal for cooling. Usually such plates are rigid and heavy enough to ensure that the print lies absolutely flat. Also, they are excellent conductors, meaning that they pull the heat out of the mount and transmit it away. The thicker the metal plate, the faster the print will cool, but do not get one that is so thick that you end up dropping it onto the print.

STEP-BY-STEP TECHNIQUES: THREE WAYS

The variety of dry mount materials from which to choose presents an embarrassment of riches, so that it is sometimes difficult to decide. Most are good for at least some specific applications.* Rather than try to become expert in the application of many different materials, we suggest using one or two of the three proved materials described here. This will provide the groundwork for trying other types.

The three products discussed below illustrate three basic types of dry mounting. The tissue is Seal Colormount; adhesives are represented by Fusion 4000 Plus; and, as an example of pressure-sensitive adhesive we have 3M's Scotch Brand Positionable Mount-

* Dry mount materials to avoid are the prebacked cold mount types, in particular the ones that use chipboard or pressed wood backing. All heat-activated materials have a neutral pH or nearly so, within a few tenths of a point of 7.0.

ing Adhesive (PMA). Spray-on adhesives have been omitted because they are not suitable for the preservation of photographs.

Seal Colormount

This is excellent and easy-working dry mount tissue that will hold resin-coated papers as well as conventional fiber-base papers. Colormount has a porous paper base with a 6.9 pH and a neutral pH adhesive that bonds permanently as it heats. The adhesive is activated at 190°F, and the recommended working press temperature is 205°F. Its thickness is about 2 mils; to help visualize this better, that is about as thick as the average heavy-duty garbage bag.

Colormount can be purchased in precut sheets and in rolls of different widths up to 100 yards long. The bond is permanent, so be very careful when positioning the print. The procedure is as follows:

1. Preheat the press to 205°F.
2. Predry the print and the mount board.
3. Tack the tissue to the back of the print and trim to size.
4. Tack the tissue to the mount board so the print is fixed in position.
5. Place the mount board on a sheet of kraft paper and cover it with release paper. In case some adhesive comes out around the edge of the print, it will not stick to the release paper.
6. Slide the entire package into the press. Close and lock the press.
7. Slide the board, print, and release paper out of the press onto the level surface next to it and place under a weight to cool.

When mounting in a window mat, have the mat already hinged before tacking. Close the mat and position the print. Use a weight to hold the print in place; then open the mat and lift one end of the print to tack it.

To protect the finish on high-gloss resin-coated prints, Seal recommends the use of its Cover Sheet rather than release paper; if you use this, trim the tissue extra-carefully because any excess will stick to the Cover Sheet. Do not use Cover Sheets with any other kind of print.

Time in the press varies with the thickness of the mount board and the real temperature of the press, which can fluctuate. Ninety seconds will provide a good trial for most boards. If bubbles show under the print after cooling, simply reheat the print in the press for a longer time.

Another tissue that you can use as an alternative to Colormount is Seal Archival Mount tissue. In general, you can follow the same procedures as with Colormount, with three important differences.

While the adhesive is a neutral 7.0 pH, the paper base of the

Archival Mount is 7.5–8.5 pH. It has an alkaline buffering agent, which makes it unsuitable for mounting color photographs.

Archival Mount is a thermoplastic material that bonds when cooling. In order to get good adhesion, when the mounted print comes out of the press you have to cool it under pressure, usually a flat weight. Cooling under pressure is critical. The recommended working temperature for the press is 175°F–200°F, but activation of the tissue occurs at 160°F.

Archival Mount also lets you remove the print from the mount by reheating. You may have to do this in stages. Slide the mounted print into the press, heat it, insert a silicon sheet behind the part of the print that can be lifted before it cools to prevent readhesion, and insert back into press. Repeat the process as many times as necessary.

Seal Fusion 4000 Plus

This is an all-purpose dry mount adhesive film with no paper core and a neutral pH. Fusion 4000 Plus can be self-trimmed so that mounting irregularly shaped prints is easy. It is activated at 160°F, and the recommended press temperature is 180°F to 190°F. It bonds as it cools, so it must be placed under a weight to ensure a complete bond after removal from the press.

Seal claims that Fusion 4000 Plus can be completely reversed by placing it in a hot press, closing and locking the press, and then peeling the print off the backing board. In our experience, this does usually work, although repeated insertions in the press may be required. You have to work fast.

However, a coating of the adhesive remains on the back of the print and must be removed with a solvent such as acetone. Also, Fusion 4000 Plus cannot be considered completely reversible without danger of damage to the print because of the severe mechanical stress created by peeling the print away from the still tacky adhesive.

We judge this material to be one of the best on the market for all-purpose work. Use it as follows:

1. Preheat the press to 210°F for trimming.
2. Predry the print and mount board.
3. Tack and trim the adhesive to the back of the print using the technique described for adhesives.
4. Take the package out of the press and peel off and discard the kraft paper. When the print has cooled, pull it off the release paper.
5. Reset the press thermostat to 185°F and allow the press to cool.
6. Position the print on the mount board. Lightly push down on it to create a pressure tack that will temporarily hold it in place. It is best to use some kind of protective sheet to avoid getting finger oils on the print.

7. Put the mount board on clean kraft paper and cover with the release paper. Slide the assemblage into the press while taking care not to disturb the position of the print. Close and lock the press.
8. Again, time in the press will vary. Try 60 seconds for a bonding time; experience may indicate that a shorter time can be used.
9. Slide the mount out of the press, still taking care not to disturb the position of the print. Put it under a weight immediately to cool.

Because the procedure for self-trimming requires a hotter press than is used for mounting, it is expedient to tack and trim all prints at the same time that are going to be mounted in one session.

Scotch Brand Positionable Mounting Adhesive (PMA)

A cold-mounting adhesive without a paper core that has a mildly acidic pH rating of 5.4, this adhesive has excellent strength, great enough to mount Cibachromes. With its plastic substrate, the Cibachrome will probably remain unaffected by the weak acidity of the PMA (although the mount board will react to this acidity). We urge that this acidity be considered when mounting other kinds of prints with this adhesive.

PMA can be bought only in rolls now, since the sheets have been discontinued. As a roll unwinds it exposes one side of naked adhesive backed by release paper. PMA is called a "positionable" adhesive because it has a relatively light initial tack. Once the adhesive has been put on the back of the print and the release paper removed, you can slide the print into position without it sticking to the mount board at first contact.

Application can be accomplished with a burnisher. 3M also makes a C-35 roller press (Figure 5.4) to speed volume work. The

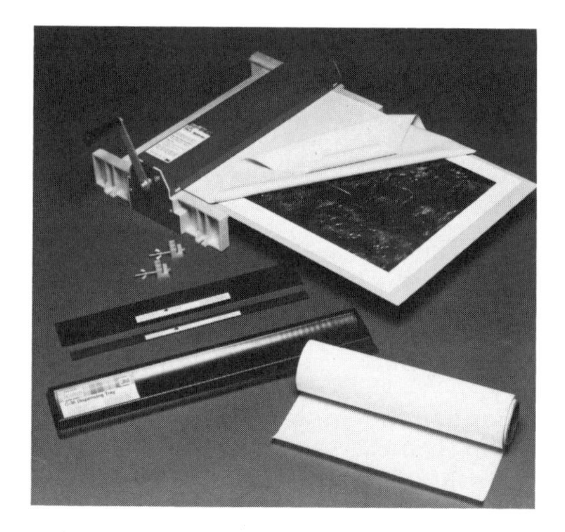

Figure 5.4
The C-35 Applicator is a tool for doing dry mounting without heat. It speeds up mounting large prints.

press provides a much more uniform and better-lasting adhesion between print and mount. Instructions come with the machine. To burnish by hand,

1. Cut a piece of adhesive from the roll. It should be slightly larger than the print. Place the back of the print firmly in contact with the adhesive.
2. Trim excess adhesive along the edge of the print; use a steel ruler to protect the print border.
3. Cover the face of the print with kraft paper and burnish firmly on a smooth, dry surface. Work from the center out toward the corners.
4. Pull off the carrier sheet from the back by lifting each corner in turn; position the print on the mount board.
5. Cover the face of the print with kraft paper and burnish firmly in place. Once this burnishing is done, the print is permanently mounted and can no longer be moved.

MOUNTING GLOSSY CIBACHROME PRINTS

Cibachrome print materials with the Deluxe Glossy surface (Cibachrome II and Cibachrome-A II) have an opaque white polyester base 7 mils thick coated on the back with a matte gelatin layer. The polyester makes the prints dimensionally stable, and the gelatin layer helps make them lie flat. Although prints made on the glossy Cibachrome give excellent color stability, they require special care in mounting. (Cibachrome prints made on the Pearl surface material can be mounted in similar fashion to other resin-coated color photographs.)

Relative Humidity

According to the manufacturer, Cibachrome prints should never be dry mounted in a room where the relative humidity is above 30–35%. Mounting in a more humid environment will affect the stability of the image. To prevent red spots, when the print is to be mounted in a press where the temperature reaches 200°F–210°F, Ilford also recommends predrying at temperatures of 175°F for 15–30 minutes in a room where the RH does not exceed 50%. This strikes us as an arduous proposition.

We suggest three alternatives: cold mounting; farming out the work to a commercial house that does dry mounting on a regular basis; or using a tissue such as Seal Archival Mount that works at 175°F.

Handling

Cibachrome prints pick up oil from fingers; the fingerprints are difficult to remove and show up vividly on the mirrorlike surface. Use clean cotton gloves to handle.

Substrates

The recommended substrates for mounting Cibachromes are:

- ☐ Aluminum — dimensionally stable; excellent smooth surface.
- ☐ Lexan — flexible with a smooth surface.

Most other materials present problems for dry mounting Cibachrome prints.

- ☐ Mounting on paper products such as mat board is likely to cause the appearance of an "orange peel" texture. This occurs when the glossy surface of the polyester exaggerates the texture of the underlying material.
- ☐ Fome-Cor shows orange peel texture.
- ☐ Acrylic sheeting requires extensive and difficult predrying. Problematic to work with during mounting.
- ☐ Gatorfoam, Ryno-board, and tempered Masonite all contain formaldehyde and similar chemicals that make them unsuitable for archival use.
- ☐ PVC sheets are unsuited for archival storage; they cause orange peel texture.

Cold Mounting

Cold mounting techniques protect the glossy surface of Cibachrome prints, and their use is recommended by Ilford. 3M's Positionable Mounting Adhesive makes a good choice, but to ensure that the photograph adheres properly when using this film, you must feed the mount and print through the roller system. Use of a squeegee to make the bond does not work effectively.

It's so difficult to mount Cibachrome prints and still achieve conservation-grade quality that it seems easier and safer to mount them in acid-free mats with photo corners, unless it is absolutely necessary to do otherwise.

UNMOUNTING

Dry mounting with tissues that are not removable is like riding the tiger: getting on is not nearly so difficult as getting off. Some prints can be unmounted, but not without risk.

Materials needed include a metal tray larger than the mount board; acetone; a sheet of glass taped along the edges for safety in handling; rubber gloves; a thin metal spatula; and an extremely well-ventilated work area.

Spot test a small area of the print for reaction to acetone. Pour approximately a ½-inch layer of the acetone into the tray and insert the mounted print face down. Cover the top of the tray with glass.

After the print has soaked for about 45 minutes, lift the board out. Wear gloves to avoid getting the acetone on your hands. Try to insert the edge of the spatula under one corner. If the print lifts easily, proceed as far as possible. If it does not, reimmerse and come back later. Repeat until the print has been removed.

As a conservation practice, this method is not recommended. It removes the print from its original milieu, and there is always the chance that things may go wrong with the procedure. Make sure that the print has been copied before starting.

Acetone is a strong solvent and nasty stuff. It may cause cancer and it will certainly cause a fire if exposed to a flame. Work safely.

Do not use this procedure with anything other than fiber-base paper prints. Resin-coated and Cibachrome papers are made of plastic, and acetone is a plastic solvent.

REFERENCES

1. James M. Reilly. *The Albumen and Salted Paper Book: The History and Practice of Photographic Printing, 1840–1895.* Rochester, NY: Light Impressions, 1980.
2. 3M, Professional and Commercial Products Department, X-PISA. "Product Information: Scotch Brand Adhesives and Tapes Aging Properties." St. Paul, MN: 3M Center.
3. Thomas Maffeo. *How to Dry Mount, Texturize and Protect with Seal.* Naugatuck, CT: Seal, 1981, p. 27.

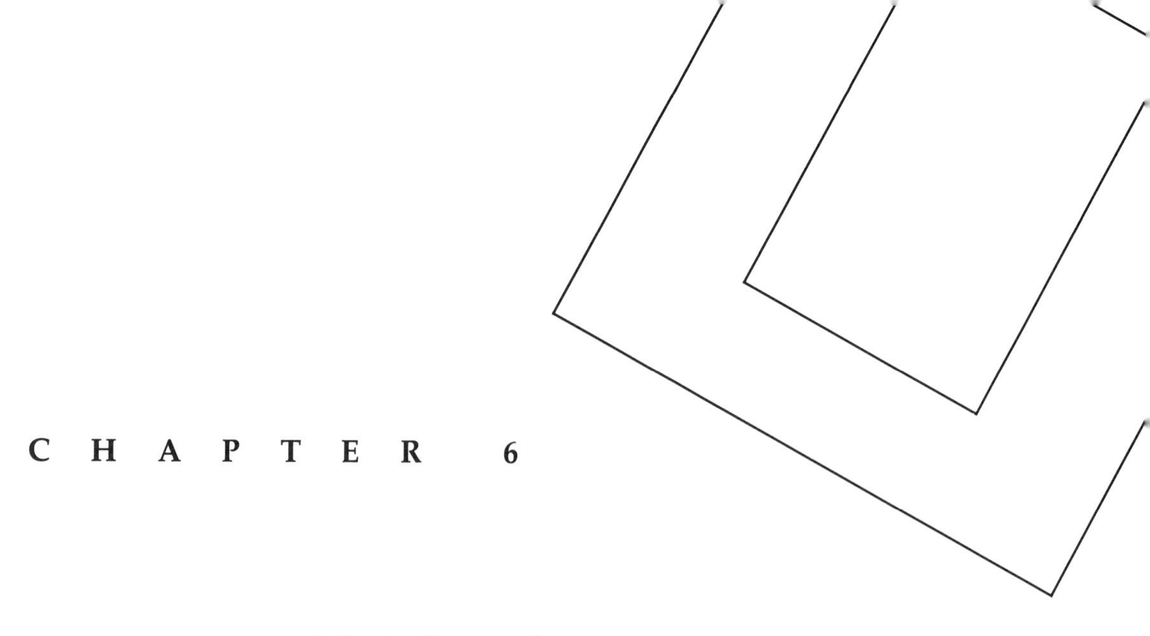

CHAPTER 6

Mat Board

THE STRUCTURE OF PAPER

To understand what makes certain types of mat board preferable for use in matting prints, it helps to have a basic understanding of the composition of paper. When examined closely a sheet of paper reveals a fascinating and complex structure.

Its key ingredient is glucose, a form of sugar. At the atomic level, glucose is made from carbon (the chemical symbol for which is C), hydrogen (H), and oxygen (O), the atoms of which are all linked together in a ring by electron bonds. Oxygen atoms located between the glucose rings link them together into long chains. These chains are molecules of the substance known as cellulose.

Placed side by side, the long ribbons of pure cellulose form bonds between one another. (They have hydrogen and oxygen atoms strategically located so that they lock together like matching parts from a jigsaw puzzle.) In this way the cellulose forms itself into sheets that are one molecule thick.

These sheets of cellulose, only a fraction as thick as a sheet of paper, stack themselves into layers. An attraction called the Van der Waals force, similar to static electricity, holds them tightly in place. These stacks of sheets are known as microfibrils, and they in turn stack in overlapping fashion into bundles, which then overlap into fibers. It's rather like the little fish that got eaten by the big fish that got eaten by the bigger fish and so on up the chain.

HOW ACID DAMAGES PAPER

Cellulose gains its strength from the length of its molecules. The more glucose units in the chain, the greater the strength of the cellulose. Acid causes the chemical deterioration of paper by breaking these chains. In the simplest terms, an acid can be described as a proton (+) in search of an electron (−). The proton that makes up the hydrogen nucleus is so hungry for an electron that it can shoulder in on most other electron bonds, and push aside any carbon or oxygen atom that gets in its way. In this way it severs the links between glucose units and weakens the cellulose.

A single proton in paper does damage way out of proportion to its size, because as a side effect of splitting the cellulose chain it generates a carbonium ion, which in turn links up with the oxygen in water and frees up another proton. This proton goes on to break more bonds, again and again, in a process called hydrolysis. This cycle can go on for only a short while before the damage becomes major and irreversible, because only a small fraction of the bonds in the cellulose need be broken to make a sheet of paper too fragile to handle safely.

A number of impurities and contaminants cause paper to deteriorate, and it is only the highest grades of paper that have anything like the purity of the substance described earlier. The following list describes substances that have been identified as affecting the strength and durability of paper.

Lignin A naturally occurring organic molecule found in paper made from wood. Lignin binds the cellulose together in the tree, and is found in great quantities in kraft paper, corrugated cardboard, and low-quality mat board. Over time it breaks down into acids and peroxides. One possible indication of its presence is a brown color.

Hemicellulose Polysaccharides (multiple sugar units) with numerous branches that prevent the microfibrils from stacking neatly. Making paper in the presence of hemicellulose is like trying to make your bed with a Christmas tree in it.

The desirable form of cellulose is usually called alpha-cellulose to distinguish it from hemicellulose.

Hydrolyzed cellulose Once the process of hydrolysis has started in a piece of paper, it continues. Each new break in the bonds between the glucose units of the cellulose releases another proton.

Alum-rosin sizing An additive from the paper-making process that breaks down into sulfuric acid.

Sulfur dioxide An atmospheric pollutant that forms sulfuric acid in the presence of water.

Nitrogen dioxide An atmospheric pollutant that forms nitric acid in the presence of water.

Glucuronic acid An acid that occurs especially as a constituent of mucopolysaccharides.

Ozone A highly reactive oxidizing and bleaching form of oxygen with three atoms instead of the usual two. It is created by a photochemical reaction with ultraviolet sunlight in the upper atmosphere or by silent electric discharges such as are generated by copying machines.

Peroxides Lignin can break down in a number of ways. One of these is to form peroxides, which are oxides with a high concentration of oxygen in them and which are particularly dangerous to photographs.

Copper and iron Metal ions absorb energy from incident light and then release it into cellulose, breaking the cellulose bonds. In addition, traces of oxidized metal have been implicated in discoloration of photographs.

Migratory acids Even when paper is initially pure, it can be contaminated by acids that migrate from adjacent materials.

THE RIGHT KIND OF MAT BOARD AND HOW TO BUY IT

To be safe for photographic prints, the paper in your mat board should meet these criteria:

- ☐ High in alpha cellulose fiber (minimum of 87%)
- ☐ No groundwork and no lignin
- ☐ No alum-rosin or other acidic sizing
- ☐ No waxes, plasticizers, or other ingredients that may transfer to the photographic materials during storage
- ☐ Maximum 30 ppm iron, and 0.7 ppm copper
- ☐ No transferable dyes or colors
- ☐ Buffered or neutral pH

Standards for papers to be used for the enclosure of photographic materials during storage are outlined in ANSI Standard *IT9.2-1988*, "Photography (Processing) Processed Films, Plates and Paper-Filing Enclosures and Containers for Storage." It outlines two standards for alkalinity and buffering in board, which we discuss at the end of this chapter.

Don't despair. You don't need to be a paper chemist to tell whether the board you are buying meets these standards. Knowing what you need is the first step. The second step is to find a reputable supplier that publishes the specifications for its mat board. A number of such sources are listed in the back of this book. The market for conservation supplies is thus far limited enough that firms in the field do not survive without a reputation for probity.

Composite Board

Knowing something about the structure of poor-quality, composite mat boards can help you appreciate the excellence of good-quality conservation mat board. You can find samples of the former almost anyplace: the board sold in most art supply stores, or the mats on many walls, will suffice as examples.

Look closely at the edge of the composite board (Figure 6.1).

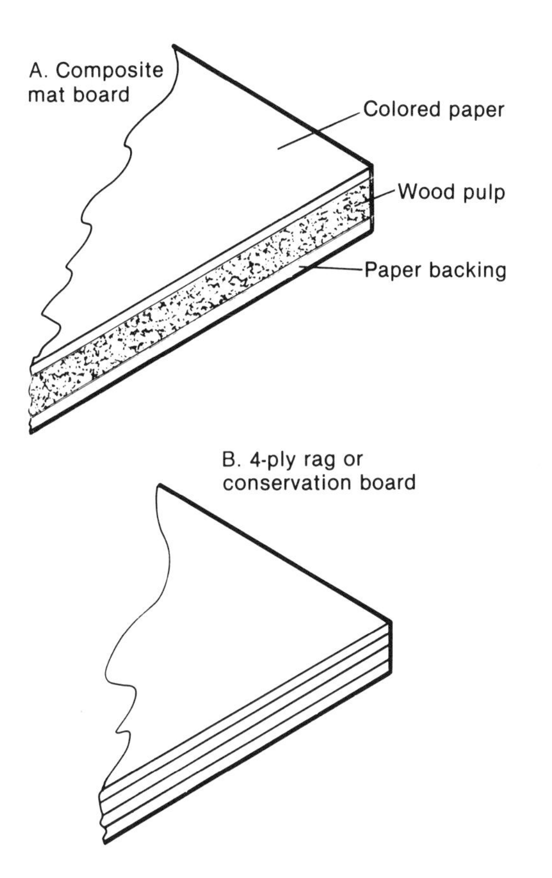

Figure 6.1

You will see that it is composed of three layers: a thin top sheet of paper, usually colored; a thick core layer inside; and a third sheet, often white or gray, on the back.

The core layer creates most of the problems. To make this chipboard material, chunks of wood are cooked in a slurry of chemicals and then mechanically ground into pulp. Bleaches, sizing, and adhesives may be added, but the substance is primarily a mixture of lignin and cellulose.

The chemical composition of the resulting product greatly resembles newsprint, a type of paper made by the same process. If you would like to measure for yourself the chemical stability of newsprint, lay a sheet of newspaper in strong light with some opaque object like a hammer on top of it. Within a few days or less, the outline of the weight should become visible as the newsprint yellows and darkens. (A less obvious effect is that the paper has become more brittle as the cellulose chains are hydrolized because of photochemical reactions.)

So, the core of low-grade, composite mat board deteriorates. You might ask, does this matter? After all, maybe you could just replace the mat from time to time.

But it does matter, a great deal, for reasons that it pays to consider. Assume that the paper on the front and the back of the board meets accepted standards of purity, though probably this is not true. When you cut a window out of such mat board, you set the exposed layer of chipboard right next to the surface of the print. The chipboard core is a veritable potpourri of acids and chemical compounds that will break down over time into more acids. As this happens, minute amounts of water vapor in the air will carry the acids onto the surface of the print.

The effect of these vapor-borne chemicals does not appear as quickly as the outline of the hammer on the newspaper. But often within a year or two, it can start to affect the print. This damage often passes unremarked for a long time. The only certain way to spot the fading, discoloration, and other changes that it creates is to lift up the window mat and look at the entire print surface. The print surface directly *behind* the mat may be unaffected, while a brown or yellow tint discolors the edges of the exposed part of the image. The interior of the image may remain unchanged.* In the long term the paper next to the bevel may turn deeper yellow and then a dark brown, and eventually it can become brittle enough to crumble. The ultraviolet radiation from sunlight or fluorescent lamps can accelerate this chemical reaction.

A related phenomenon occurs when a print has been backed by corrugated cardboard. Parallel light and dark lines, corresponding to the corrugations of the cardboard, appear on the surface of the print. This results from the migration of substances, usually sulfuric acid, formed by the slow breakdown of the adhesive used to glue the corrugations to kraft paper. These acids penetrate the print from the back, and by the time they make their presence visible on the surface the damage has been done. Prints blemished in this way may be treated by a conservator, who will deacidify the paper, but this cannot reverse the damage already done.

Conservation-Grade Mat Board

If you look closely at any of the several types of conservation-grade mat board, you will see that their structure is distinctly different from board with the chipboard core. First, it is the same color all the way through. Second, it is built up in thin layers, called *plies* in the paper industry. You may not at first be able to see the plies. Take a small piece of board and light one corner with a match. Let it burn briefly and blow out the flame, and you will see that the board separates into discrete layers. These are the plies. Most board has 2, 4, 6, or 8 plies.

Conservation-grade mat board can be made from 100% cotton fiber or from high alpha-cellulose made from wood pulp.

The board made from cotton fibers is called rag board, because it used to be made from old cloth that had been macerated by huge water-powered hammers. Modern rags contain too many synthetics, and paper mills today bypass the intermediate stage.

* Fading or yellowing of the exposed part of an image may also indicate a photochemical reaction. In this case the discoloration generally happens over the entire exposed surface.

Instead of rags they purchase cotton fibers direct from the textile manufacturer.

The board made from purified wood pulp is called conservation board. All the lignin and other contaminants are removed by a bleaching process, so that what is left is long-chain alpha-cellulose.

In both cases, the start of the actual papermaking process is a water-and-fiber slurry. When it is poured onto a moving screen similar to a conveyor belt, the water flows through and the fiber stays behind. It used to be that these screens were stretched on frames, called molds. When the water had drained out, the screen would be turned over and the sheet couched on a sheet of felt. Thus the front and back of the sheet, with their slightly different textures, came to be called the *wire* and *felt* sides.

All paper used to be made of rags. It had excellent strength and generally a laudable absence of chemical impurities. Commercial printing for mass audiences and the industrialization of paper-making changed all that. Now, most paper costs very little and will at best last a couple of decades. Pure rag paper has become a rare and relatively expensive commodity.

When to Avoid "Acid-Free" Board

A few years back, one major board manufacturer introduced a new line of mat board widely touted as "acid free." Technically, this was correct — when you checked the board for acidity, it would in fact test out at pH 8.0 or higher, which meant that it was mildly alkaline. What the manufacturer had done, however, was take regular wood pulp (the kind used for chipboard) and dose it heavily with alkaline calcium carbonate. However, as time passes, all the lignin and other impurities in that board deteriorate into acids and peroxides and will overwhelm the ability of the buffering agent to neutralize them. The end result is highly acidic board. The point is that the term "acid free" is no panacea. Conservation-grade board must meet the other specifications described earlier. It does help, of course, to understand how the acidity of paper is gauged, because this is still *one* of the criteria for judging quality.

The pH scale runs from 0 (very, very acidic) to 14.0 (extremely alkaline). The midpoint, or area of neutrality, is 7.0. Like the Richter scale for earthquakes, the pH scale is logarithmic, so the increase in acidity from pH 5.0 to 4.0 is much greater than the increase from 7.0 to 6.0. As measured over time, the pH of paper inevitably shifts, usually toward the acid end of the scale. This can be counteracted by *buffering* with calcium carbonate. (Calcium carbonate is legitimately added to many conservation-grade boards, and not just used to mask unpurified pulp.)

Choosing Between Rag and Conservation Board

Two major factors favor conservation board over rag (cotton fiber) board. One is money. Conservation board costs far less and offers the same conservation quality. It also has a more homogeneous arrangement of fibers and gives a more consistent cut for making window mats. It has less tendency than rag board to produce ragged bevels, and so cutting goes faster with less waste. Rag board has the advantage of extremely brilliant whites and a soft, velvety feel and look that conservation board cannot match. It also has the weight of tradition behind it.

Until recently, both types of board came in two colors: white and cream. Only composite boards, the "painted ladies" of mat board, offered the option of color. A black rag board that was on the market only briefly showed a regrettable tendency to transfer its dye to the surface of the print merely by touching it. In 1979 Bainbridge brought out the Alphamat line, which met the need for colors by facing a conservation-grade board with a colored sheet.

Before using any colored board, test for color fastness. Place a piece of damp blotter on a sample and allow it to sit for 24 hours under a weight. Any color that shows up on the blotter is reason to reject that type of board for use next to the print. For applications that require a specific color it is possible to use composite board in a multilayer mat. Instructions for this type of mat are given in Chapter 8.

Full Sheets Compared to Precut Sizes

Most manufacturers cut board to standard sizes: 30 × 40 or 32 × 40 inches and 40 × 60 inches. If you buy direct from a wholesale distributor these may be the only sizes offered, which can be something of a problem. They are expensive to crate and to ship. Small quantities in these sizes can get damaged when you stuff them into the hatchback of your car. You also take a risk of making costly mistakes in cutting them to smaller sizes. However, institutions that do a great deal of matting should definitely buy the full sheets because the savings outweigh the costs.

Many individuals find that precut sizes offer an economical alternative. Mail order sources listed in the back of the book sell different kinds and grades of board in standard sizes. The most common sizes (in inches):

8 × 10	14 × 18
9 × 12	16 × 20
11 × 14	20 × 24
14 × 17	22 × 28

Suppliers specify minimum quantities, but will send samples for examination.

When you plan to mat a large show or to produce a portfolio edition of prints, purchase all your board at the same time and from the same supplier. Purchasing in quantity saves money and assures uniformity of presentation. Board color varies from batch to batch even at the best mills. Paper technicians tell us that at best they can hold color within a certain range during the course of a run. There is no guarantee that one batch will match the next. Differences may not appear when you are looking at the board during the matting, but when the framed prints go on display a variety of shades may suddenly become apparent.

Standard Mat Sizes

It makes good sense to use standard mat sizes. The dimensions above serve quite well to hold most prints, and they confer some important benefits in addition to reducing waste.

- Uniform-sized mats simplify storage requirements. They make it possible to reduce a collection of prints from diverse sources into a set of uniform sizes.
- There is less slippage in print storage boxes when all the prints fit snugly into the box. This reduces abrasion and corner damage. Archival-quality print boxes can be made in custom sizes, but the standard-sized ones are considerably less expensive.
- If prints are all in standard-sized mats, overhang is eliminated when they are stacked. This prevents the print from getting bent.
- Standardizing your mat sizes also means you can reuse standard-sized aluminum sectional frames. This saves on framing costs. Since prints can be stored in mats alone, it also reduces the need for storage space.

WHETHER TO USE BUFFERED OR NONBUFFERED BOARD

Photographic conservators do not yet fully agree on the benefits of matting photographs with board that has a buffering agent added.

Our daily environment is mildly acidic. Rainwater becomes mildly acidic even when it absorbs carbon dioxide out of the air. Sulfur dioxide, a common atmospheric pollutant, forms a stronger acid in rainwater. This acid rain has devastated areas where the soil has no buffering agents; in numerous lakes it has wiped out all the fish and plant life.

An alkali can neutralize an acid. The two combine to form a compound called a *salt*. (Sodium chloride is the most common but not the only salt; we use it as a seasoning.) Limestone is a milk alkali found in many watersheds, and when present it helps neutralize some effects of acid rain. It is composed almost entirely of calcium carbonate extracted from sea water by ancient, microscopic creatures for their shells.

During the 1950s and 1960s, William Barrow investigated why certain old paper was still in good condition. It turned out that these papers had all been made in mills that used water from limestone aquifers. The residue of calcium carbonate in these papers had buffered them against acid hydrolysis of the cellulose.

When Barrow's discovery became widely known, makers of mat board for museums started to add calcium carbonate to their products to give them an extra margin of protection. The thinking was, and is, that a slightly alkaline quality counteracts the natural drift toward acidity.

The ANSI Standard *IT 9.6-1988* specifies that board for *black and white photographs* should have a pH between 7.0 and 9.5, with a 2% alkali reserve. A pH of 9.5 seems rather extreme and may eventually be modified.

Some conservators questioned the effect of the alkali buffers on photographs. Their research indicated that *color prints* could be affected by the presence of alkalis, and that these should be mounted on nonbuffered acid-free board rather than the buffered board. As a result, the ANSI Standard reads: "Paper that is in direct contact with processed color photographic material should have a similar composition to that used for black-and-white except that the pH should be between 7.0 and 7.5 and the 2% alkali reserve requirement shall not apply." This refers to chromogenic prints like Ektacolor and Fujicolor, and to dye transfer prints.

The *cyanotype* (or blueprint) is another photographic medium definitely known to discolor in the presence of buffering materials.*

No research published to date suggests any problem resulting from using buffered board with gelatin-emulsion black and white photographs. For economy when matting black and white photographs, buffered board may be the better choice. Paper mills clearly identify buffered board, and any dealer who does not specify the presence or absence of a buffering agent will supply the information on request.

* See the identification flow chart in James Reilly, *Care and Identification of 19th-Century Photographic Prints* (Rochester, NY: Eastman Kodak Company, 1986) for assistance in identifying older media.

CHAPTER 7

Hand Cutting Conservation Mats

A FEW WORDS OF ENCOURAGEMENT

Anyone who can draw a straight line and use a few simple hand tools can make a mat that not only protects a print from chemical and physical damage, but also enhances the appearance of that print.

By doing the job on your own, you make certain that the materials and methods used meet conservation standards. It is a sad fact that many commercial shops still work ignorantly — using acid-core composite board, mounting prints with rubber cement, and backing them with corrugated cardboard.

How much will you save by cutting your own mats? We compared the current prices for tools and mat cutting in our city (Rochester, New York), and we calculated that by cutting five or six mats the average person can recoup the cost of tools and basic materials used in a lifetime of mat cutting (excluding mat board, of course). If you do your own framing as well, these savings increase.

WHAT IS A WINDOW MAT?

A window mat has become the accepted standard enclosure for prints. It differs from other types of print enclosures in that it allows the image to be seen without opening the housing. Most window mats are made of paper or paper mat board, although for housing photographs made in certain media like daguerreotypes they are also made of metal and other materials.

Prints are mounted in window mats both to protect them during storage and to provide a border while they are in a frame. The mat in the frame also serves to separate the print surface from the glass and to avoid adhesion of the emulsion to the glass.

Framers have developed many elaborations of the window mat, but the basic concept remains simple. It consists of two layers of mat board larger in both dimensions than the print. A tape hinge

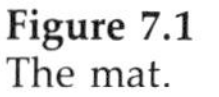

Figure 7.1
The mat.

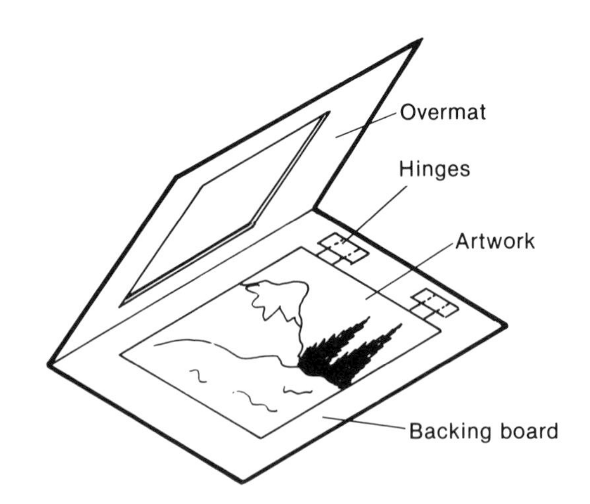

along one side attaches these two boards to one another. The upper board, or overmat, has a window with beveled edges cut in it. The print is positioned on the lower board, called the mount, in such a way that the image shows through the window. The print is then attached to the mount (Figure 7.1).

THE CONSERVATION MAT

Certain materials and techniques of construction help protect the print from deterioration. Any mat designed to enhance the longevity of a print can be called a *conservation mat*.

The use of conservation matting techniques does not need to detract from the tasteful and imaginative display of a photograph; in fact, we feel that an understanding of the principles of conservation matting adds to the repertoire of choices you have in exhibiting prints.

Conservation mats must satisfy these criteria:

- ☐ The mat board and mounting materials contain no chemical impurities that could harm the print.
- ☐ The print can be removed without damage. If the mat is damaged or becomes dirty, little effort should be required to move the image to a new housing.
- ☐ The mat can be opened for inspection without cutting or tearing. This makes possible early detection of fading, insect attacks, and damage from fungus and mold. It encourages periodic examination.
- ☐ The mat provides separation between the print and the glass of the frame (if any). In the case of gelatin emulsion prints, the presence of minor amounts of moisture can cause a phenomenon known as local ferrotyping. The emulsion adheres to the glass and can be pulled off the print.

☐ The mat, when possible, is cut to a standard size to allow for safe storage in controlled conditions.

START WITH A WORKSPACE

Planning a work area for cutting mats and mounting prints requires the same forethought and attention to detail needed to build a darkroom or any other studio work site. Working in a pleasant and convenient environment will enable you to avoid the temptation to take shortcuts. Depending on the volume of work planned, set aside a separate room or area of a room exclusively for this purpose. If this cannot be done, at least set aside a specific area for associated multiple uses of which matting and framing will be one, and store all tools and materials there to speed up preparation.

When planning your workspace, you will want to take these factors into account.

Lighting

Three types of lighting should be available, and you can probably get them for a minimal investment of around $20. *Daylight* can come from a nearby window or skylight. A couple of lamps or a ceiling fixture with conventional light bulbs will supply *incandescent* illumination. For *fluorescent* lighting, watch the local papers for sales at major discount store chains, which regularly hold sales on four-foot fluorescent work light fixtures complete with the tubes. These put out good light and they are made for the home handyman, so they are easy to install. Just put two hooks in the ceiling, attach the chains on the fixture to hang it, and plug it in.

You need three kinds of light to assess the effect of a mat on a picture. Daylight has a blue cast, incandescent is orange, and fluorescent is usually greenish. Use the same kind of light as will be present in the area where the prints will be displayed. If in doubt, use daylight.

Work Surface

Build your work surface sturdy and free from sway (an especially important factor for mat cutting), make it between 36 and 42 inches high for comfort, and keep it uncluttered. If it is too high, you cannot reach everything, and if too low, you will strain your back, so try different heights for comfort before putting it permanently in place. A usable work surface measures at least 30 to 36 inches deep and 42 to 96 inches long.

Covering

Cover the work surface with a protective sheet of corrugated cardboard, chipboard, or foam core, and overlay the sheet with kraft paper taped in place under the counter edge. This surface protects the print from imperfections in the counter top and can be replaced when it gets nicked and cut up. A disposable covering also encourages cleanliness, a must when handling prints.

Storage
Have a lower shelf or separate area in which to keep all liquids when prints are on the work surface. Hang tools on a peg-board arrangement for easy access, and store mat board flat on shelves, covered to protect it.

Cleanliness
Dust and dirt can harm prints. Vacuum frequently. Put everything away when you have finished working. If you absolutely must set up in a basement or other space where dirt comes down from the ceiling, staple sheets of plastic or garbage bags to the joists overhead.

Lip
At one end of the counter you should have a raised lip. You can bolt or nail a 1 × 2-inch board to the counter to make this lip. It will hold wood frames securely while you put in the backing.

GATHER TOOLS AND MATERIALS

Assemble everything needed before you start. This speeds up the work, and gives you the important psychological advantage of not interrupting the continuity of your effort. Among the items you need are:

☐ Mat cutter

☐ Supply of mat board

☐ Guide bar

☐ Graduated ruler

☐ Pushpins

☐ Double-edge razor

☐ Masking tape

☐ Mat-marking scribe or T-square

☐ Artist's utility knife

☐ Linen tape

☐ Art gum eraser

☐ Hard-lead pencil

Some of these items may be unfamiliar if you have not cut mats before, so we will discuss them in more detail.

CUTTING A MAT

- ☐ Assemble all tools.
- ☐ Clean the work area.
- ☐ Tape down the cutting surface.
- ☐ Measure both image and sheet sizes.
- ☐ Sketch the dimensions of the mat.
- ☐ Transfer the measurements to the back of the mat board.
- ☐ Adjust and test the mat cutter blade.
- ☐ Cut the window with guide bar and mat cutter.
- ☐ Free the corners with a double-edge razor.
- ☐ Finish the beveled edges and erase the guidelines.
- ☐ Tape the backing board to the long side of the window mat.

Mat Cutter
A hand-held mat cutter like the Dexter, which sells in art supply stores for around $15 will do just fine. Logan, EZ/Mat, and X-acto make similar models in the same price range. You can spend hundreds of dollars for a heavy-duty production mat-cutting machine, and the speed and ease these offer make them well worth the price if you plan to cut thousands of mats. But remember, the cheapest tool that can do the job is the best, and a hand-held mat cutter used properly does just as good a job as more elaborate equipment.

We chose the Dexter for demonstration purposes because, in our experience, most people who work in the visual arts always have one tucked away in a back drawer. It is a fine tool, but the lack of adequate instructions from the manufacturer causes many people to set it aside in disgust after trying a few abortive mats. Our advice is pull it out, dust if off, and use it.

No matter what kind of mat cutter you have, make sure to lay in a plentiful supply of fresh blades. Nobody can cut a good window mat with a dull blade. Replace the blade as soon as it starts to drag, or if the point gets chipped.

Supply of Mat Board
Have more than enough on hand, because there is usually some waste. Try to have enough for at least 10 to 15 mats for the first learning session, and do not forget to include backing board. This translates to about 20 sheets of 4-ply board for windows, and 15 sheets of 2-ply board for backing. As time goes on, build up a

working stock of varied colors, thicknesses, and finishes. You also need board for cutting surfaces, for sample corners, and for practice. Remember, you are going to need more than you might think.

Guide Bar

This item is indispensable. You simply cannot cut a straight line with a hand-held mat cutter without using a guide bar. Light Impressions sells a real gem, but if you cannot buy it, make your own. Get a solid piece of metal with a straight edge. It should measure about ½ by 2 inches, but dimensions are not critical. However, a straight edge *is* critical, so look down the length of it to check for curves or irregularities. Draw a line on a piece of paper along the edge, turn the bar over, and draw the same line. The two lines have to overlap or the bar is not straight. Length is not critical, but a bar approximately 3 feet long is nice. Make sure that the metal does not have any dirt or grease on it, and file off any sharp points. Next, make a nonskid bottom by attaching a strip of thin cork or foam rubber along the bottom with double-sided tape.

If you are thinking of substituting a yardstick, forget it. Yardsticks are not heavy enough.

Graduated Ruler

Almost any kind will do, so long as it measures in at least ¹⁄₁₆-inch increments. C-Thru makes an excellent clear plastic one sold widely in art stores. Try to get one with the measurements marked clear out to the edge. You will find that this makes it easier to measure the distance between guidelines and the guide bar.

Pushpins

This is a convenience item. Pushpins hold the mat in place while you cut the window.

Double-Edge Razor Blades

These may soon go the way of the straight razor, but get some now for finishing cuts at the corners of the windows. Cover one side with tape so you can hold it safely. Single-edge razor blades have a metal strip that makes them too thick for this purpose.

Masking Tape

Never use masking tape to mount a print. Use it to hold a cutting surface in place on your workspace top, and to make a positioning mark on the cutter for setting the blade.

Mat-Marking Scribe or T-Square

Either one of these tools does the job, but the mat-marking scribe makes it easier. To make one, buy a carpenter's scribe at the hardware store. It is a short wooden rod about ¾ inch square, marked in inches and with a nail at one end. This rod goes through a squarish block of wood. The block gets locked in place by a set screw. Carpenters use it to mark cuts parallel to a given edge for cutting. Pull out the nail, drill a hole large enough for a pencil, and wedge the pencil in place. If this sounds too complicated, Light Impressions sells a modified one for a few dollars. The Logan mat cutter has a marker already built in.

If you already have a T-square, make sure that the T is firmly set at right angles (*exactly* 90°) to the crossbar, and that the crossbar does not wiggle. Sometimes it can be tightened simply by setting the screws deeper; if not, discard it and get a new one. The board you ruin will cost more than a replacement T-square.

Artist's Utility Knife

This is another convenience. Utility knives with heavy handles are needed to cut mat board down to size. For general work, Olfa makes a fine line of knives in all weights that have disposable break-off tips and a very sharp edge.

Linen ("Holland") Tape

This is not needed to cut the mat, but you use it right afterwards to attach the backing board to the window mat. For the safety of the print, get tape that has an acid-free adhesive. Some "acid-free" linen tapes have an acidic glue, so buy what you know and can trust.

Art Gum Eraser and Hard-Lead Pencil

You use these to draw guidelines and later to erase them. A soft-lead pencil will smear all over the place, so use one that is graded 3H or harder.

CUTTING A BASIC WINDOW MAT

Prepare the Work Area

Clear out all the clutter. Tape down a piece of scrap board so that there will be a soft surface that the mat cutter blade can penetrate. Make the board longer than the biggest mat to be cut, but it does not have to be more than 6 inches wide. Put tape just at the corners, so that the board can be pulled off and discarded when it gets sliced up.

Calculate the most economical use of your mat board. If you bought precut sheets and plan to use just those sizes, this will not be a problem. If you did not, sit down with pencil and paper and draw some sketchy rectangles. Label the long and the short sides of the rectangles with the measurements of the sheets you are going to cut from. Sketch in different arrangements of the mat sizes you plan to cut, until you find the most economical arrangement. Use this final sketch as the plan for cutting down the board.

Use the utility knife and guide bar to cut the board down to size. Always measure from a factory edge to get straight cuts. Do not draw guidelines all the way across the sheet; all that is really necessary is two marks on opposite sides to position the guide bar. Cut all the way across from side to side because otherwise you end up with an L-shaped remnant, and it will not be of any use unless you are matting L-shaped pictures. Save leftovers for smaller mats and for scrap.

Look carefully at the mat board to decide which side is the front and which the back. We mentioned before that board has a felt (rough) and a wire (smooth) side, which creates a slight difference in texture. For individual pieces, it does not really matter which side is out, but be consistent when doing a large batch. Look at the board for scuffs, dirt, and embedded particles. Planning to cut around these imperfections will save board.

On a related topic, if you have not already washed your hands, do so before you handle much board.

Calculate the Size and Position of the Window

Finally we get to the actual process of making a mat. It boils down to figuring how wide the borders will be — that's all (see Figure 7.7).

A natural question that occurs to many people is, "Do I measure the width of the borders on the front or the back? Because of the bevel, the two widths are different." The answer is: on the back. When making a mat, do *all* measuring, marking, and cutting on the back of the mat board.

Measure the picture. Be exact. Note the following information: paper size (both dimensions), image size (both dimensions), and the width of the borders.

For the sake of convenience, let us assume that you already know how big a mat will go on the picture. Your mat should be bigger than the art. If it is not bigger, make a bigger mat. Do not ever trim a picture to fit a mat or a frame. There is more detailed information in Chapter 8, but for now all you need to know is that you should make a bigger mat if the borders come out to be less than 2 inches wide.

Sketch a rectangle on scrap paper. Measure the mat board *exactly* and enter the dimensions next to the long and the short sides of the rectangle. Precut mat board can vary as much as ⅛ inch in either direction. Now sketch another rectangle inside the first. Note the picture dimensions on the long and the short sides of it. Do not worry about scale, because you need only a rough idea. Subtract the width of the picture from the width of the mat board, and divide by 2. The result is the width of the side borders. Subtract the height of the picture from the height of the mat, and divide by 2. The result is the width of the top and bottom borders. Enter the border widths on the sketch.

At this point, you can transfer the measurements to the back of the mat board. But first, stop and think.

Are the top and side borders nearly the same size? Ideally, they should be pretty close, unless you have a specific visual reason for doing otherwise.

Should the bottom border be thicker than the top and sides? It is a common practice in European and American matting to have the

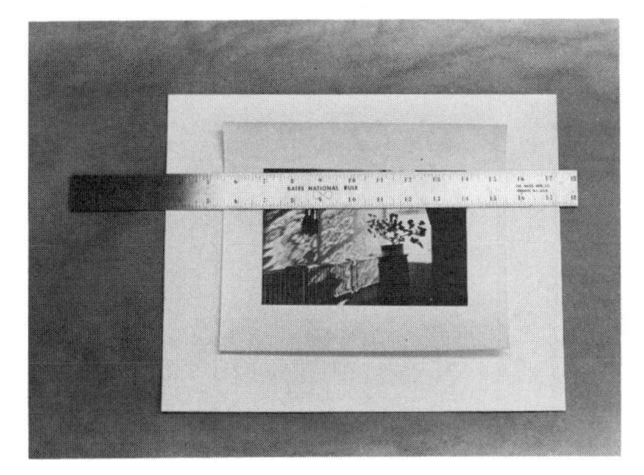

Figure 7.2
Measure both the image and sheet size of the print. Plan the mat so that its exterior dimensions are larger than the size of the print sheet.

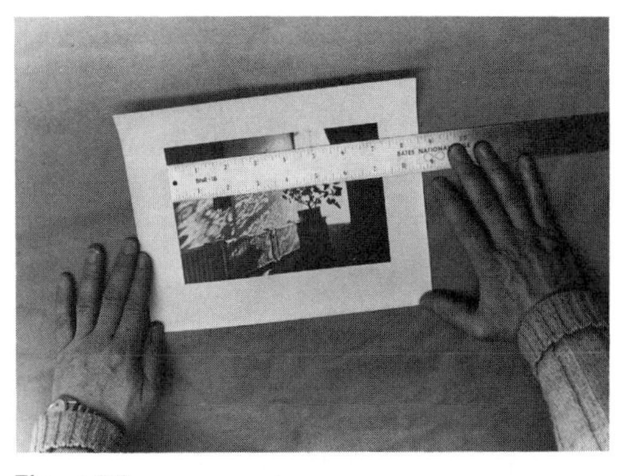

Figure 7.3
Measure the image size exactly to decide the size of the mat window.

Figure 7.4
When transfering measurement to the back of the mat board with a T-square, allow the end of the board to overhang the countertop so that the crossbar of the T-square moves freely.

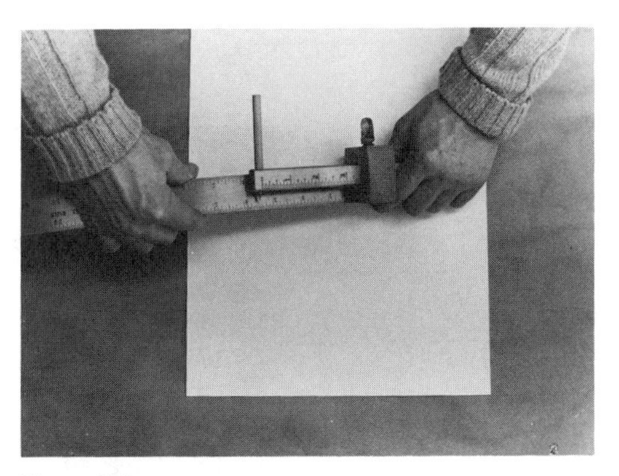

Figure 7.5
Use a ruler to check the exact distance set on the mat-marking scribe when using it to transfer measurements to the back of the mat board.

Figure 7.6
Slide the mat-marking scribe along the edge of the mat board when the measurement has been set. Note that the guidelines overlap at the corners.

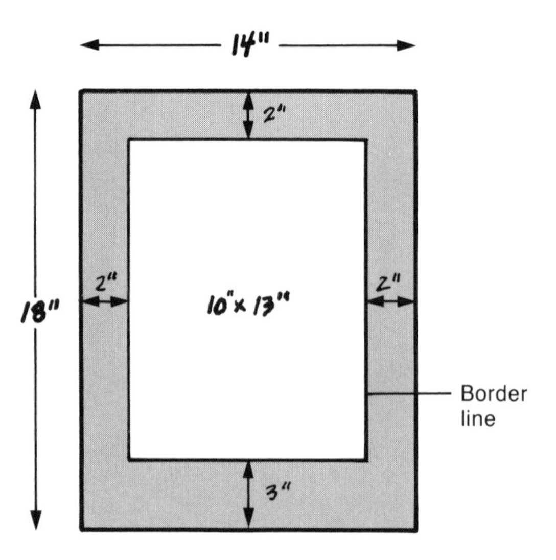

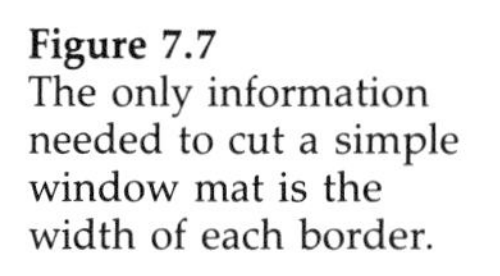

Figure 7.7
The only information needed to cut a simple window mat is the width of each border.

bottom border wider than the others, rather than having the picture at dead center. If the mat size stays the same, any width added to the bottom must be subtracted from the top border.

Will the mat cover enough of the border of the picture? At least ⅛ inch of the sheet should be overlaid by the mat on each side. Otherwise, the print can pop through the window. Paper changes size depending on how much moisture it gets exposed to, and an 8 × 10 sheet can expand and contract ¼ inch.

After making adjustments, you can transfer the measurements to the back of the mat board. For this, either the mat-marking scribe or the T-square will be needed.

Mat-Marking Scribe

Set the block of the scribe so that the edge facing the pencil stops on the mark that corresponds to the side border width of the mat. Set the screw to hold it in place. Measure the distance from the block to the pencil point to be sure it is right. If not, adjust it.

Put the mat board face down on the counter with the side extending over the edge. Slide the block along the mat board so that the pencil enscribes the first guideline. Go nearly to the end of the board. Turn the mat board 180° and repeat on the opposite side. You have now marked both border widths.

Reset the mat-marking scribe for the width of the top border. Mark that, and then do the same for the bottom border. On the back of the mat board you should now have four guidelines that look like a big square hash mark (#). Make sure that these lines overlap by an inch or two at each corner.

Once you get the hang of it, you can mark up mats very fast with a scribe.

T-Square

Say you have a T-square and want to use it. That's all right too, because many people do use T-squares. Working with a T-square is almost self-explanatory, but here are a few tips you might not have thought of. Use the T-square only on the top and one side. This compensates for the board not being exactly square. When measuring with the ruler, make the starting marks right against the edge. This also helps keep things square. Hang the edge of the board over the side of the counter so that the crossbar of the T-square does not get twisted by any surface other than the edge of the mat board.

Set the Blade of the Mat Cutter

Now that the window is measured and marked, it is time to cut it out. Set the marked-up mat board aside for a moment, and look at the mat cutter.

The cutter blade needs constant attention. If it gets dull, if the point snaps off, if it is set too deep, if it is too shallow, if it is loose, you will waste time and materials. When you find that you are having trouble cutting a mat, look to the blade.

To set the blade, hold a piece of mat board on the bottom of the mat cutter — right against the blade. Move the blade up and down until its point extends just a very little bit beyond the bottom of the board. Tighten the set screw.

Now test it. Lay a piece of scrap board on the cutting surface. Gently push the blade through the board and slide the mat cutter a couple of inches in a straight line. Put the mat cutter aside and look at the bottom of the scrap. Did the blade cut through all the way? If it did, look at the cutting surface. There should be a very faint score mark just barely visible. Adjust the blade if more than the slightest score mark appears on the cutting surface, or if the blade does not cut through.

Some advice about handling the mat cutter may help. Always rest the cutter on its back with the blade up when not holding it; this protects the tip of the blade. Use a light touch when cutting, because with a properly set blade you do not have to plow through the board. Never cut on a hard surface like metal or wood; this breaks the blade. Protect the blade by cutting on a surface of mat board. Replace the cutting surface frequently, so that previous cuts do not drag the blade out of the intended line. Always use a scrap from the same lot of board that you are using for the mat when making the test cut; different batches of board have slightly different thicknesses.

Use this checklist to diagnose problems with the blade.

- ☐ Cutter pushes too hard — blade set too deep.
- ☐ Cutter difficult to set in board — tip is broken.
- ☐ Rough edges on the bevel — blade is dull, hacking and chopping through the board; or failure to put protective cutting surface under mat.
- ☐ Cutter "dives" or flares out at corners — blade is set too deep. This is also called "hooking."

Most mat cutters except the Dexter hold the blade at a preset angle to make the beveled cut for the window mat. On the Dexter, put the top end of the blade firmly against the wall of the inset to get a uniform bevel. Slight variations will not be apparent, so do not worry about them.

Cut the Window Out of the Mat Board

Retrieve the premarked mat board and lay it face down on the cutting surface. For ease and greater control, cut directly away from yourself. The guideline for the first cut should be at right angles to the edge of the counter.

If you like, set a couple of pushpins through the center of the window area to hold the mat firmly in place. At the near left-hand corner, set the cutter blade in the mat board. Slide it in at an angle, rather than pushing straight down. The blade should go in right on the guideline, on the *near* side of the crossline.

Remember that the cut is beveled. To go all the way through at the corners, the cuts on the back have to overlap slightly. With experience you will be able to judge how much in advance of the crossline the blade needs to be set. When you have found this point later, you can mark it on the side of the mat cutter with a piece of masking tape.

Do not make the cut yet. First, lay the guide bar parallel to the cut line, and about ¼ inch to the *left* of it. Push the near end of the guide bar against the flat side of the mat cutter. Now measure the distance between the line and the bar at a couple of places to make sure that they are exactly parallel. The distance should be the same at each point, and the near end of the guide bar should rest solidly against the side of the mat cutter.

After a little practice, you will not need to measure the distance between the guide bar and the cutting line. But keep this rule in mind: the guide bar always goes on the left side of the line. That is, it always rests on the outer margin of the mat.

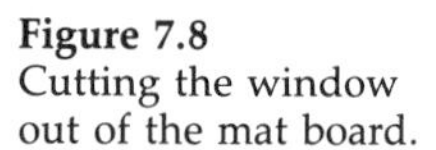
Figure 7.8
Cutting the window out of the mat board.

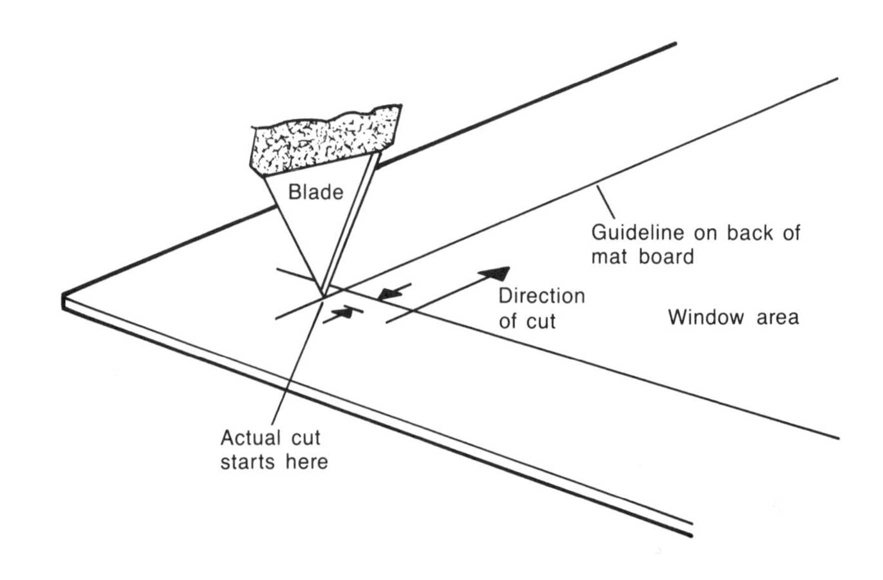

Figure 7.9
Work clockwise around the window when cutting a mat.

Now for the moment of truth. Push the cutter firmly but gently along the guide bar until the back of the blade crosses the crossline at the far side. Make the cut in one smooth motion, without stopping or jerking. Hold the guide bar firmly with the left hand.

Different people hold the mat cutter in different ways. We find it easiest to push with the heel of the palm, using the fingers to push the cutter laterally against the guide bar. Other people prefer to push the cutter with their fingertips. Do whatever is most comfortable.

The cutter glides easily across the board when the cut goes well. If you find it necessary to push down hard, adjust the blade.

To make the next cut, rotate the mat board 90° counterclockwise. The pushpins have to come out for this. Repeat the same operation for the last three cuts.

The Dexter may show a slight tendency to snag at the end of the last cut. Correct this by filing the leading edge of the cutter.

If your cut has gone well, the blank will fall out of the window when the mat is turned over. Often a few corners remain attached. So lift up the mat, holding the blank in place, turn it over, and see if the cut is complete all the way around. If not, slip the double-edge razor into the cut and finish the job. Do not just rip out the blank. It is a sign of poor workmanship to have ragged corners.

Occasionally there will also be a few attachments in the center of a cut where the blade has lifted momentarily. Use the razor for these, too, rather than trying to recut with the mat cutter.

For Left-Handers

With some mat cutters, like the EZ/Mat, the tool can be reversed for left-handed use. In this case, start at the right-hand corner and reverse all the directions as appropriate. The Dexter contour fits the right hand, but it too can be used by left-handers. The secret is to insert the blade on the corner of the mat away from you, and to *pull* the mat cutter toward you. Otherwise, follow the directions as given.

Figure 7.10
Adjust the blade on the mat cutter by holding a sample of mat board against the bottom of the tool. The tip of the blade barely extends beyond the thickness of the board.

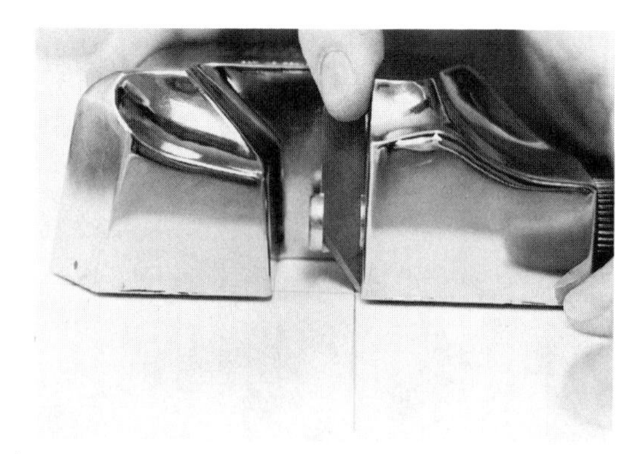

Figure 7.11
Start the cut by inserting the blade on the guideline just in front of the crossline at the corner. Cuts must overlap on the back of a window mat with beveled edges in order to go all the way through at the front.

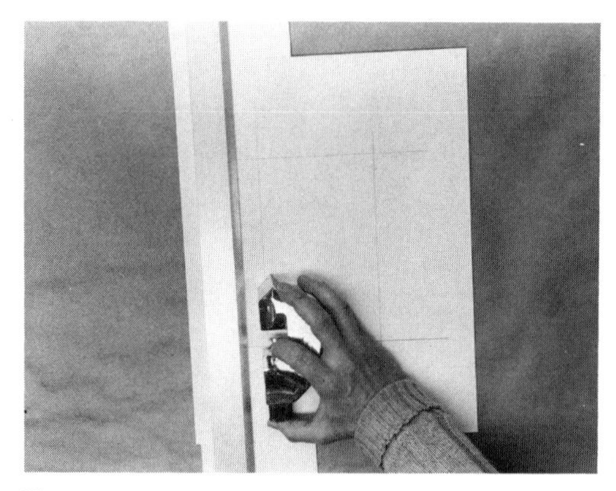

Figure 7.12
Set the blade first, then bring the guide bar over against the edge of the mat cutter. The guide bar always rests on the outer margin of the mat, away from the window.

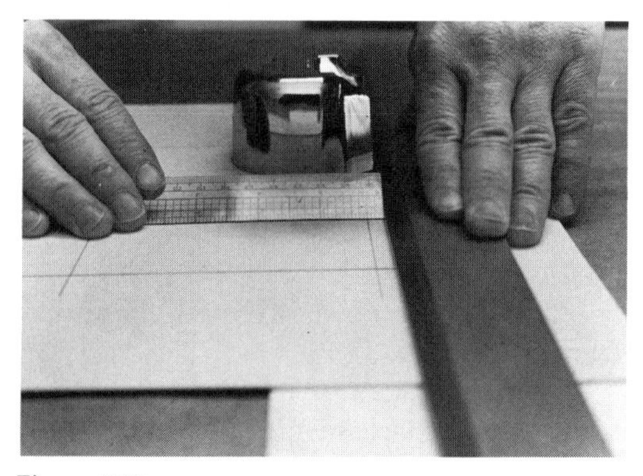

Figure 7.13
When the guide bar is snug against the edge of the mat cutter, measure the distance between the bar and the guideline for the cut at several points to make sure they are parallel.

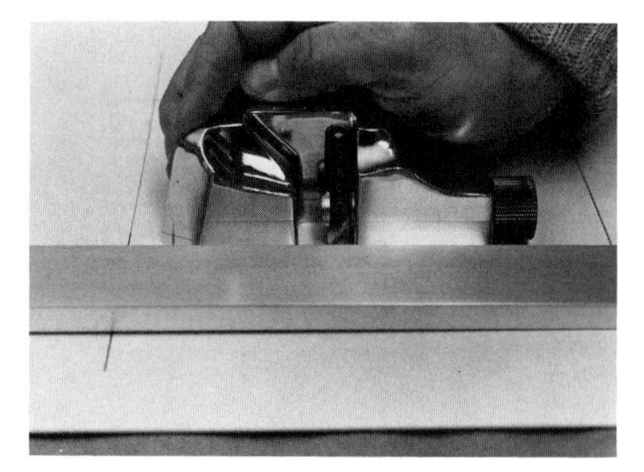

Figure 7.14
Push the mat cutter along the guide bar with the heel of your palm to make the cut. If the blade has been set properly, the cutter wil glide across the surface of the board.

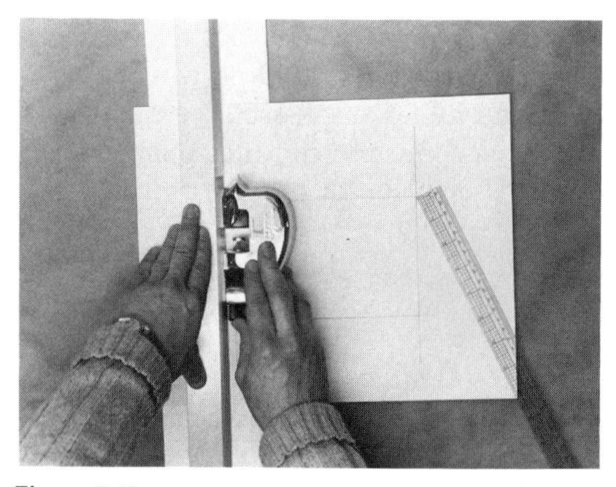

Figure 7.15
Some people prefer to push the cutter with the fingertips to get a more precise feeling of control.

Figure 7.16 (left)
A pencil mark on a piece of masking tape can help determine the precise degree of overcut needed at the corner. Set the cutter blade in the board using this pencil line to position it on the crossline of the cut. This mark will be useful for only one particular thickness of mat board.

Figure 7.17 (below left)
Left-handers can pull the mat cutter instead of pushing it. The guide bar in this demonstration has been hooked through a nail at the far end to prevent sideward movement.

Figure 7.18 (below right)
After setting the blade, use the heel of the palm to exert lateral pressure, while pulling with the fingertips. The right hand holds the guide bar.

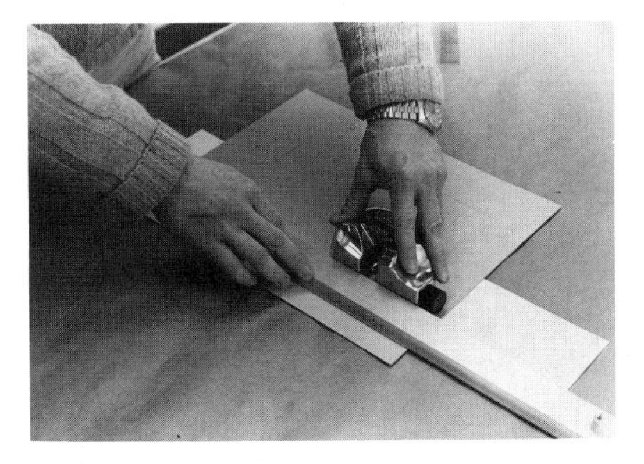

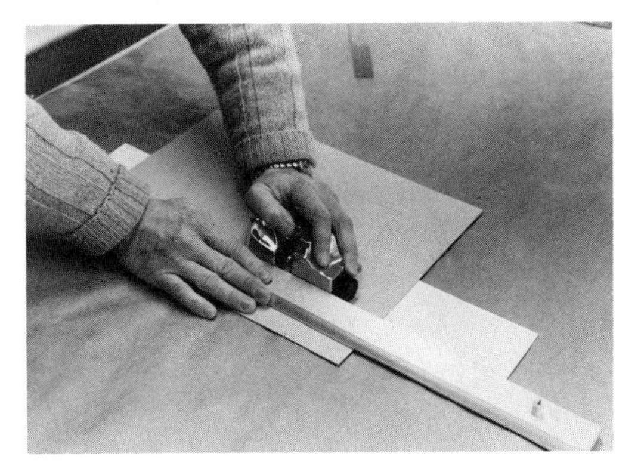

Finish the Edges

Lightly run a burnisher or thumbnail along the inside edge of the cut to make it look less raw; this also reduces the chance that a sharp edge will cut the print. Some workers use a piece of very fine sandpaper for this.

Overcuts on the front corners can also be burnished lightly to make them less visible. If you do not overcut, the result can be a dented or unsquare corner. If a dent is not too severe, rubbing a piece of mat board into the corner helps disguise it.

Erase All Marks from the Back of the Mat

Your cutting marks can transfer to the print. Never get careless or rushed into using a ballpoint pen to make the guide marks. One very fine Matisse print we saw was brought in for reframing because the original job was sloppy. When the print was opened for rematting, it turned out that the original framer had used a ballpoint pen and the marks had transferred to the print. Needless to say, this reduced its value.

Practice

To cut a mat takes skill, not talent. We have tried to include every detail that might help a beginner do things right from the start,

Figure 7.19
When the corners of the window blank remain attached to the mat after the cut, use a single-edge razor blade to finish the cut without leaving ragged corners.

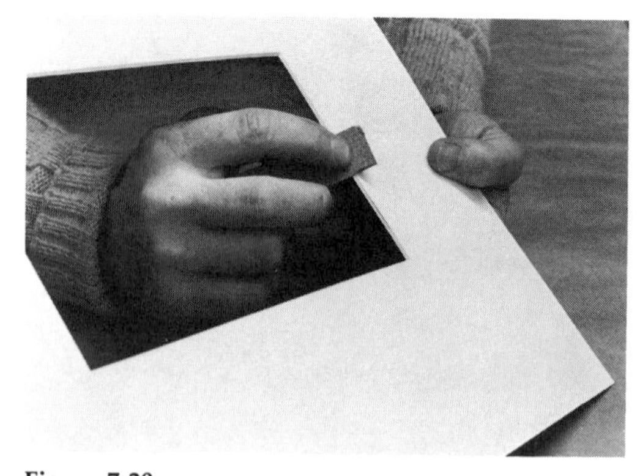

Figure 7.20
Sand the bevel lightly with fine sandpaper to soften knife-sharp edges and to remove any slight blemishes.

Figure 7.21
A burnisher can smooth out the bevel for a smoother look. It will also help remove any traces of the overcuts on the front, but excess pressure can make the surface noticeably shiny.

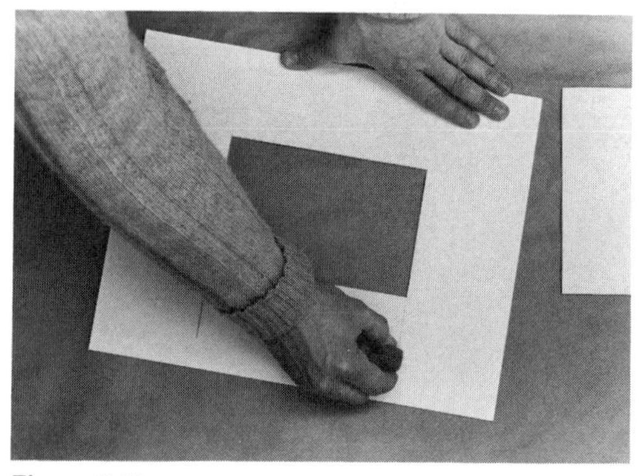

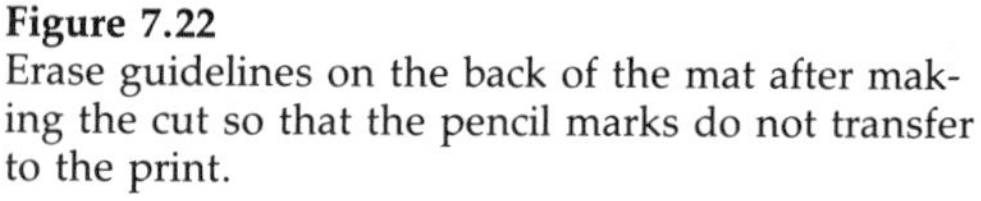

Figure 7.22
Erase guidelines on the back of the mat after making the cut so that the pencil marks do not transfer to the print.

sometimes to the point of being tedious. But only practice creates skill. It might help to buy some cheap board and practice making windows until you get the knack. If you do practice on cheap board, use something like a nice lewd purple so it will not be tempting to use your practice mats on good prints.

While practicing, why not make some sample corners? L-shaped corner samples can be held up against a corner of a print to determine the visual effect of different shades and grades of mat board. Make the samples about 2 inches wide and 1 foot long in both directions; note the kind of board on the back in pencil

If the guide bar seems to have a tendency to travel during the cut, try hooking the far end over a nail. This assumes, of course, that your bar has a hole in the end. Secured at one end, the guide bar swivels on a pivot and can be controlled more easily. With experience, however, you will probably find that the guide bar handles easily enough when loose.

Assemble the Mat

The completed conservation mat has a backing board attached to the window. The two pieces of board are held together along the long side with linen tape, so that they can open like a book (Figure 7.23). For this step you need the linen tape, the mat and backing board, some kind of burnisher, and a means to moisten the tape.

Put the backing board and window mat face down and butt them on the long side. Use a ceramic tongue or sponge to moisten a strip of linen tape that is about an inch shorter than the long side. If you use a sponge, put the water on with a patting motion rather than wiping off all the adhesive.

Wait 30 seconds or so for the adhesive to get tacky. Make sure that the two edges of the boards are butted tightly together, and run the tape down the seam, so that half of it sticks to each side. Run the burnishing tool down the seam to crease the tape.

Close the mat like a book. Check that all four corners of the window and backing align perfectly while there is still time to adjust them.

If you are too slow, there is still hope. Cut down the crease and grasp the top of one piece of the tape. Pull it inward toward the center of the mat and down, until it all comes off. Do not worry about some of the board coming with it, because this area will be concealed when the mat is closed. Do the same on the other side, and try again.

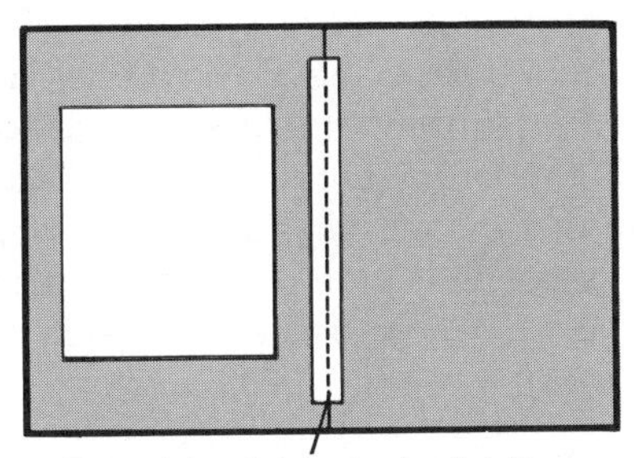

Tape (about 1 inch shorter than the long edge).

After creasing the tape, close the mat like a book.

Put the tape on with mat and backing side by side.

Figure 7.23
Assembling the mat.

Figure 7.24
Crease the linen tape when joining the backing board to the window mat. The mat closes like a book, with a tape as a hinge.

Backing Material

Either 2- or 4-ply rag or conservation board will do for the backing. Two-ply costs less and takes up less room in storage.

Always make the backing board the same size as the front mat. Small backing boards are false economy; the mat bows when backed or put under pressure in the frame.

Never use anything other than conservation-grade board. Foam core, chipboard, and corrugated cardboard cost less and cannot be seen, but they can destroy the print. Remember that the backing board is in even more intimate contact with the print than is the window mat.

While on the subject of substitutions, let us mention masking tape. Do not substitute it for linen tape. While the print lies closed in the mat, usually in a sealed environment like a storage box or frame, vapor-borne contaminants from the tape can affect the print. Anyway, within six months the masking tape gets so brittle that it breaks when the mat is opened. Linen tape stays flexible. If you have to substitute, make a Japanese tissue and starch hinge like the ones described in Chapter 4.

Why Tape the Long Side?

Most matted prints spend their lives in storage boxes. If you establish early the convention of hinging all mats on the long side, you already know which side is attached when lifting them out of the box. In that way mats do not go every which way, like items in a badly packed suitcase.

A reminder: make all mats so they can be opened for inspection of the print. If the window mat and backing board are attached along more than one side, inspection is difficult — and hence more unlikely.

THE SINK MAT

The sink mat is a variation on the basic mat developed by the Library of Congress staff to house pieces that are too thick, too bulky, and too heavy for a conventional mat [1]. Albumen prints,

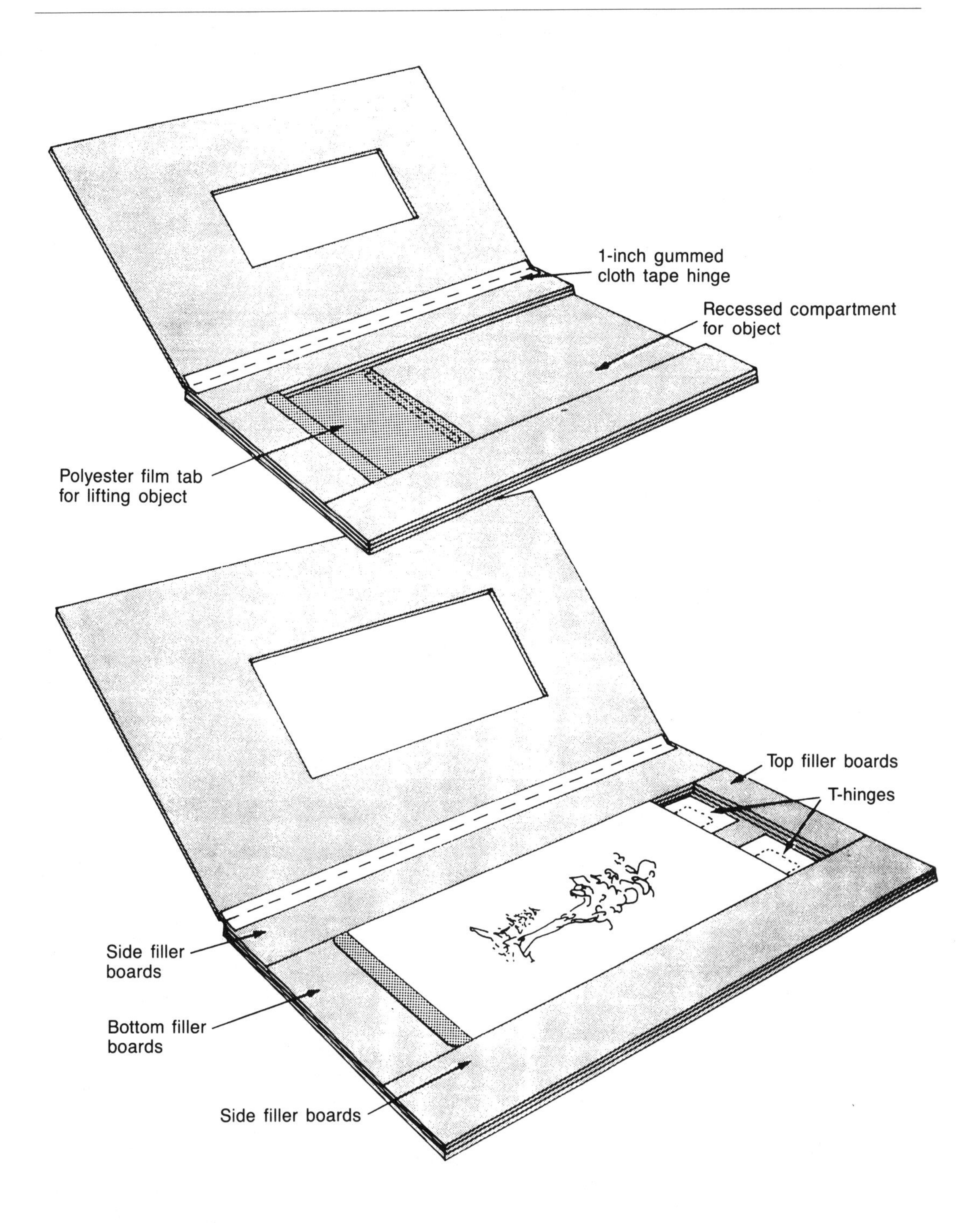

Figure 7.25
The sink mat.

which are mounted on thick card stock, would be a typical candidate for housing in a sink mat.

The structure of the sink mat resembles that of the basic mat, except that it has filler boards attached to the backing board around the outside of the print being matted. These filler boards are built up to a thickness equal to that of the mounted print. The overmat is hinged to the top of one filler board.

Why a Sink Mat?

The sink mat simultaneously solves a number of problems.

- ☐ *It protects the mat.* In the basic mat, a print on a thick mount causes the overmat to droop when the mats are stacked. This distortion of the mat can both damage the mat and put stress on the print.
- ☐ *It supports the print and saves the hinges.* When a heavy print in a mat is placed in an upright position, as it will be in a print framed for display, the weight can put extra strain on the hinges. This creates a dilemma: you have to either put extra heavy hinges on the back of the print with the chance that they will be difficult to remove, or else take the risk that the print may tear loose and slide down in the frame.
- ☐ *It makes it safer to move prints.* Filler boards on the sides keep the matted print from swaying side to side and possibly breaking the hinges.
- ☐ *It protects warped and brittle prints.* With a sink mat, it is not necessary to put great pressure on a possibly delicate print to get it to lie flat in a frame.

Cutting the Filler Boards

The filler boards must be cut precisely to make them flush with the outer edges of the mat and with one another. They should not be so close to the sides of the print as to restrict its movement when it is lifted on its hinges, but they should be as close as possible otherwise. Most prints do not have corners that are exactly square, so it is wise to cut each set of filler boards individually. Each sink mat should be approached as a custom fitting job.

Notes on Making the Mat

The illustration (Figure 7.25) gives you a good idea about how to modify the basic mat.

You need to cut the window on the overmat so that it overlaps the print by at least ⅛ inch on all sides, because the print cannot be floated in a sink mat.

When you attach the hinges to the back of the print, do it at places where it comes into contact with the backing board. This keeps the hinges from being stressed by the natural tendency of the print to pull away.

Decide how many thicknesses of mat board are needed to equal the thickness of the mounted print, and if the print is warped, how many are needed to match its highest point.

After cutting the filler strips for one side, attach them to one another with Scotch Brand Double-Sided Tape No. 415 and then attach them to the backing board. Do the same for the filler strips on the other side of the print, measuring to make sure you have a precise fit.

Attach the filler strips along the bottom in the same way. These run clear across the width of the backing board.

On the backing board where the print will lie, use the double-sided tape to attach a piece of polyester film that extends over the bottom filler strips. This provides a tab for lifting the print out of the sink.

The filler strips on the top run between the two filler strips on the sides, and do not carry any weight (unless the mat is stacked, in which case they serve to keep its shape intact). They should be high enough to leave room for the hinges.

Hinge the print in place in the accepted manner and attach the overmat to the front of a side filler strip with linen tape.

THE POLYESTER SLING

The name suggests a drink served in a pineapple with an umbrella on top, but the polyester sling is a method of putting a print into a mat without using hinges.

A glance at Figure 7.26 reveals the basic idea of the polyester sling. A piece of polyester film behind the print supports it in the window without being attached at any point to the back of the print. This support, called the sling, is attached to the back of the window mat with Scotch Brand Double-Sided Tape No. 415, the same as was used in the sink mat. The sling eliminates mounting attachments, holds the photograph securely, and can be fully reversed.

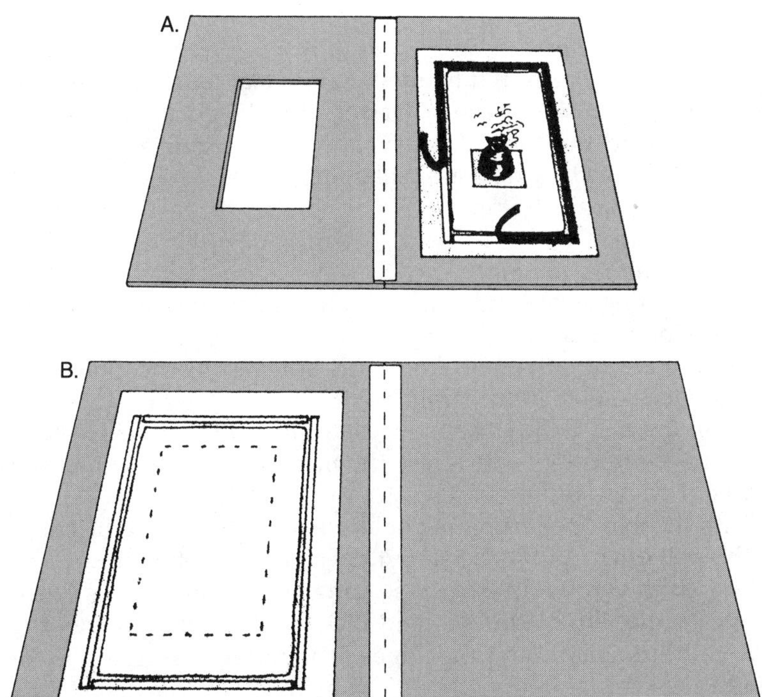

Figure 7.26
The polyester sling.
A. Step one.
B. Step two.

What situations might call for using a polyester sling? Three come immediately to mind.

- ☐ *Prints stored outside of mats.* Some collections store their holdings unmatted. Mounting and dismounting take time (and therefore money) and increase the risk of damage. Putting on and taking off hinges can create unnecessary wear on the top edge of the print.
- ☐ *Fragile prints.* Prints in a deteriorated state or with a delicate or friable support may be at increased risk from mounting for display, and may even be too weak to be safely supported on hinges.
- ☐ *Photographs on resin-coated (RC) paper.* These resist the application of conventional adhesives.

We can think of possible problems with the polyester sling, but they are minor. Heavy prints might slide down and pick up some adhesive along the edge. (The conservators at the Library of Congress say this has not proven to be a problem.) The design of the mat does not let you put a slipsheet between the window mat and the print. And finally, when you lift the window mat, the backing board does not protect the back of the print.

Print clips will not work for this mat, so you need a print weight. You can make one with a piece of lint-free cloth filled with BBs or lead shot, gathered together and taped securely at the top with duct tape.

Critical Dimensions

The print borders covered by the window mat must be at least ¾ inch on all four sides. The polyester film should be 3 to 4 mils thick, more in the case of heavy prints. The double-sided tape should be ¼ inch wide for smaller prints and ½ inch wide for large, heavy, or thick prints.

Notes on Making the Mat

Start by cutting a conventional mat and hinging the backing board to the window mat.

Cut a piece of polyester film that is at least one inch longer in both dimensions than the print.

Open the mat and put the polyester film on the backing board and the print on top of it. Close the mat, position the print, and put your weight on top of a small piece of blotter on it.

With the mat open, run the double-sided tape around the edges of the polyester film. Make sure it does not touch the print.

Pull off the brown backing tape and close the mat. Press down to secure the film to the mat and pick up the weight. Open the mat and burnish around the edges of the film to secure it more firmly.

To take apart the mat, cut the polyester film carefully around the edge of the print with a razor knife. Any traces of adhesives from the tape can be taken off with a "pickup" of dry rubber cement.

CONSERVATION MAT WITH WRAPPER

Some prints have such a delicate surface that they cannot withstand any type of physical contact, even with a protective slip sheet. An extra board hinged to the backing board of a standard conservation mat protects this type of print from dust and abrasion without touching it directly. This wrapper board can be swung behind the mat when the print is framed for display, and it keeps the window mat clean outside the frame.

You will need some starch-filled buckram cloth reinforcing tape of the kind used by bookbinders to make this type of mat. You will also need a strong, nonvolatile adhesive. A good choice is polyvinyl acetate (PVA), thinned to the consistency of heavy cream.

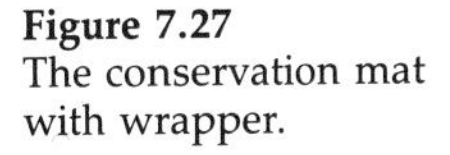

Figure 7.27
The conservation mat with wrapper.

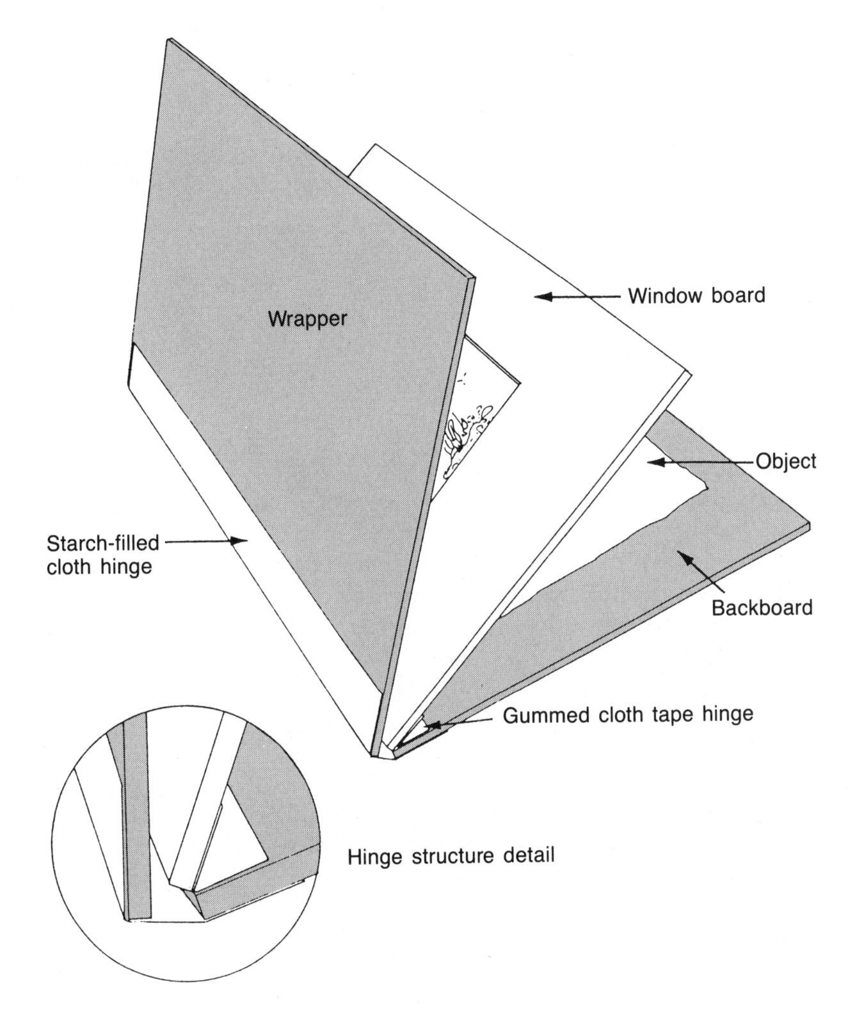

Notes on Making the Mat

Cut three pieces of mat board to the same size and cut a window mat from one in the standard way. To keep this wrapper from touching the print, the board for the window mat has to be — at a minimum — a 4-ply board. Either 6- or 8-ply may be needed for large or wavy pieces. The print can be overmatted on the borders or floated in the center of the opening.

Hinge the backing board to the window mat.

Put the wrapper board on top of the mat and measure the thickness of the spine they form. Add two inches to that and cut a strip of buckram that wide. (For example, if the spine is 3⁄8 inch thick, the strip should be 2 3⁄8 inch wide.) Trim this strip to the height of the side of the mat.

The overlap of the cloth hinge will be 1 inch, front and back. Lightly pencil a guideline on both pieces of board 1 inch from the edge.

Brush the PVA onto the cloth hinge and lay it on a clean work surface.

Put the backing board of the closed mat onto the hinge, so that the edge of the hinge is aligned with the guideline, and gently press down.

Put the wrapper board on top of the window mat and clip it in place. Use a bone burnisher to turn the hinge up and over the window mat. Rub carefully to prevent the cloth from getting puckers and use a sheet of scrap paper to keep burnisher pressure marks from showing.

Turn over and burnish the hinge on the back.

Open the wrapper immediately so that the PVA on the cloth dries to the edges of the boards. The cloth hinge should stick only to the front of the wrapper board and to the back side of the backing board.

SLIP SHEETS

To protect against dirt and abrasion, conservators at the Library of Congress suggest using a slip sheet of smooth, archivally permanent material when prints are stored in mats. Specifically, since 1972 the Library has used clear, uncoated polyester film. The film must have no plasticizers, UV inhibitors or absorbents, and must be guaranteed nonyellowing, dimensionally stable, and resistant to moisture, abrasion, and chemicals. Polyester films that meet government standards are Mylar Types D, A, or S; Melinex Type 516; and Scotchpar Industrial Grade Polyester Film.

REFERENCE

1. *Matting and Hinging of Works of Art on Paper*, a National Preservation Program Publication compiled by Merrily A. Smith. New York: Consultant Press, 1986.

CHAPTER 8

Decorative Mats

The rather simple designs presented in the last chapter have a number of things to recommend them. Chief among them is the ease with which they can be created. For many prints, a plain mat enhances their appearance most effectively. The mechanistic aesthetic that started at the Bauhaus coincides with a purist tradition in American photography in demanding a clean and simple style of matting. Respect for the inherent elegance of photographic media makes this viewpoint acceptable to connoisseur and conservator alike.

A case can also be made for the occasional use of more elaborate and decorative mats. This chapter gives you information on how to make a number of decorative mats simply and without departing from sound conservation practices.

The design of even a decorative mat calls for restraint. Everything about the mat should call attention to the image, rather than the other way around. A mat maker should consider himself or herself the assistant, not the partner, of the artist. Any other approach leads to a clash of wills that is painfully apparent to a sensitive observer. When in doubt, it is better to err on the side of simplicity.

MULTIPLE-WINDOW MATS

To make a mat with more than one window requires no more advanced techniques than those used to make a basic window mat. The important difference lies in the amount of planning you need to do.

Laying Out the Mat

Starting with a sketch is the best way to assure a well-designed mat when more than one window is planned. A convenient way of making one to scale is to use draftsmen's quadrille paper, and assign a scale of one square to the inch. If a quadrille pad is not available, a rougher sketch can serve as well.

Sometimes the relation of two windows will be determined by the material itself, such as the need to display a caption printed on the mount board of the print. In other cases, you have to decide the proper relation of the imagery. It helps to visualize the final result if you lay the prints on a piece of board the same size as the final mat, and move them around until they feel right. Measure the results and transfer them to the sketch.

In most cases these two principles help assure a pleasing result:

- ☐ Borders *between* windows should be narrower than the openings themselves.
- ☐ The *exterior* border around all the windows should be wider than the borders between them.

This approach makes the images cohere into a single group.

You might be tempted, when matting two equal-sized prints in the same mat, to simply divide the mat in half and lay out each half separately. In this case you would get a border between the prints at least twice as wide as the side borders. This distances the prints from one another rather than making them cohere.

Even though the overmat might conceal it, print borders should not physically overlap. Leave room so that the prints do not touch one another, and never trim a historic or potentially valuable print to fit your personal conception of how the mat should look.

You might need to make a second sketch after looking at the first. Write down all measurements on the sketch.

Cutting the Mat

Transfer your planned measurements to the back of the mat board with either a T-square or a mat-marking scribe. Erase all crosslines to avoid overcutting through confusion.

Draw an X in the center of each window. Reversed bevels give the mat maker the biggest problem with multiple-window mats. The X's solve this problem: simply keep the mat cutter on the inside of the window and the guide bar on the outside; that is, the mat cutter always goes closest to the X. In case the bevels get reversed, the mat has to be recut.

Plan your sequence of cuts carefully. If possible, treat all interior openings like one big window with interruptions. Cut all the tops of the windows in the upper row, then go down the side, move onto the bottoms of the windows in the bottom row, and thus up the final side. If the outside borders are laid out flush with one another, you need only lift the cutter quickly while skipping over the interior borders.

Cut the interior window sides next. While cutting the bevels on the inside edges of the window, always check to make sure that the cutter goes in from the side with the X, and that the guide bar rests on the window's border.

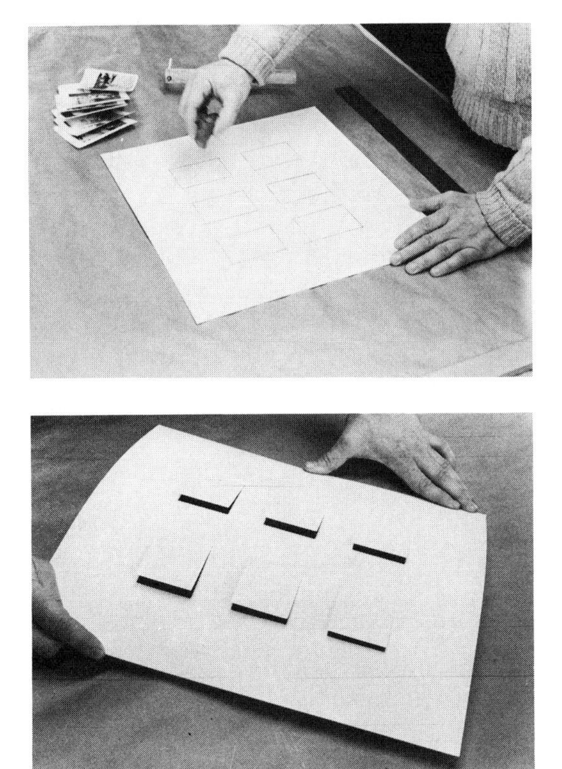

Figure 8.1 (above left)
Multiple windows: Erase all lines except the window lines before starting to cut, to avoid possible overcutting due to confusion.

Figure 8.2 (above right)
Make all cuts that line up directly at the same time to ensure that they form a straight, continuous edge.

Figure 8.3 (left)
Three sides of each window have been cut, and only the bottoms remain.

When two or more windows have sides that run in a continuous straight line, make certain that the mat cutter gets set true to the bar each time. Even if the bar does not move, a sloppy cut can result from starting wrong. When a couple of windows line up perfectly the mat looks great, but the eye can pick up the slightest discrepancy.

Finish the bevels as described in the last chapter. Hinge the backing board as for the basic mat. Check the size of the windows. You are cutting more, so the chances for a mistake increase. Finally, mount the prints and close the mat.

DOUBLE OR FILLET MATS

A double or fillet mat consists of two (or possibly more) mats with successively larger windows laid one on top of the other.

The term *fillet* refers to the exposed part of the undermat that makes an interior border around the picture edge. The fillet's width is ordinarily uniform around all four sides, though this can vary.

When you master the technique of cutting a fillet mat, an endless variety of advanced matting methods opens up to you. Mat boards of contrasting colors can highlight or subdue image colors, and by using a rag or conservation board undermat as a barrier you can draw on stocks of other colored board that would not otherwise be safe to use. Two or more mats create a feeling of luxurious

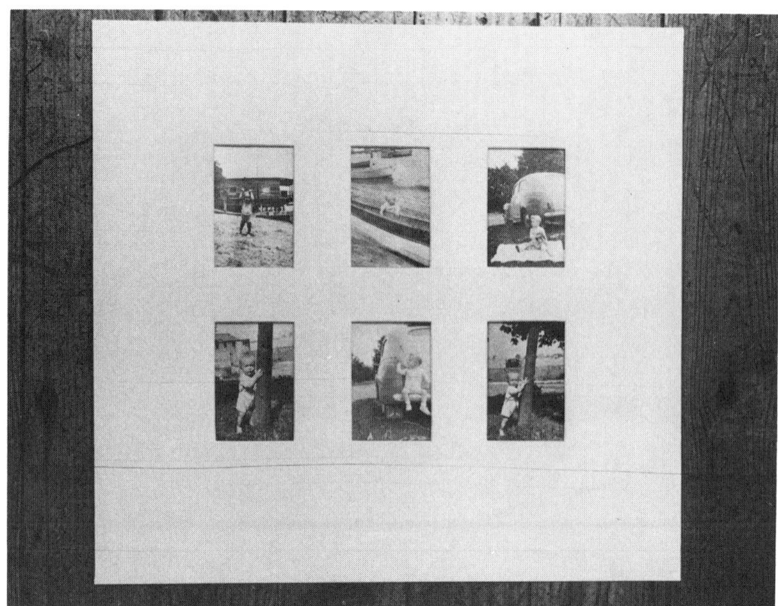

Figure 8.4
A collection of snapshot images on the theme of one child makes good use of a multiple-window mat.

depth even when the boards are the same color, and especially when they are an extra-white rag. Double mats put extra space between the print and the glass of the frame. A fillet gives the printmaker a convenient place to sign the work, as well. When you learn the technique of covering the mat with fabric or paper, you will find that a fillet mat can provide a useful way of creating a slight distance between the image and the texture of the overmat.

The necessary tools and cutting skills, despite the more accomplished look of a fillet mat, have already been acquired in learning how to cut a basic conservation mat. The only significant problem you face in cutting a fillet mat lies in making the fillet exactly the same width all the way around the picture. The relatively narrow width of the fillet makes minor discrepancies immediately obvious.

A common practice is to make the undermat a darker color than the overmat to frame the image and draw the eye to it; contrasting colors further accentuate any variation in the fillet's width.

Cutting the Mat

Cut both boards to exactly the same size. Put them together and trim any excess from the edges with a razor-sharp knife.

The cutting sequence is overmat first, undermat next. The overmat is used as a template to mark the window on the undermat.

Measure the image area you want to show. The window on the overmat will be larger than the image. To determine its size, add twice the fillet width to both horizontal and vertical dimensions of the image.

To do this you need to know the width of the fillet. Anything smaller than 3⁄16 inch tends to get lost. If you intend to use a composite board for the overmat, we suggest a wider fillet than that, perhaps 3⁄8 inch.

Let us take an example to show how to figure the size of the overmat window. We will assume that we are gong to use a ¼-inch fillet. If the image area is 7½ × 9½ inches, we would add ½ inch (that is, twice ¼) to both dimensions to get a result of 8 × 10 inches for the overmat window. As before you will find that a sketch helps.

Mark and cut the window on the overmat just as you would for a basic mat. Put the cut mat back to back with the piece of board you will use for the undermat. Check again to see that all edges are flush. On the backs of both, write "top" or put an X on the ends that will be uppermost in the frame. This ensures a uniform fillet when the mats are put together.

Cut a strip of scrap board about 4 inches long and to the exact width of the fillet. In the case of our example, this would be ¼ inch wide.

With print-positioning clips, clip the mat boards back to back and check that the edges are flush. From now on they should not move until you are finished marking. (If you do not have print-positioning clips, you can substitute clothespins, but put some scrap pieces of the board between their jaws and the mat board to avoid marring the board.)

Hold the spacer strip of scrap board against the edge of the window in one corner. Draw a pencil line down its length. Slide the spacer down to the next corner and repeat. Do this around all four edges. Connect the lines with a ruler.

Separate the mat boards and use the guidelines to cut the window in the back of the undermat.

Erase the pencil marks from the back of the undermat, turn it over, and hold the two mats together for a visual check.

You have a choice of adhesives for joining the two mats together: rice paste, methyl cellulose, carpenter's white glue, or Scotch Adhesive Transfer Tape No. 924. Glue gives you more time to exactly position the mats, but tape has the advantage of speed.

Transfer tape has not been mentioned before. It consists of a layer of adhesive on a strip of release paper. You put it in place just like ordinary tape, peel off the release paper, and get, in effect, a double-sided adhesive tape without the tape.

From this point on you follow the same techniques outlined in the section on basic matting. Hinge the backing board in place with linen tape and mount the print.

INLAID MATS

An inlaid mat has two or more colors of board butted flush against each other so that they give the appearance of being a single piece of board. The effect is at once subtle and striking.

An inlaid mat is not easy to make without the proper tools. In fact, for the first time in this book, we are describing a mat that you cannot make with a hand-held mat cutter. Our justification for including it here is that it represents a high point in the mat-

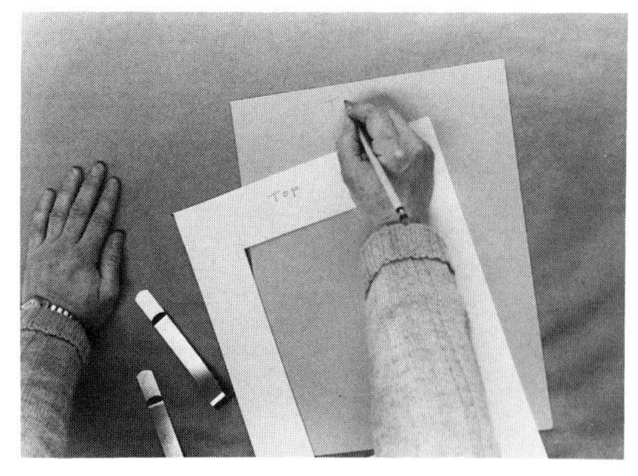

Figure 8.5
Fillet: Mark "top" on both pieces after squaring edges and cutting the first window for the overmat. Assemble both pieces with print positioning clips (shown).

Figure 8.6
With the two pieces clipped firmly together, use a precut spacer as wide as the desired fillet to mark the position of the second window.

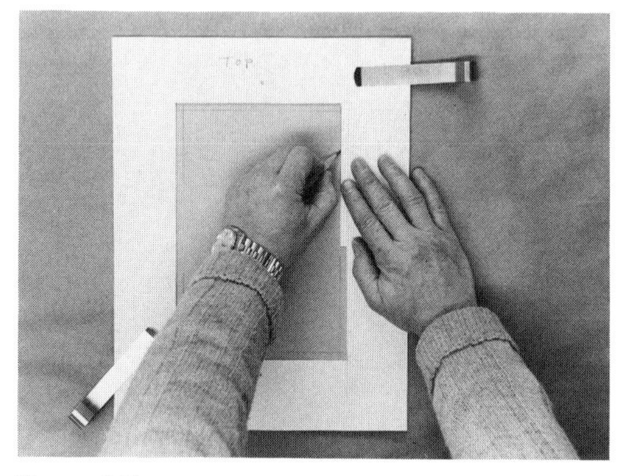

Figure 8.7
Use a pencil to mark the outlines of the fillet mat window along the spacer, and then connect the lines all the way around the window.

Figure 8.8
Use a ruler to check the position of the two mats after finishing the cutting.

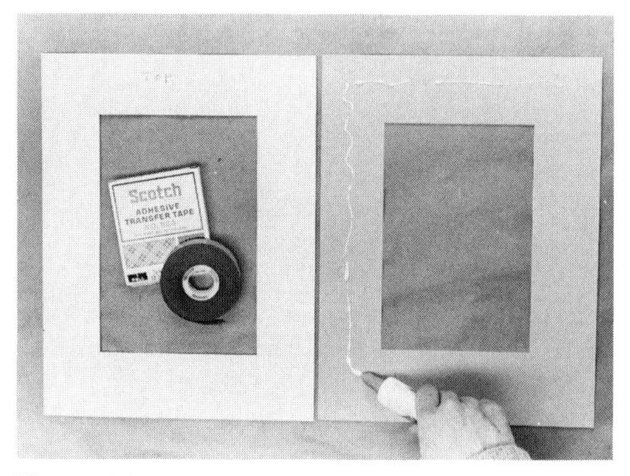

Figure 8.9
Either white glue or adhesive transfer tape can be used to put the two pieces of board together before hinging.

making craft, requiring a degree of proficiency comparable in pottery to fitting a tight lid onto a hand-thrown jar.

We tried repeatedly while taking photos for this section to cut an inlaid mat with a Dexter, and, sad to admit, we failed just as often as we tried. The difficulty lies in the fact that all four outer bevels of the inlay have to butt *exactly* flush with the bevels of the outer mat that surrounds it. The most minuscule deviation, even less than 1/64 inch, spoils the effect.

The simplest tool that will cut a satisfactory inlay is Alto's EZ/Mat Cutter (they're not expensive, but they are a step up from simple, hand-held cutters). Of course, if you have access to a Keeton, a C&H, or a Logan production cutting machine, all the better.

Use either rag or conservation board. The increased number of conservation board colors makes possible combinations quite numerous. Before choosing colors, make sure that both boards match exactly in thickness. Here, too, slight deviations show up noticeably. For ease of working use 4-ply rather than 2-ply board.

You may decide to back the finished mat with acid-free barrier paper to avoid any chance of the linen tape's adhesive touching the print. Otherwise, all tools and materials remain the same as for basic mats.

Cutting the Mat

There is a simple trick to getting a perfectly butted inlay. You use the edge of the outer mat board as your reference for cutting *all* bevels. Tape two pieces of board together to use the edges of just one to position the cutter blade. To do this effectively, you need mat boards of different sizes.

Begin by cutting a piece of board to the exterior dimensions of the mat. This will become the outer panel of the mat. We assume that you have already measured the image and know how big a mat is needed.

Next, cut a smaller piece of board. Exact dimensions do not matter, but they have to be larger overall than the inlay and smaller than the outer mat. This board will be the inlay.

You have to decide the inlay width as a matter of taste. For practicality make it at least 1/4 inch wide; anything smaller is hard to handle. We think that a good ratio of outer panel width to inlay is around 4:1, but take that just as a general guide. Usual practice makes the inlay panel appreciably narrower than the outer panel.

Do not forget to add double the inlay width to both image dimensions when figuring the size of the second board. That is, if you have a 5 × 7 image with a 1/4 inch inlay, you need a second board *larger* than 5 1/2 × 7 1/2 inches. Allow some margin.

Tape the smaller board to the back of the larger, face to back. Use masking or drafting tape. Drafting tape is better because it has less tack, but either will do. This tape gets removed later.

So much for the easy part.

Use a pencil and the EZ/Mat cutter to mark a window outline

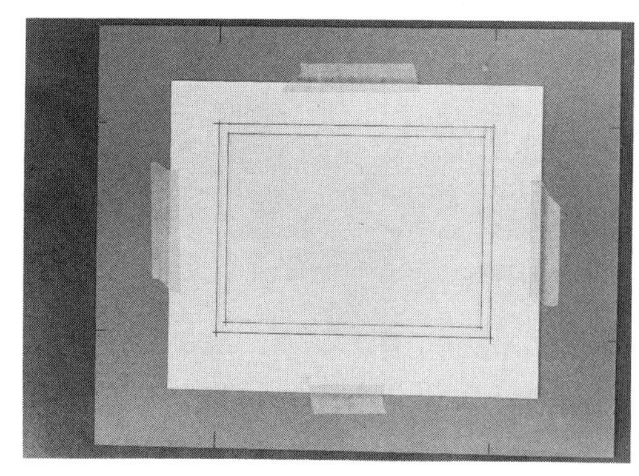

Figure 8.10
Inlaid: An assemblage taped and marked prior to cutting a gray mat with an inlaid white strip to run around the image. The two lines on the back of the white board indicate the width of the white inlay. Notice that the outside line's position has been marked along the edges of the gray board, for later extension to cut the window in the gray mat.

Figure 8.11
This photo shows the assemblage after the last cut has been made to complete the inlay strip. See how the edge of the gray board rests against the side stop of the EZ/Mat cutter, and how it's not possible to check the completeness of the cut until the entire strip is finished.

Figure 8.12
The assemblage has been taken apart to reveal the inlay strip (right). The remainder is scrap.

Figure 8.13
Cut the window on the outer mat (gray) as marked; fit the inlay strip into the opening and check for fit. It must be exact.

Figure 8.14
Tape the two pieces together with linen tape. Scotch Magic Transparent Tape can be used as a substitute.

Figure 8.15
This close-up shows the corner of an inlaid mat to demonstrate the close fit that can be achieved.

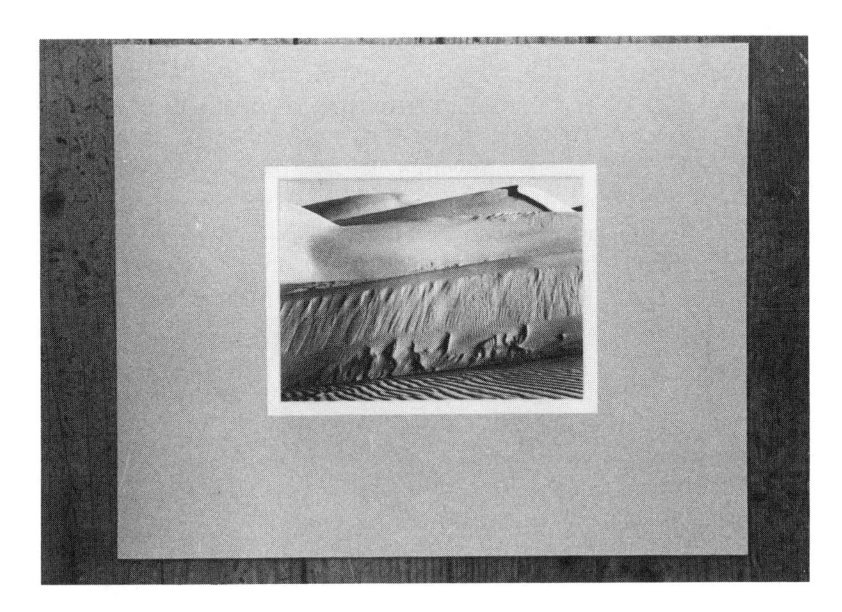

Figure 8.16
The finished piece with image in place, surrounded by the white inlaid border.

on the back of the inlay. It makes things easier to think about if you pretend that the two boards taped together are just a single piece; mark and cut as though making a regular window.

Change settings on the EZ/Mat and mark a second, bigger window on the back of the inlay board. This bigger window will become the outer edges of the inlay panel and the inner bevel of the outer panel. If you want an inlay that is ½ inch wide, make this window 1 inch bigger along both dimensions.

While marking this second window, extend the lines past the edges of the inlay board so that they run onto the back of the outer mat. This saves time later when you cut another bevel along exactly the same line.

Now reset the EZ/Mat for the first window. Make the cuts through the back of the inlay board, just like for a conventional window mat.

Cut through only one layer of board doing the inlay, but the blade has got to go all the way through it. There is no easy way to check at this stage if your cut is good, so test the blade setting on a scrap of the same board before doing the window.

While you do all this cutting, be certain that the board's edge is pushed tight against the guide bar of the cutter so that the board cannot move.

Next, set up the EZ/Mat for the larger window (the outer edges of the inlay). Make this cut all the way around. Now you can untape the boards. If the cut was perfect, you have three parts from the inlay board: a window blank (discard), an inlay (keep), and an outer margin (discard).

The usual precautions apply here: do not rip the sections apart. Check for incomplete cuts and finish them off with a double-edge razor if necessary.

Do you have any serious overcuts at the corners? Remember that the inlay attracts attention. Overcuts on an inlay consequently get attention, too, and your standards should be stricter for them than for a conventional mat. If necessary, do it over.

Now we go back to the outer mat. Extend those guidelines you made all the way across. You have now marked the position for the inlay window on the back of this mat.

Cut the inlay window just like a basic window. Remove the blank. Drop the inlay into the window of the outer mat. Check for fit. If the inlay does not fit exactly, try rotating it 180°. It should fit now. It helps to get a quicker and more accurate match if you mark an X or "top" on the corresponding sides of the outer and inlay mats while they are still taped together, so you can put the same two sides together.

Erase all pencil marks from the back. Run strips of linen tape on the backs of the two mats down the seam.

At some time in the future, adhesive from the linen tape might transfer to the print surface. Now is the time, if you decide to do so, to put a sheet of barrier paper on the back of the inlay mat; you may instead want to cut a second, hidden mat of 2-ply conservation board. Make its window a trifle larger than the image window and adhere it to the overmat. Hinge the mat to a backing board and mount the print.

Inlaid Mats: A Panel Variation

A more elaborate version of the inlaid mat incorporates a panel of contrasting color into the body of the mat itself.

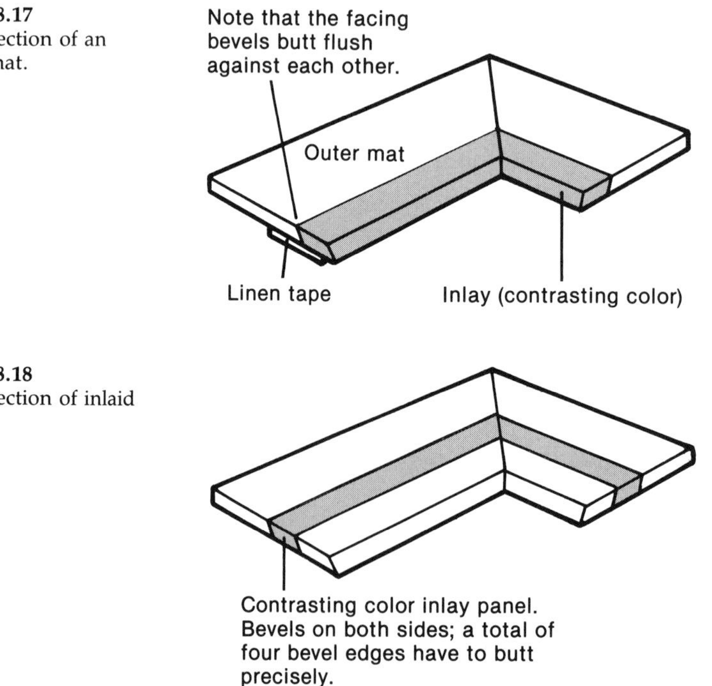
Note that the facing bevels butt flush against each other.
Outer mat
Linen tape
Inlay (contrasting color)
Contrasting color inlay panel. Bevels on both sides; a total of four bevel edges have to butt precisely.

Figure 8.17
Cross section of an inlaid mat.

Figure 8.18
Cross section of inlaid panel.

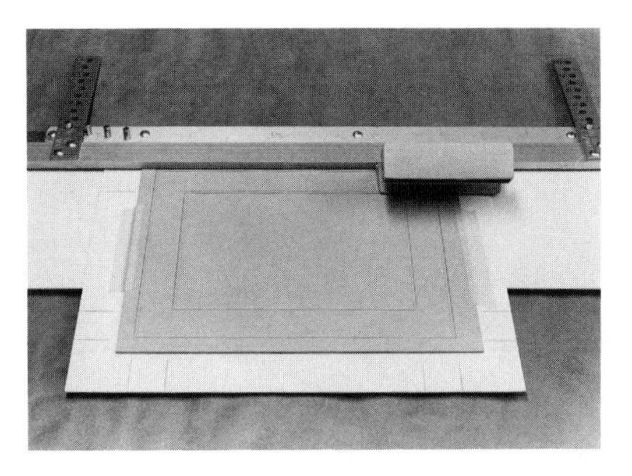

Figure 8.19
Inlaid panel: The entire assemblage ready for cutting a mat with an inlaid gray panel. This time both the inner and outer edge lines have been marked on the larger mat board.

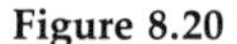

Figure 8.20
After the panel has been cut, the outer mat must be cut all at one time. This photo shows, going from the outside in, the various panels remaining: the outer border, the inlay panel (scrapped from this piece of board), the innermost panel surrounding the image, and the window blank (also scrapped).

Make the inlaid panel a different width to avoid the appearance of boxes inside boxes. A thin panel of a dark color will serve to constrain an otherwise open print, and a wide panel of white or ivory on a dark mat has the same general effect. Before investing all the effort to cut an inlaid panel mat, try drawing it to scale and shading in the dark areas to get some sense of the impact of different widths. Calculate all dimensions before starting.

Basically the same procedure as for cutting an inlaid mat will be followed except for the order of the cuts. Tape the inlay panel board face to back on the larger board after cutting them to size. Draw the guidelines for the inlay panel on the back of the board. Extend both sets of lines onto the back of the other board

Cut both inner and outer bevels of the inlay. Remove it and the blanks from the back of the other board after marking both with an X or "top."

Mark and cut a window in the other mat board to fit the print, just as you did with a basic mat. Do not remove the blank yet.

Use a straightedge and pencil to extend the guide marks for the inlay. Cut the inner, then outer, bevels along these lines.

At this point, you have made a total of 20 cuts, of which 16 have to match exactly. Finish the corners with a razor and assemble the mat.

Join the pieces on the back with linen tape, hinge, and mount the print.

These instructions describe the basic version of the panel inlay. You can experiment with possible variations like inlaying a couple of colors. There is no reason for the panel to run all the way around the border, so you might try putting panels in along the sides or the top and bottom. Or use the inlaid mat as the top in a fillet mat — your imagination will suggest a range of possibilities.

Figure 8.21
The completed inlaid panel mat with a gray panel.

FABRIC-COVERED MATS

There is a long tradition of using fabric to wrap mats, and without question cloth enhances certain prints. An old sepia-toned portrait gains added warmth with a watermarked silk mat, or you can accentuate the textures of a salted-paper print with a coarse-weave linen mat.

Any type of cloth can be used to wrap the mat, but not all materials may be safe from a conservation point of view. Little testing on fabrics has been reported in the technical literature, and the wide variety of fabrics gives us pause. With so many types of cloth available, we cannot make valid recommendations about the safety of all kinds of material. A tremendous number of additives are used to improve the looks and handling characteristics of the raw fabric: dyes, both waterfast and otherwise; sizing and starch; inks or paints on silkscreen fabrics; and fluorescent brighteners. Cloth with a heavy nap, like velvet, causes static electricity that attracts particles to the print surface.

If you construct a fillet mat with fabric covering only the outer mat, there is no need to forgo any of the cloths you can buy. For mats touching the print, choose only natural-finish fabrics like linen, cotton, or silk that have not been dyed, printed, or colored. Launder these fabrics to take out starches and sizing material. For these mats, a concealed sheet of barrier paper or a 2-ply undermat gives the print extra protection.

Tools and Materials

A little more in the way of supplies is needed than for most other mats; here is a list.

☐ Window mat of acid-free board *

☐ Fabric, larger than mat

☐ Methyl cellulose adhesive †

☐ Utility brush 2 inches wide

☐ Sharp knife

☐ Kraft paper

☐ Water sprayer

☐ Electric clothes iron, *or* dry mount press and tacking iron

☐ Linen tape

☐ Burnisher

Covering the Mat

Spread kraft paper on the work surface to protect it from the adhesive. Brush methyl cellulose on the front and bevels of the mat. If it is going into the "frameless frame" described in Part III, do the outside edges too. Cover with a thin, even layer and make sure that all surfaces, especially the bevels, get a good coating.

Let the methyl cellulose dry to the touch on a fresh piece of paper. The extra water speeds the drying, so it should be ready in a few minutes. Do not let it get bone dry, because that makes reactivating the adhesive more difficult.

Iron the fabric to take out any creases. Lay it on the front of the mat and line up the weave square with the mat. Cover overall with a fresh sheet of kraft paper. (Make sure that you do not accidentally use the first piece.)

Spray this cover sheet lightly with water, and then iron it and the fabric onto the front surface of the mat. The combination of heat, pressure, and moisture reactivates the methyl cellulose and bonds the cloth to the board. The cover sheet keeps the glue and fabric from scorching.

You can substitute a dry mount press for the iron. Use release paper to keep adhesive from getting on the platen.

After the mat cools, turn it over and make four cuts, each 1 inch long, in each corner at a 45° angle extending toward the center of the window. Leave a little space at each corner so that an uncut piece can cover the corner bevel, rather than starting the cut right

* Cut the mat to the same size as if making a basic window mat. You might decide to reverse the bevels to get an easier wrap, but we give instructions for the more attractive mat with bevels in the ordinary fashion.

† Dilute premixed methyl cellulose adhesive with water, 2 parts methyl cellulose to 1 part water. See "Adhesives" in Chapter 4, for the methyl cellulose recipe.

Figure 8.22
Coat the face of the mat with adhesive and allow to dry.

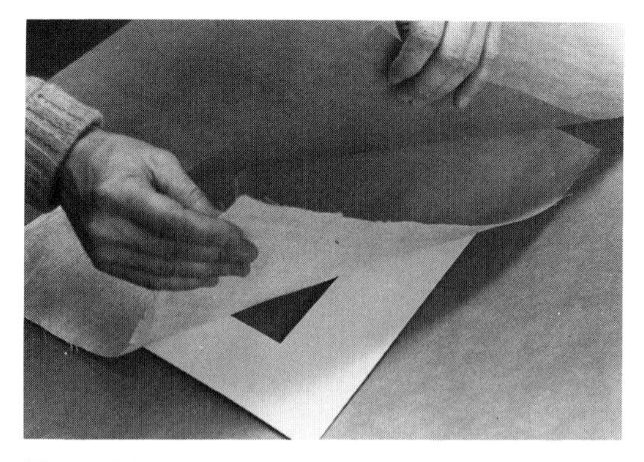

Figure 8.23
Cover the face with an oversized piece of fabric and cover sheet of kraft paper.

Figure 8.24
Spray the cover sheet with water and use an iron to adhere the fabric to the mat face.

Figure 8.25
Make angled cuts at each corner and cut out the center section of fabric, leaving enough to wrap around the bevels.

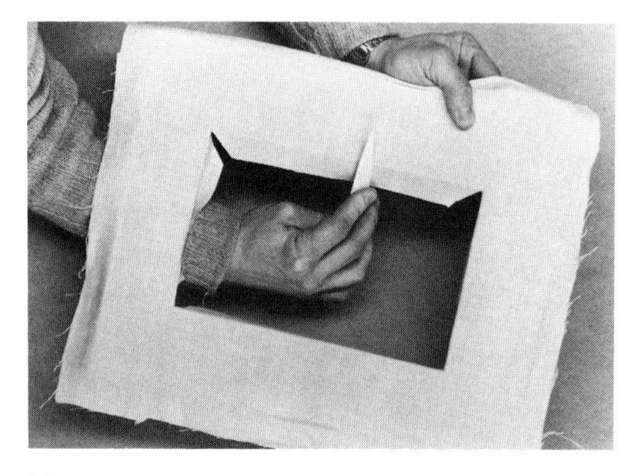

Figure 8.26
Crease the fabric along the edges of the window.

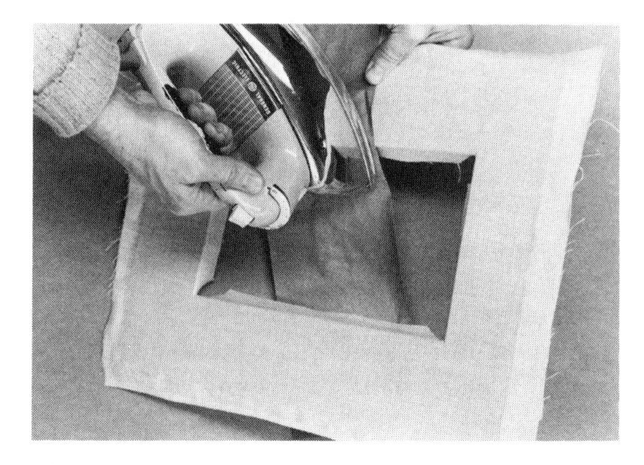

Figure 8.27
Join the fabric to the beveled window edges with a cover sheet and iron.

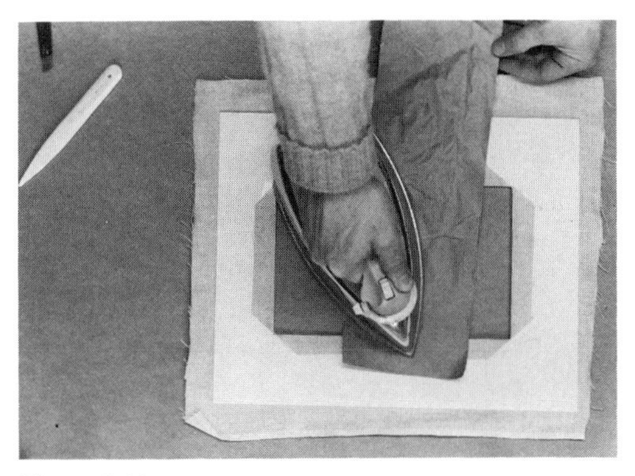

Figure 8.28
Join the folded-over fabric flaps to the back with the iron.

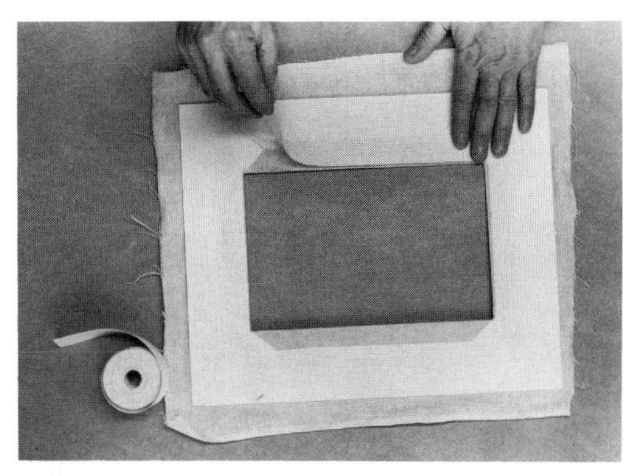

Figure 8.29
Cover the folded-over strips on the back with strips of linen tape along their edges.

Figure 8.30
A hidden undermat with a window only slightly larger than the fabric mat will provide extra protection against adhesive contacting the print surface.

Figure 8.31
The completed piece ready for framing.

at the corner. Cut out the center of the window, leaving an unattached strip on each side for wrapping around the bevel.

The next step is to join the fabric to the bevels. Fold the cloth where it touches the inner line of the bevel. Make this fold on all four sides before going on. A burnisher produces a sharp crease; run it along the back of the mat several times with the cloth folded around.

Join the cloth to the bevel with the tip of the clothes iron or tacking iron. Again, a cover sheet sprayed with water provides the needed moisture. Make certain that the cloth sticks firmly to the bevel, because sloppiness will show up later.

Spread methyl cellulose on the part of the mat back that will be covered by the fabric wraparound. Take care when spreading the adhesive that you do not coat parts of the back that will be exposed, especially if the part in question comes in contact with the print. Let the adhesive dry, and iron the wrapped-around strips.

To finish the wrapping in a pleasing style, run linen tape strips along the edges of the cloth on the back. This gives the wrapping extra strength and prevents fraying.

The outside of the mat still has extra material hanging loose. For a conventional frame with the mat edges concealed, trim the excess by hanging the mat edge over the side of the table and running a knife along it. For frames that show the mat edge, wrap and iron the strips around the outside of the mat as you did on the bevel.

If the fabric mat is not backed by a fillet, we suggest a concealed mat of 2-ply acid-free board fitted between fabric and print. Make the concealed mat exactly the same size as the fabric mat but without beveled edges, and join the two. This keeps adhesive from contacting the print, and gives extra protection in case the cloth contains any chemical contaminants.

PAINTED BEVELS

A painted bevel on an ordinary mat transforms this edge into a strong image boundary, adds a note of color, and increases the polished appearance of the framing. Painting a bevel on a fillet mat attracts added attention to the fillet and can complement its color scheme.

You have a wide choice of paints, none of which need be expensive. Suitable ones include watercolors in tubes, acrylic artist's paints that come in tubes like oil colors, and colored drawing inks. Do not buy the palette type of watercolor, because it does not deliver enough pigment for this kind of work. As for inks, get ones that are waterfast, like those made by Pelikan or Hunt Speedball.

A size 00 watercolor brush of the kind made by Windsor & Newton will satisfactorily apply paint or ink. While it is not necessary to have an expensive brush, red sable bristles make the application more accurate.

Start by cutting a mat of conservation-grade board. The cutting procedure stays the same as for a basic mat, except that the corners cannot be overcut. Capillary action draws the color into the overcuts, which show up as hairlines extending into the mat surface. Undercut the corners and finish them with a razor. Save the window blank for practice painting.

You can make a colored bevel wider and more obvious by cutting the bevel at a shallower angle. To do this with a Dexter, snap off the rounded *top* of the blade before putting it into the mat cutter. This gives more clearance so that the blade can be set in the cutter at a greater angle. Butt two pairs of pliers along the line where you make the break. Protect your eyes from flying metal shards by wearing safety glasses. Set the top of the broken blade snugly against the pocket side of the mat cutter, and do not move it between cuts on the same mat. Otherwise, you will get different angles and different-width bevels.

Mix a small amount of paint with water and stir thoroughly to get an even dilution. Paint should be thin and watery so that it flows freely from the brush. Inks can also be diluted with water for better control of the shade of color.

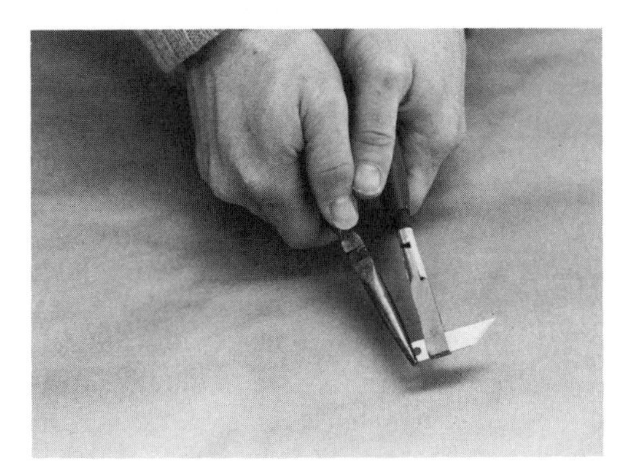

Figure 8.32
Painted bevel: To make a wider bevel on the window with a Dexter mat cutter, snap off the top of the blade . . .

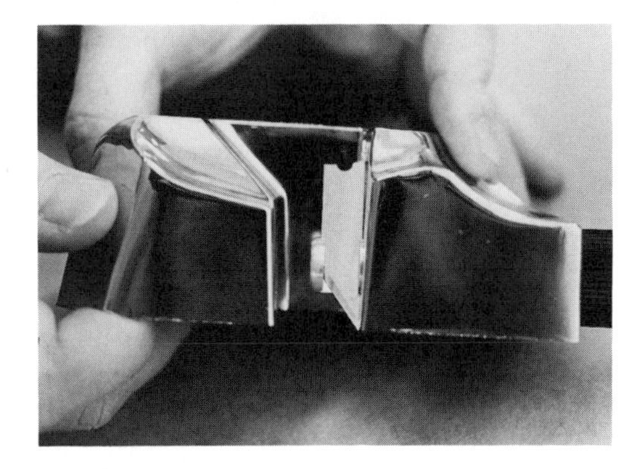

Figure 8.33
. . . and push the top of the blade against the side of the inset. This makes the blade cut at a more oblique angle.

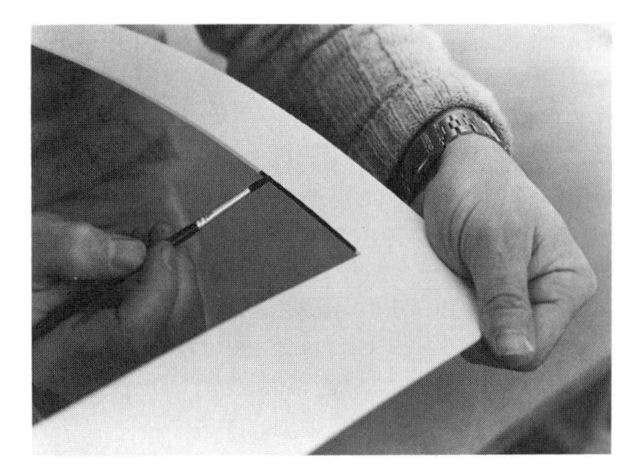

Figure 8.34
Start painting the bevel a little distance from the corner . . .

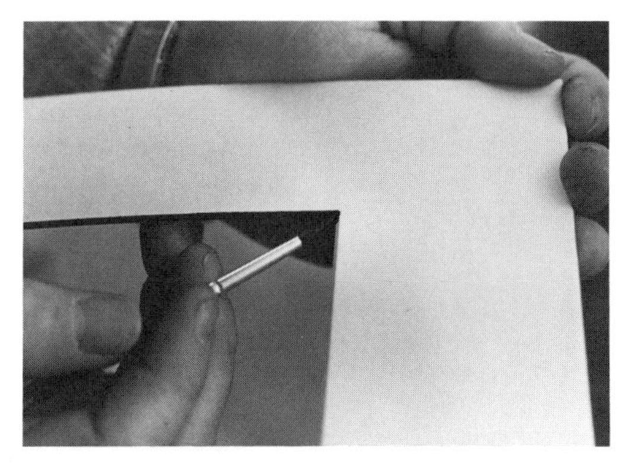

Figure 8.35
. . . and then go back to paint in the corner.

Do not try painting on the mat right away. Practice on the cutout scrap to get a feel for how the paint reacts to the board and to check the color intensity.

The brush should not be heavily loaded with paint. It is better to go over the same area several times to build up the right color rather than risk slopping some on the front of the mat.

When you do start to paint the bevel, work from the back by extending the tip of the brush through the window. Start painting at some little distance from the corner, and then go back to fill in the gap. Work continuously around the edges of the window.

Small amounts of paint that get onto the front of the mat will chip off with a sharp utility knife or razor blade after it has thoroughly dried. When you make a real botch of it, start again instead of trying to salvage the mat.

A Few Other Tips

Vivid colors make the most attractive outlines, because pale colors look like a discoloration on the mat board. For extra print protection, a thin mat recessed under the window mat can be made from 2-ply board. Joined to the back of the window mat, it holds the painted edges away from the print surface and prevents any possible transfer of color.

FRENCH MATS

In basic form a French mat consists of panels of pastel hues painted on the mat face and outlined with thin lines of more intense color. It lends an aura of elegance and refinement to the right kind of picture. Not every kind of photographic print goes well in a French mat, but the work and skill needed to make one are an excellent guard against abuse of the idea.

Because a French mat in traditional form suggests a certain nostalgia and romance, it is particularly appropriate for landscapes, portraits, and genre images dating from photography's Pictorialist era and before. You might well put one around a cathedral interior by Frederick Evans, a Woodburytype, or a Hill and Adamson salted-paper portrait.

Modern photographic print makers have since the 1960s taken up again many of the colorful printing methods used by the Pictorialists, though contemporary aesthetic concerns are naturally different. Diazo, platinum-palladium combinations, and hand coloring, to name just a few techniques, are often done on richly textured paper with an exquisite range of subtle colors not matched by any modern photo papers. This kind of imagery can also look good in a French mat, but care should be taken to appeal to modern sensibilities in your choice of colors and layout.

Tools and Materials

This suggested list includes the least expensive materials:

- ☐ Watercolors (tube or dry)
- ☐ Mixing pans or dishes (white preferred)
- ☐ Drafting pen
- ☐ 2 brushes: ½- and ¼-inch red sable watercolor
- ☐ Hard-lead pencil
- ☐ Art gum eraser
- ☐ T-square, *or*
- ☐ EZ/Mat cutter base, *or*
- ☐ 45° French Mat Marker (from C&H Manufacturing)
- ☐ Window mat (rag or conservation board)
- ☐ Cork-backed ruler

Two methods of color application are currently in vogue. One uses watercolors, and instructions on this method follow. The

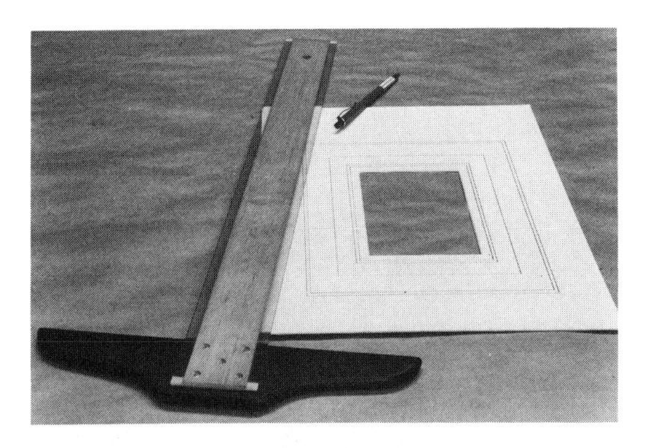

Figure 8.36 (left)
French: The layout of a French mat done in pencil before painting in the panels with watercolor.

Figure 8.37 (below left)
After the lightly colored panels have been painted with watercolors, each panel is outlined with color using a ruling pen. Note that the progression being followed is from the innermost lines to the outer.

Figure 8.38 (below right)
An extreme close-up shows how the color applied on the outlines can bleed into the panels if they aren't dry enough.

other method uses dry colored powders that are brushed onto the mat; a complete kit for this technique is sold under the trade name Mat Magic French Matting Kit.

Many frame shops choose Mat Magic for rapid production of custom mats. It is fast and easy and it needs only a little bit of skill. It comes with an excellent set of instructions that work well when followed exactly, and it does forgive minor mistakes.

Either rag or conservation board can be used with watercolors. Our example was done on conservation board, but 100% rag board closely resembles the paper that most watercolor painting is done on.

You will find that in the beginning you waste lots of board, so do not sit down with just one board and think that you are going to make a French mat.

Either kind of watercolor, the type sold dry in a metal palette box or the kind in tubes, works equally well. Use a good-quality brand of dry color, something like Windsor & Newton, because the less expensive brands suitable for grade school children do not mix nearly as well and often contain insoluble flakes.

Our experiments with colored inks convinced us that they should be avoided for French mats. The colors showed a marked tendency to separate when cut with water, and after drying they left irregular outlines on the outside of the panels.

Colored inks work well in the drafting pen for making the panel

Figure 8.39
The finished piece actually has a rich array of colors to complement the print.

outlines, however. You do not *need* them to make the strong border colors; you can just as well mix strong shades from watercolors.

Designing the French Mat

If you can find them, examine some old French mats to get a feel for the effect you want to create. Sometimes a design can be lifted directly from an old mat.

An excellent design to start with is a single outlined panel of color, with a thin rule between the panel and the picture and another outside it. Do not attempt too complex a design at first. The discouraging results may lead you to abandon the whole project, and really, it is not *that* hard.

Once again, let us suggest a sketch before you start. If you make it to size, measurements can be transferred directly to the mat.

Some general principles to follow when starting to design your French mat:

☐ Avoid equal widths *between* panels and lines.

☐ Make panels of different widths. Two panels of equal width, one inside the other, create an optical interference effect that distracts from the image.

☐ Vary the width of the lines so that they are not all the same. You can adjust a drafting pen to get this effect. But, for most applications, each line should have the same width around its entire length.

☐ Either confine yourself to very thin colored panels close to the window, *or*

☐ Extend panels almost to the edge of the mat. Anything in between leaves the edges looking noticeably bald.

Once you get the hang of things, you can break all these rules whenever you like. Try using them for starters.

The traditional look in French mats consists only of rectangular panels, but for more modern motifs it is possible to liven this up with varied geometric shapes. Be aware that this takes considerably more skill with the brush, and if overdone it can look gaudy.

Executing the French Mat

Draw the design lightly on the face of the mat in pencil. Use the T-square, EZ/Mat Cutter base, or the C&H marker as a guide. Measure the spacing of each line rather than trying to eyeball it. Inaccuracy mars the precisely finished appearance essential to this kind of mat. Make all measurements from the top of the window bevel. To avoid confusion later, erase overdrawn lines that extend past the corners.

Now you can start to color in the panels. Progress from the inner to the outer panels, and leave each panel to dry before starting on the next. Work as follows:

☐ *Mix plenty of color in the pan.* You cannot stop to mix fresh color in the middle of a panel, and besides, it is impossible to match shades exactly.

☐ *Test the color on a scrap of board.* It should be very pale, because even pale colors gain intensity when surrounded by an outline of similar color.

☐ *Load the brush with color and start in one corner of the innermost panel.* Work around the mat, trying to keep a bead of liquid moving ahead of the brush. Work rapidly once you start. If a line of color dries, it leaves a mark that cannot be covered.

☐ *Let the mat dry between panels.*

☐ *Complete any remaining panels.*

- ☐ *Let the mat dry thoroughly.* Any moisture left on the panels causes the outlines to bleed into the interior of the panels and ruins the mat.

Lines (more appropriately called *rules*) are applied with a drafting pen, a cheap and truly wonderful tool. You can vary the width of the lines it draws by turning a little screw on the side. The pen holds a reservoir of color between its points by capillary action. To avoid damaging the rather delicate points, do not dip the pen into an ink bottle or mixing dish; instead, use a fully laden brush to transfer liquid to it. We suggest that you buy the kind that has a little hinge just above the set screw. The points can be fully opened for thorough cleaning.

Load the pen with ink or strong watercolor. Adjust and test the width of the line on scrap board. Use a cork-backed ruler as a guide for the pen. The guiding edge of the ruler has to be off the mat surface, or capillary action will draw the color under the ruler and smear the work. Keep the flat side of the pen against the ruler, and draw each line all the way around before going on to the next. To avoid smearing already drawn lines, work from the bevel out.

The most attractive effects in a traditional French mat come from using a deeper shade of the panel color for the surrounding rules. Free-standing rules can be the colors of your choice.

A variation you may wish to try is the use of gold and silver Mylar-backed tapes in place of some of the rules. Apply them after all the other colors. Overlap the tapes where they meet, and then cut through both of them with a razor-sharp knife at a 45° angle to get a mitered corner. Gently pull off the excess tape.

Finally, put away all paints, inks, and other liquids before bringing out the print. When the mat has thoroughly dried, the print can be mounted in the regular fashion.

OTHER MATERIALS AND METHODS

Obviously, in a work this size you cannot hope to cover all the possible varieties in mat design and decoration, nor would it be appropriate to try doing so, because what we are especially concerned with is the preservation of photographic images.

Our discussion of matting techniques will be incomplete, however, if we fail to include some reference to methods that can be used to customize a particular presentation.

Oval- and Circle-Cut Mats

Production of a mat with an oval window or with circular cuts as part of the window motif requires the use of a specialized mat-cutting machine. These devices currently cost in the neighborhood of $1600 and up, require a fair amount of valuable space, and produce beautiful results. To get the full benefit of their versatility requires the additional use of an industrial-grade mat-cutting machine like a C&H.

We did extensive searching and experimentation in the attempt to find a less expensive way of producing the same results, and concluded that there is no way to do it. A hand-held mat cutter seems inevitably to produce an irregular and sloppy job. However, there are a number of practical alternatives if the volume of matting you do will not justify purchase and maintenance of an oval cutting machine.

The first alternative to consider is farming out the work to a frame shop. Most of these shops have an oval machine or at least access to one, and their personnel are skilled in its use. Either specify the type of board you want them to use, or supply it when you place the order. To save money and to ensure that the work gets done right, order just the mat cutting from the shop and hinge and mount the print yourself.

For a simple oval you need to specify *two* dimensions for the window: the vertical and the horizontal. For a purely circular window, only the diameter need be specified, but make certain to specify where on the mat the center of the circle should fall. It is a common practice to cut more than one oval window in a mat. In this case, draw up a sketch beforehand with all measurements indicated to show exactly where you want each window to fall.

If you supply the mat board and precise cutting instructions, it may be possible to negotiate a good price with the frame shop, particularly if you plan to do a large volume. Most mat-cutting prices are set by frame shops on the basis of size, but for a simple cut done in volume a shop is often willing to talk about a single fee.

When your work requires more elaborate variations, the price will probably go up. A particular image, for example, might require a rectangular window with an arched top or with circular corner cuts. When you find yourself in this situation, talk over your needs with the shop owner and try to work out the most economical means of accomplishing the job.

Another alternative is to dispense with an oval window. For example, it might seem necessary to remat a photo that has come into your possession in an oval frame with a deteriorating mat. You could cut a window out of a piece of board larger than the

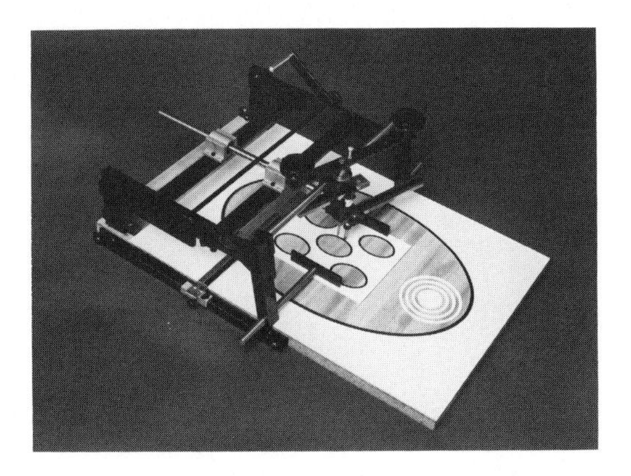

Figure 8.40
An oval mat cutter is an elaborate and expensive piece of machinery. It is also the only good way to cut an oval mat.

frame. Set the frame on top of the board and move it until the window is properly positioned. With a pencil, trace the outline of the back of the frame rabbet on the mat board. Remove the frame and cut along the tracing line with a utility knife. When the mat goes into the frame, the front lip hides the roughness of this cut. To make the visual transition from the oval frame to the straight-sided window less abrupt, you might cut inset corners at 45° angles to one another so that you have a window with eight sides instead of four.

Paper- and Foil-Decorated Mats

Wallpaper, gift wrapping, and marbled papers can be added to the surface of a mat as accent motifs. One easy-to-use and attractive technique is to cut thin strips of gold or silver foil wrapping paper as outline for one or more panels of a French mat. Apply transfer adhesive tape to the back without removing the release paper from it, cut several thin strips of uniform width (an EZ/Mat or production mat cutting machine makes this easy), peel off the release paper, and apply to the mat. Overlap the corners and cut through them at a 45° angle with a razor-sharp knife to make mitered corners. Burnish down firmly.

For a "decorator" look, you can also use wallpaper — for example, as a mat outline next to the frame rabbet. Apply a positionable adhesive to the back of the paper, such as 3M's Promount, and cut out the desired shape with a sharp knife. Burnish firmly to the front of the mat. This kind of effect can easily be overdone, so experiment first, and then use restraint and discretion in deciding whether the final result merits framing.

PART THREE

Frames and Framing

C H A P T E R 9

Metal Frames

The aluminum section frame meets all criteria for the display of valuable prints in accord with sound conservation practices. It can be considered "state of the art" in framing and whenever it can tastefully house a print it should be used. Lest this be considered an overly strong recommendation, let us enumerate the positive features of metal frames.*

A metal frame is unaffected by decay, insects, or fungal growth. It is chemically inert and gives off no destructive vapors. It is physically strong. The flexibility of the assembly hardware used in the better types allows them to be opened easily for inspection of the print. And one last, major advantage: it is cheap. Stock sizes that correspond to stock mat sizes can be used over and over.

The metal frame has much to recommend it from an aesthetic standpoint as well. Its clean and simple molding is devoid of any decoration that might clash with either the print or with the surrounding environment. In addition to the standard silver (chrome, actually) and gold, metal molding now comes in a wide variety of colors that can subtly or boldly complement the print colors.

By now more than two decades old, the aluminum frame business has become an industry in its own right. The origins of the concept are difficult to document with precision, but it seems that the first aluminum frames were used at the Museum of Modern Art in New York City to display contemporary art. These prototype frames had welded corners. The print and mat were held in place by a wooden strainer like the stretcher used to hold a canvas taut

* The term *metal frame* is used in this chapter to refer to aluminum section frames. There are not any comparable types on the market made of other materials. Stationery stores and novelty outlets sell cheap desk-top models of die-pressed metal (often with snapshots conveniently provided!), but we assume that you will not confuse these with sectional frames.

for oil paintings. Screws driven through the sides of the molding held the strainer in place.

Each welded frame was a hand-crafted object intended solely for one work of art. The metal frame represented a triumph over the confining conventions of the carved wood frame, and it introduced the sleek functionalism of industrial fabrication into the museum world, as a complement to works of art that had already incorporated this new beauty.

Practical improvements followed initial use of the new design. The chief obstacle to widespread use of the new frames was their cost. Fabrication of individual custom units by a skilled artisan made them expensive. The low volume of production further restricted their availability. A demand, however, had already been created by the taste-setting influence of major museum shows.

A parallel development in the use of extruded aluminum molding for storm windows provided the breakthrough that made mass-manufactured metal frames possible. Aluminum storm windows became a major industry in the United States following World War II, in part due to the creation of fasteners to join sash molding around the glass. An adaptation of these fasteners made possible the changeover in framing from welded corners to angle cuts joined by a simple device of several screws and an L-shaped plate. Instead of needing a highly paid expert and expensive equipment, framers could simply put together the molding sections with a screwdriver.

Welded frames have remained in the market, and deservedly so, because they have some features not found in the screw-plate frames. For example, the beautiful finish of a welded corner has a strong visual appeal, and the extra physical strength of a strainer can compensate for the lack of flexible application it entails. For general application, however, the welded frames have been largely supplanted by the new type of aluminum section frames.

BUYING

For the purpose of framing, aluminum is aluminum, and one need not worry about such fine points as temper or gauge. The critical factor is hardware design. By *hardware* we mean the corner fasteners, print-securing clips if any, and hanging apparatus. You want to take a careful look at this before buying, because each type of brand name frame has its own design specifically made to fit the particular molding that distributor uses. No matter whether you buy lengths of molding and cut them to fit your needs, or whether you utilize a chop service of some sort, the hardware determines how well that particular style of molding will work.

Before buying a quantity of molding, examine a sample of the corner hardware sold with it. Look at these three points:

- ☐ *Strength:* is the hardware adequately strong so that the frame cannot pull apart from its own weight on the wall?

- ☐ *Ease of assembly and disassembly:* does the frame go together quickly, and can it be taken apart to allow for inspection and the insertion of different artwork?
- ☐ *Fit:* do the corners butt tightly without leaving gaps?

A well-made system meets all these criteria, while some of the cheaper ones do not meet any of them. The closeness of the corner fit is a good indication of whether the other two points are also covered. Many cheap frames will not go together at the corners no matter how you wiggle them around.

One of the earliest and most prestigious suppliers, the Nielsen Frame Company, overcame this problem at the start by ingeniously making its hardware so that the hardware tilts the molding ever so slightly toward the inside. The face corners butt tightly, but leave a slight and virtually unnoticeable gap on the sides. The Nielsen frames are so strong that it would be virtually impossible to separate the corners without bending the metal — and if your frame gets into that kind of situation somehow, you can probably forget about the print.

Avoid frames with pop-in plastic braces at the corners. Not only do these abominations never fit tightly, but they are also weak. By definition they cannot be any stronger than the pressure used to put them together. From time to time they pull apart just from the strain of hanging on the wall. If you have to use them, be certain to rig a 4-point hanging harness on the back.

Some aluminum frames have "one-way" corner hardware. In other words, once you put these frames together, they will not come apart. Avoid this type too.

Another aspect of hardware design to examine with a skeptical eye is the means used to hold the print mat in place. None of the aluminum section frames we know of uses a strainer. The best ones have a set of spring clips, which are bowed metal strips with much springiness. You slide them behind the back of the mat, between it and the rear lip of the molding. They supply enough pressure to hold the glass, print, and mat firmly against the front lip. To release the pressure, you simply pop them back out again. Makers of cheap frames often do not supply anything for this purpose, so you have to wedge wadded pieces of cardboard in the channel behind the mat — an awkward procedure at best.

Once you have satisfied yourself that the hardware on a certain kind of frame meets your requirements, it is necessary to consider the different types of molding offered in that particular line. Most distributors of metal frames have a variety of molding styles that are suitable for different purposes. Molding can be bought in precut, paired sections for top and sides, usually in increments of one inch, or custom chopped to specific measurements. Custom-chopped Nielsen molding can be bought through the mail from chop services advertising in framing and photography magazines, or even from local framers.

One important factor to consider in choosing a particular style of molding is the width of the face lip. A very narrow lip looks good and provides adequate security for pieces as large as 11 × 14 inches. For pieces larger than that, a wider lip is required. Any long section of aluminum has a certain degree of flexibility, and even a small amount of force causes the lip to twist slightly. Force exerted from the back can then cause the glass (which is also slightly flexible) to move forward and pop out of the lip. If the lip is too narrow, the problem persists no matter how many times the frame is taken apart and reassembled.

Another thing to consider when looking at molding is how the print will be separated from the glass. For a while, it became common to put prints in metal frames without any mat at all, so that the print is pushed against the glass itself. This practice, needless to say, should be discouraged. However, the same effect can be achieved by making a hidden mat. This can be made of either four narrow strips of conservation board, or an actual window mat with very thin borders that is concealed behind the face lip. Making either of these two alternatives gets easier, the wider the face lip.

Another alternative is to use a double-channel molding that has separate channels for the glass and for the print. Because the glass is held apart from the print surface by the molding itself, no mat at all need be constructed.

As a guide to evaluating different molding styles, here are cross sections of typical extrusions with some notes on their application.

COMMON EXTRUSION PROFILES

Standard

This is the most commonly used profile, and an excellent general-purpose style. Face-lip inside dimension measures about 3⁄16 inch, wide enough to prevent popping on any but the largest frames. (See Figure 9.1.)

Thin Face

This severe profile is more sharply angled than the standard and with a narrow face lip no more than 1⁄8 inch wide. It provides a strikingly modern surround for small and delicate prints. Not for use on large frames. (See Figure 9.2.)

Deep Box

This type has the same face width as the standard, but the deeper frame gives greater corner strength at the price of less resistance to flexing. Do not use it to carry more weight than a standard molding frame despite the appearance of greater substance. Originally the deep box was designed to hold oil paintings on stretchers. The extra depth makes it possible to put in spacers easily, so that glass and print are more effectively separated. (See Figure 9.3.)

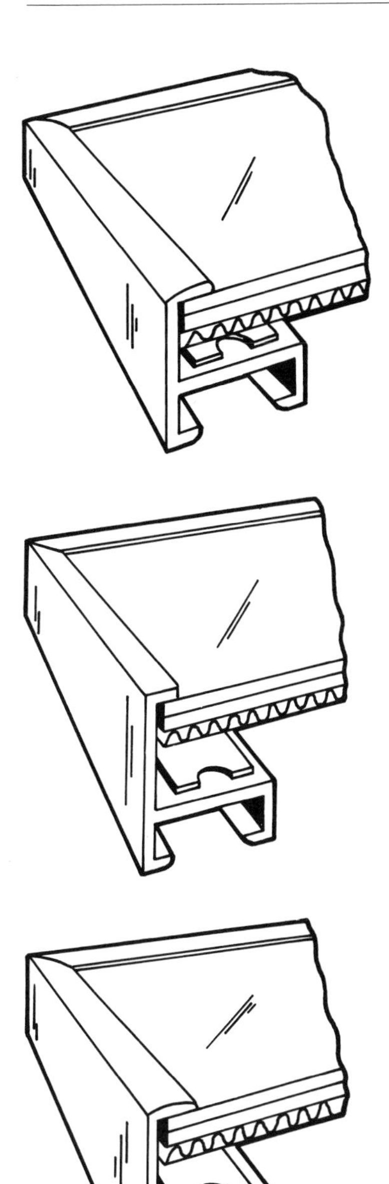

Figure 9.1
Standard profile.

Figure 9.2
Thin face profile.

Figure 9.3
Deep box profile.

Shadow Box with Two Channels

This model has separate channels for the glass and for the print. It is a good choice for graphics like posters or large-screen prints that need an unmatted presentation. Manufacturers like Nielsen make the glass channel in a choice of widths: one for grade B single-weight glass and another for ⅛-inch-thick Plexiglass. The glass is slightly loose in the channel to allow for easy insertion. That makes this molding a questionable choice for traveling exhi-

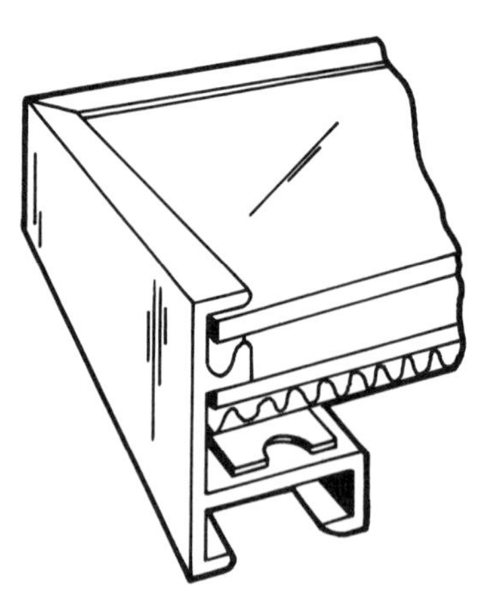

Figure 9.4
Shadow box profile.

bitions, where small chips of glass might crumble off and migrate into the print area. (See Figure 9.4.)

FITTING ALUMINUM SECTION FRAMES

Because the assembly of aluminum section frames is essentially a simple matter, these instructions may seem overly long. Training new framers, however, has taught us that even the simplest things can go disastrously awry and common sense can be totally forgotten when you attempt your first fitting.

The standard fitting method calls for glass, mat and print, and backing to be stacked on the work surface and then slid into the frame. Variations include putting a U-shaped polypropylene channel around the glass (see the end of the chapter) or making a semisealed enclosure of the glass, mat, and backing package by sealing the edges with Scotch Magic Transparent Tape. Edge taping (explained in the next chapter) protects against glass chipping, seals out contaminants, and dampens humidity fluctuations.

Clean the glass on both sides. Never spray or use liquids in the immediate vicinity of the print.

The print should already be matted appropriately. If it is to be framed unmatted, cut spacing strips of conservation board, or make a thin window mat.

Cut the backing board to size. (Any of the materials for the structural backing described in the next chapter will suffice.)

Check the size of the frame. Even if you have ordered from a supplier before, that supplier may have changed its policy on leaving an allowance for fitting. Remember that an 11 × 14 piece of glass will not fit into a frame that has interior dimensions of exactly 11 × 14 inches. Look for an allowance of at least ⅛ inch on wide-lip frames, a little less for the narrow-lip ones. Oversize glass can jam in the frame and be difficult to remove.

Assemble three sections of the frame into a U-shape. Use the two shorter sides for the arms of the U.

On the Nielsen frames, a corner is assembled by putting two L-shaped plates into the bottom channel of one piece and then

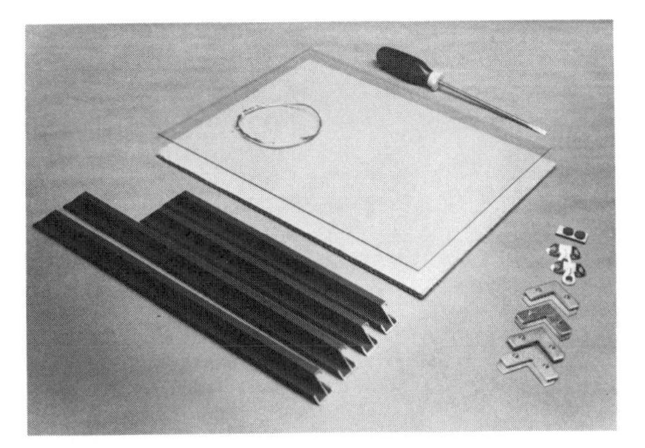

Figure 9.5
Layout showing the parts of a typical metal frame ready for assembly. The only tool needed to put everything together is the screwdriver, so we included that.

Figure 9.6
Measure the *inside* dimension of the rabbet before starting assembly, and make sure by comparison that there is some play to allow for the mat and glass to slide in the channel. Check both a top (or bottom) and side section.

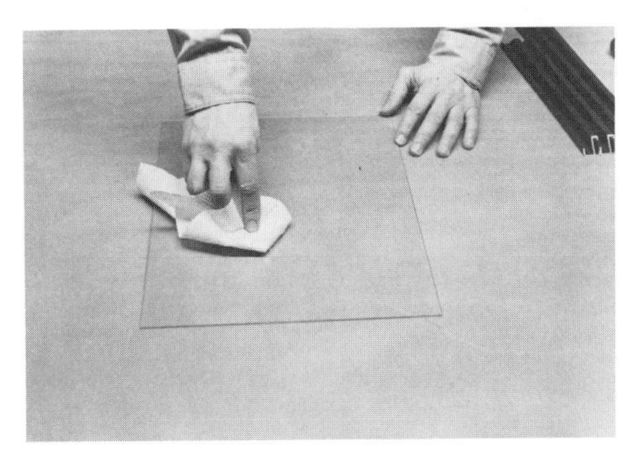

Figure 9.7
Clean the glass on both sides separately from the print before starting assembly. Make sure that it is absolutely clean. This one is — that's why it's so hard to see in the photograph.

Figure 9.8
The entire print package before inserting consists of, from top to bottom: the glass, the mat and mat backing board, and the mechanical support — in this case, archival corrugated cardboard.

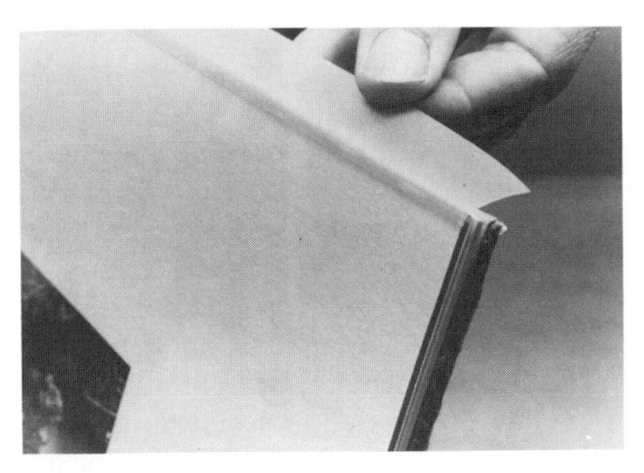

Figure 9.9
To make a sealed enclosure for the print package, Scotch Magic Transparent Tape is run around the edges of the entire package. The back edge of the tape can run on the moisture barrier, or on the structural backing.

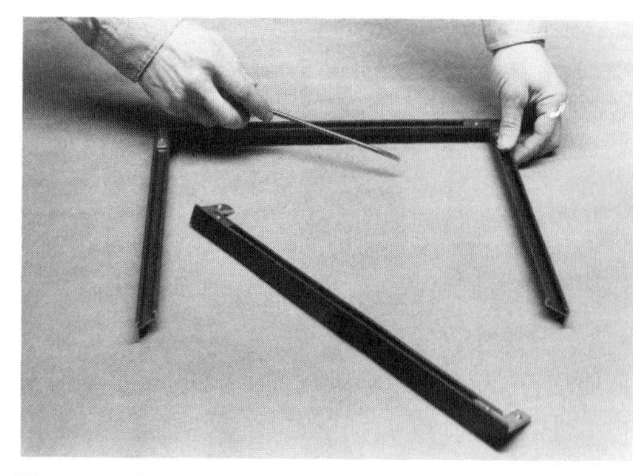

Figure 9.10
Put together a U-shaped assembly of three sections prior to inserting the print package, with the long side at the bottom of the "U."

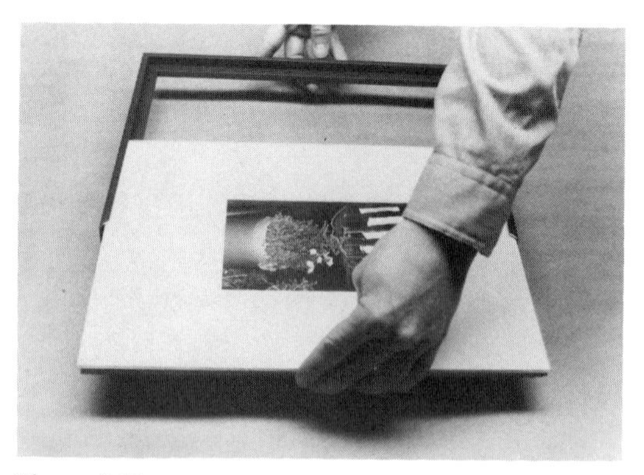

Figure 9.11
Rest the print package on a small book and slide the U-shaped section onto it.

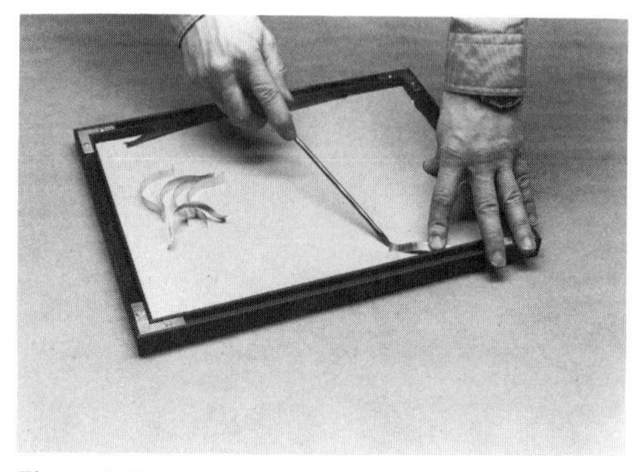

Figure 9.12
Slip the remaining section onto the frame and tighten all the screws. Check thoroughly to be sure you have a good fit. Slide the spring clips into the back of the rabbet with the tip of the screwdriver to exert pressure against the print package so that it fits snugly against the front of the rabbet.

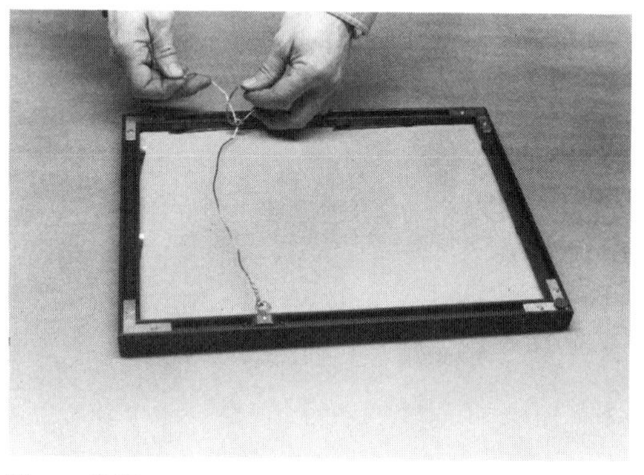

Figure 9.13
Put the hanging hardware onto the back of the frame and string the wire. If you look carefully, you will see that the wire is being looped twice through the hardware before being twisted around itself. Both bottom corners on the right have Bumpons in place to hold the frame away from the wall.

slipping the other section onto the protruding arm. One of the plates has short screws that face toward the back of the frame. Tighten these and check the fit. If necessary, loosen again and wiggle the pieces around until the faces fit. Tighten again. Most other brands require a similar method of assembly.

Do not assemble two L-shaped frame sections together and try to scoop up the glass, mat, and backing. Many beginners try this; it simply cannot be done. Remember: use the U.

Another caution: do not slide the mat and backing into the U and then push the glass in on top. The razor-sharp edge of the glass can gouge considerable material from the print surface — not a desirable effect!

The standard method is to start by piling, in this order, the backing board, mat, and glass on the work surface. Lift them all together and check for trapped dust.

For more permanent framing, make a semisealed enclosure. Pile material in this order: backing, moisture barrier of Mylar or foil, matted print, and glass. Check for dust. Run Magic Transparent Tape around the edges, and leave a small vent opening at one top corner. Make sure that no more tape appears on the front than the face lip will conceal. For a variation on this same idea, the tape can be run along the front of the glass and onto the moisture barrier, with the backing inserted separately.

Transfer the glass, mat, and backing package to the top of a book or wood block so that the edges hang over on all sides. The print should be face up. Then insert the package into the frame by pulling the U-shaped section toward yourself so that the channel scoops in the package. An alternative is to have several inches of the print package extend over the edge of the work table, and then pull it down into the frame channel. Turn the entire piece over after checking one last time for dust.

Put the corner pieces in the free section of molding, slide them into the appropriate slots of the U, and tighten the screws. Keep the frame face down and attach the remaining hardware. Put in the pressure clips (or wadded cardboard, if you are using one of *those* frames) to hold the print package against the front lip.

Put on the hanging clips. The higher on the molding they go, the more nearly vertical the frame will hang.

String the wire. (See Chapter 11 for details of wrapping the wire.) To make a reinforced 4-point wire hanger, use extra hanging clips on the bottom. They can be bought separately in most cases. They go onto the bottom track. String wire from the bottom clip to the top clip on the same side, across the back to the top clip on the other side, and down to the bottom.

Turn over and admire.

EXTRA GLASS PROTECTION

When an exhibit in metal frames is going to travel, thought must be given to the extra stresses it will undergo. Minuscule fissures on the glass edges can propagate across the entire surface as large cracks. You can provide an extra cushion for the glass by using a chemically inert polypropylene U-shaped channel on its edge. Frame Tek sells its Clear Plastic Channel for this purpose. The channel is wrapped around the edges of the glass. It keeps the frame movements in transit from crumbling the edges of the glass and provides additional space between the glass and mat.

WELDED METAL FRAMES

Welded aluminum frames were mentioned earlier as the prototypes of contemporary styles. Their attractively sealed corners still make them an ideal choice for very valuable works.

Welded frames are "back loaded," which means that the entire print packages goes in from the back instead of sliding into a retaining channel. The wood strainer, custom made for the particular frame, then goes in back to hold everything in place. It is secured with screws through the molding side.

The same precautions need to be taken with the strainer as with a wood frame. A polyester moisture barrier and archival corrugated cardboard back should go behind the mat. Use a semisealed enclosure. Finally, seal the strainer itself with artist's matte medium or polyurethane varnish, like the rabbet of a wood frame.

CHAPTER 10

Clip and Passe Partout Frames

Some occasions, like a quick show of student work, call for a way to put prints on the wall without expensive hardware and time-consuming labor. At the same time that you want to get the work up on display, however, you want to make sure that it is not going to get a thumb-tack put through it.

For occasions where expedience counts, there are two quick, cheap ways to "frame" the print after it has been matted: clip frames and passe partout framing.

CLIP FRAMES

Clip frames, sometimes paradoxically called "frameless frames," cost only a few dollars apiece. The only additional expense is a sheet of glass and a hanging hook. They can be used more than once, and they are more versatile than metal frames because they accept mats of varying sizes. Glass has to be cut separately for different-size mats.

Clip frames differ slightly from model to model, but all have some features in common. Each uses a sheet of glass or other glazing material as the primary support member. The only parts visible on the front are several minuscule fingers of metal or plastic, usually two to a side.

None of the clip frames has side members to keep out environmental intrusions. This makes regulation of the environment to conservation standards for temperature, humidity, and dust control more important than with conventional frames.

Clip frames are so minimal in construction that there is no point in giving detailed instructions about their assembly. Diagrams and notes on the package give the pertinent information. You will start, we assume, with a print matted to conservation standards, since a mat is necessary in all cases anyway.

The four kinds of clip frames described here are the most widely

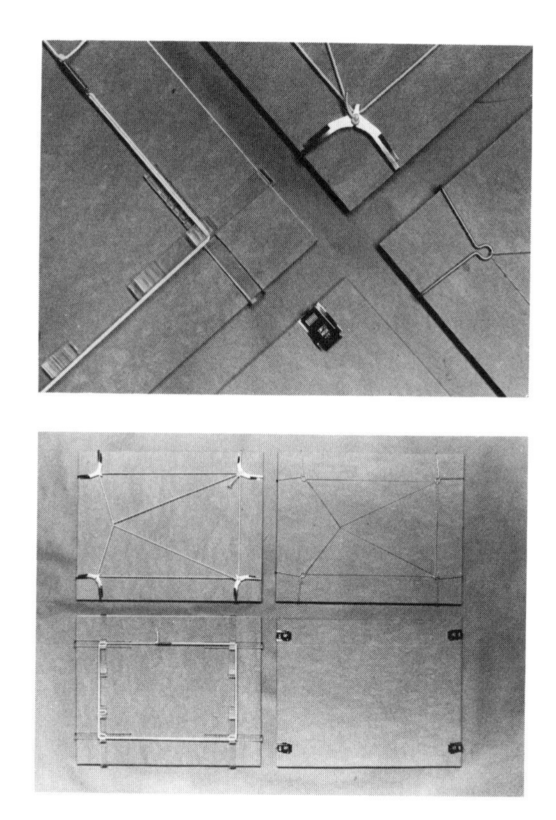

Figure 10.1
The back of four assembled clip frames. Clockwise from the left: Uni-Frame 20, Swiss Corner Clips, Gallery Clips, and Swiss Clips. Notice that the Swiss Clips are the only ones that aren't held together with a cord, and they go on only one side of the frame.

Figure 10.2
A more distant shot of the backs of four clipless frames, showing the way in which cord is strung on the back of each.

available types. If you come across other kinds, do not be afraid to experiment with them, because there seems to be little difference among the various kinds with regard to conservation qualities. The most important thing to check for is that the construction must be *strong*. Make sure that the whole assemblage will not disintegrate on the wall.

Uni-Frame-20

A Uni-Frame-20 supports pieces up to 20 inches square; a metal version called the Uni-Frame-40 is made for pieces up to 40 inches square. The Uni-Frame-20 is constructed of clear plastic.

Each corner has an A piece with two positioning cleats and a B piece that is flat. A and B parts can be positioned at right angles to each other in any of four positions, so that the clips appear on the front with the desired spacing separating them.

When the corners have been assembled, the J-shaped clips at the end of each arm slip around the sandwich of glass, mat, and structural backing to hold them together. A nylon cord (supplied) is wrapped around each of the four corners where the two pieces meet, and makes a rectangle. Its ends are tied through a small spring at the bottom of the frame to take up the slack. The cord hangs on the framing hook on the wall, and the spring allows enough play to permit positioning the frame upright.

The flat protruding faces on the backs of the cleats are a strong feature of this frame, because they hold the frame flat against the display surface.

Gallery Clips

In the days before agribusiness, farmers used to fix everything from combines to milk pails with a little bit of baling wire. They would feel right at home with Gallery Clips. These frames are sold in a package that contains nothing but four preshaped pieces of thick wire and one length of cord. The wire, formed in L shapes with semiclosed loops at the juncture of the two arms, has a small square hook at each end that wraps around to the front of the glass.

It is hard to believe that so little could do so much, but when cord is strung through the loops in the prescribed fashion, it makes a tight, solid frame. The finish on the wire is not too fancy, but so little shows from the front that it hardly matters. Gallery Clips require a thick backing, usually a material like foam board.

Swiss Corner Clips

Swiss Corner Clips are functionally identical to Gallery Clips. The difference is that the corner pieces are made of stamped spring steel with a chrome finish. An additional benefit is little spring arms on the sides of the main arms that exert pressure on the frame back so that these clips can hold any piece thinner than the $7/16$ inch maximum thickness. This does away with the need for additional padding on the back.

Cord is tied in the same fashion as for Gallery Clips. The flat steel faces on the front definitely look more attractive than wires. Maximum recommended size is 20 inches square. When solidness of construction and quality of finish are considered together, these are the best clip frames we have seen.

Swiss Clips

Do not confuse Swiss Clips with Swiss Corner Clips. They are a different type altogether. The Swiss Clips consist of four edge fasteners for the top and bottom of the frame. None goes on the sides. A separate hanger bracket, which can double for an easel back, is pounded into the structural backing before assembly. This is the only clip frame about which we have certain reservations.

To secure them, place the clips so that the clip goes around the glass, mat, and backing. Then tap them with a hammer to seat a locking pin in the backing. This does not seem to cause the glass to break, to our surprise, but we are still dubious about using this method for framing valuable prints.

If the hanger bracket is not used, the manufacturer recommends securing the clips with epoxy glue and suspending the assembly from two nails, one through each hold in the top clips.

Spring arms on the sides of the clips mean that you can use

different width backings. The seller claims they "frame any size picture," but prudence suggests nothing larger than 16 inches square. The clips have some application as hangers for passe partout frames.

Conservation Measures with Clip Frames

You do not need to take extensive conservation measures with clip frames if they are used for temporary hanging. For long-term framing, clip frames should not be used. The open sides provide absolutely no protection against insects, moisture, dust, and atmospheric pollutants. Handling over the long term will probably result in breaking the unprotected glass.

If you want to hang prints in a heavy traffic area in a building where a large population goes in and out, choose another method of quick framing, like passe partout frames. Sealing the frames is the only way to protect against the dirt tracked in by crowds, and the high humidity levels that crowds create.

Frame assembly needs to be carried out with the same rigorous care for print safety used in handling other prints. For example, clean the glass separately from the prints, and handle the glass carefully so as not to scratch the prints.

With frames that use a cord, make sure that the cord gets drawn up tight to make a solid assembly. You will not be saving any time if the whole thing comes apart on the wall.

The manufacturers of some of these clip frames call for the use of Masonite or composition board as a backing material. Disregard these instructions. The unknown components of these materials (wood resins, adhesives, etc.) can be damaging even in the short run. Substitute some kind of foam board like Artcor or several thicknesses of conservation board.

Their design is such that clip frames provide little pressure from the back — just enough, in fact, to keep the edges of the glass and mat together. As the entire unit draws moisture out of the surrounding air, print and mat are likely to buckle. Over a few months the problem can become severe, and in itself would be an excellent reason for using clip frames only temporarily.

Take the same precautions when choosing a hanging site as you would for any other framed print. In other words, do not hang in direct sunlight, on cold exterior walls, or on a wall that will not hold a framing hook securely.

PASSE PARTOUT FRAMES

Passe partout originally referred to nothing more than sticking a mat around the print and putting it on the wall, a fashion still popular in college dormitories. In more advanced circles, it evolved into the practice of taping a sheet of glass on the front of the mat, using strips of gummed paper. At present we take the term to mean the creation of a sealed enclosure with fabric, foil, or polyester tapes specially made for the purpose. It is a cheap way to make a nice-looking frame.

Passe partout frames have an advantage over clip frames in that they seal the edges against dirt, insects, and moisture. Consequently, buckling is less of a problem.

Weight plays a large part in determining how large a piece can safely be framed with passe partout. When possible, use thin, picture-weight glass or acrylic sheeting like Plexiglas. It is a dubious practice to frame anything bigger than 16 × 20 with passe partout, and even that is stretching it. Consider both the strength of the tape you use and the holding power of your hanging system in setting your size limits.

Passe partout has much to recommend it for cheap framing. It looks modern and very "artistic," if you like, and it is an excellent way to display postcards and small reproductions that do not merit a full-scale framing job.

Suitable tape is available from a number of sources. 3M makes both foil and polyester self-adhering tapes that are easy to work with. (The type in the demonstration photographs is Polyester Film Tape No. 850, Black.) Fabric tapes are made by Mystik and Filmolux in white and a variety of colors. Or you can use medical bandage tape, the type called adhesive tape. It is available only in the hygienic white. A minimum width of ¾ inch is needed to give the tape good purchase on both the front and the back.

The tools and materials used for passe partout are a miser's fantasy. A knife, mat-marking gauge, and burnisher are all the tools you need, and even the burnisher is optional. Materials are a sheet of glass, the matted print, a sheet or two of backing board like Artcor or thick mat board, the tape, and your choice of hangers.

Passe partout rings for hanging are optional. They require two sheets of backing board, one to protect the back of the print and one to hold the rings. A slit is cut in the outer backing to hold the rings before assembling the frame. Wire is strung through the rings just like through screw eyes on a conventional frame.

One may be tempted to forgo a mat in favor of the more "modern" look of a floated print. But, even when doing temporary framing, a mat is still needed to prevent contact of glass and print.

Assemble glass, matted print, and backing sheet(s) into a package. The glass, as always, should be clean; be sure to check for dust. Glass, mat, and backing need to be exactly the same size. Before you start taping, check this very carefully by standing the package on end on all four sides. If there is any discrepancy, trim it at this time. Run a sanding block along the edges of the glass to take off sharp edges that might cut the tape or your fingers.

Lay the frame package face down on the work surface, and lift off only the outer backing sheet.

The secret to a good-looking passe partout frame lies in getting the tape straight. We have seen a number of ideas suggested for this, including taping a ruler to the front of the glass, but putting guidelines on the outer backing seems the simplest way of getting the job done.

Use a mat-marking gauge of the kind described in the chapter on cutting a window mat. Rule a line about ¼ or ⅜ inch from the edge of the backing. If a gauge is not readily available, a straight-edge and pencil can be substituted, though they take a little longer and may not be so accurate.

Place the backing on the rest of the package, with the lines face up. Run the first piece of tape along one line and let a bit extend over each end. Cut the tape. Accuracy in alignment is critical, because the width of the tape on the face depends on the way it is laid on the back.

Grip the package tightly on both sides and turn it over so the back now lies on the table. To adhere the tape to the front of the glass, lift the side opposite the tape, with the taped edge still resting on the table. This creases the tape and connects it to the frame side. With your finger, apply a little pressure to tack the tape to the front of the glass in the center of the frame. Burnish out toward both corners.

Polyester tape can stretch, so burnishing this kind of tape needs to be done very carefully to avoid creases and wrinkles.

Repeat the same operation for the opposite side of the frame. Then trim the ends of the tape flush to the glass. An Olfa knife with break-off points provides a constantly sharp cutting edge. A single-edge razor will also work, but have a plentiful supply on hand. Apply tape to the two remaining sides in the same fashion and trim it.

To make a sharp crease along the sides and connect the tape to the glass, run a burnisher or the back of your thumbnail all the way around the front. The overlapped tape at the corners makes little raised squares. To remove them and to counterfeit the look of a molding frame a little more closely, cut a diagonal line from the inner to the outer edges of the corner. Peel away the excess tape to create a miter joint.

As with any sealed print enclosure, moisture buildup can create problems. Vent the passe partout frame by puncturing it at a few points along the side.

Hanging Passe Partout Frames

Passe partout rings make it possible to hang the frame in a conventional manner. Insert them into the back before taping, and use a second sheet of board to prevent the clasps from making an impression into the print surface over time. Linen tape over the clasps on the inside will give extra strength.

Clip frames will hold passe partout frames, with the advantage that the tape seals out the environment. The chances of breakage are also cut down. Passe partout framing is also recommended for prints shown in a temporary exhibition with pushpins on homosote board (see Chapter 13).

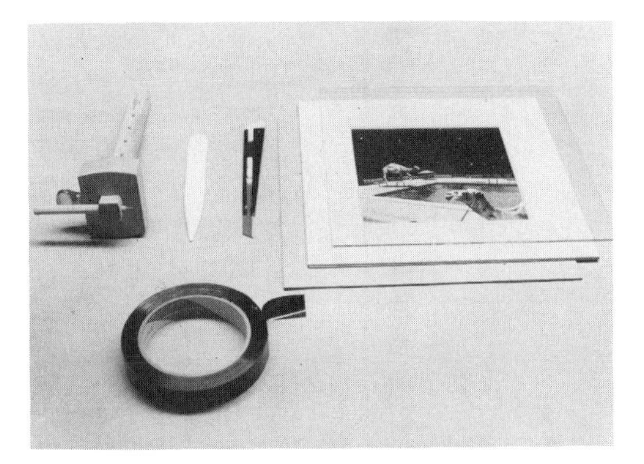

Figure 10.3
Everything needed to make a passe partout style frame. But where's the frame? It's on the roll in the foreground. Hardware for hanging isn't shown, because it's up to you to choose.

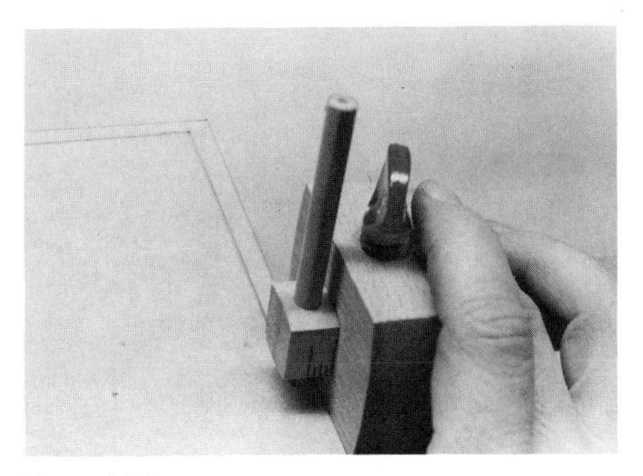

Figure 10.4
Mark a guideline on the outer backing board to indicate where the edge of the tape should be placed. Run the guideline around all four sides.

Figure 10.5
Assemble the entire package; run the tape along one side with overlap at each end. Crease the tape by lifting the opposite end . . .

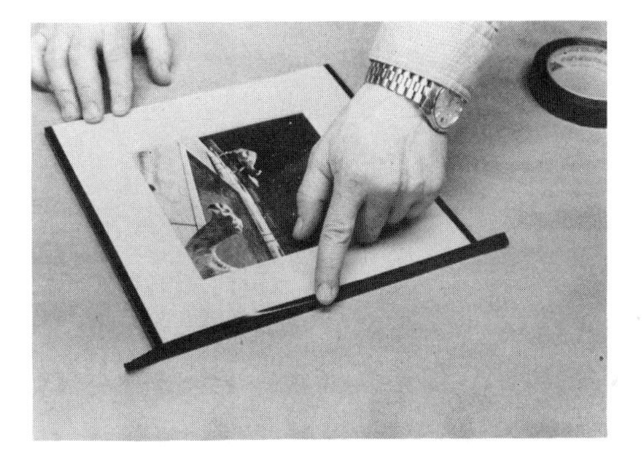

Figure 10.6
. . . and press down in the center of the front and then burnish out to the edges. (The careful observer will notice that we have this step slightly out of sequence, because tape already has been applied to two other sides. Nobody's perfect.) Before putting on the side pieces of tape . . .

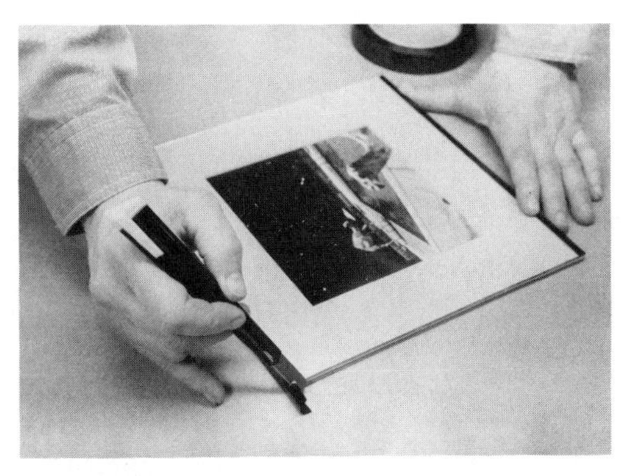

Figure 10.7
. . . you have to trim off the overlapping tape at the top and bottom. Cut it flush with glass edge.

Figure 10.8
To get rid of those ugly little overlaps at the corners, make a miter cut all the way through the tape and pull off the excess.

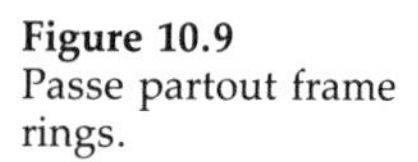

Figure 10.9
Passe partout frame rings.

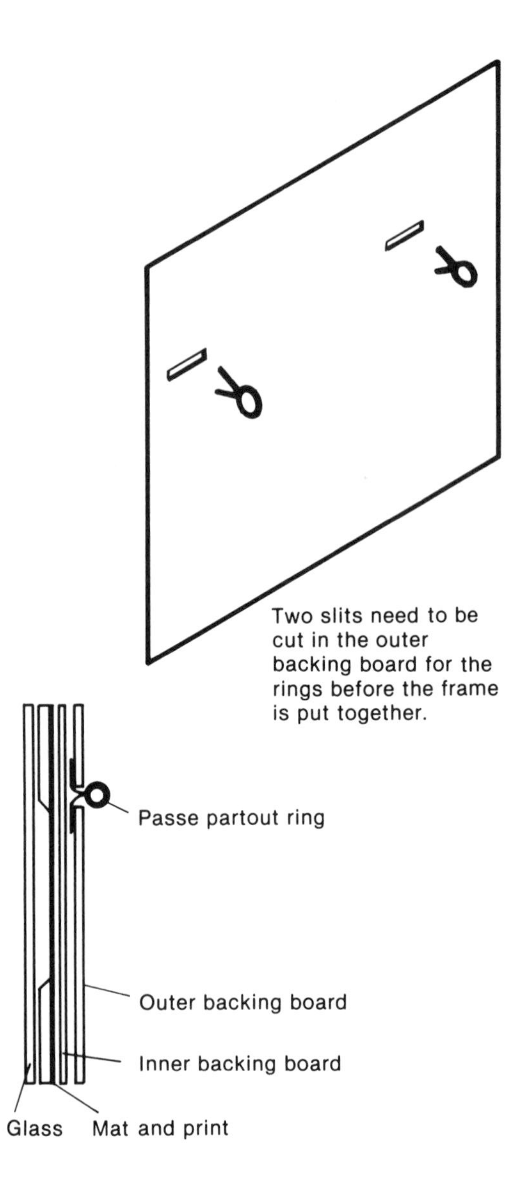
Two slits need to be cut in the outer backing board for the rings before the frame is put together.
Passe partout ring
Outer backing board
Inner backing board
Glass
Mat and print

PLASTIC BOX FRAMES

If you are not satisfied with any other quick, cheap method of framing, try plastic box frames. They have a clear plastic box, open on the back, that fits snugly over a chipboard box on the interior. A pressure fit holds the print against the front of the frame.

These can be a safe method of framing, especially in the short term. Probably the greatest potential problem comes from the chipboard box in the interior. The frame is so tight that it holds only the thinnest of mats, so for protection it would be a good idea to insert a moisture barrier behind the print. An alternate idea would be to duplicate the interior box with acid-free mount board, but in that case the effort might be better spent on a more elaborate frame.

Discount-store versions of this frame can have a needle sharp injection sprue mark right in the exact center of the face. It is hard to believe that anyone could be so thoughtless when it would take so little effort to put the sprue onto one of the sides. Look before you buy; the jagged edge is as dangerous as a vandal's knife.

The plastic in these frames has an affinity for static electricity, so you will not want to use them for charcoals or pastels, or for photographs like hand-tinted daguerreotypes where the static might lift pigment off the surface.

A few things that are not problems: condensation and plasticizer compounds in the plastic. But do not hang box frames for long periods in direct sunlight, because there is the chance that ultraviolet light might cause some outgassing onto the prints.

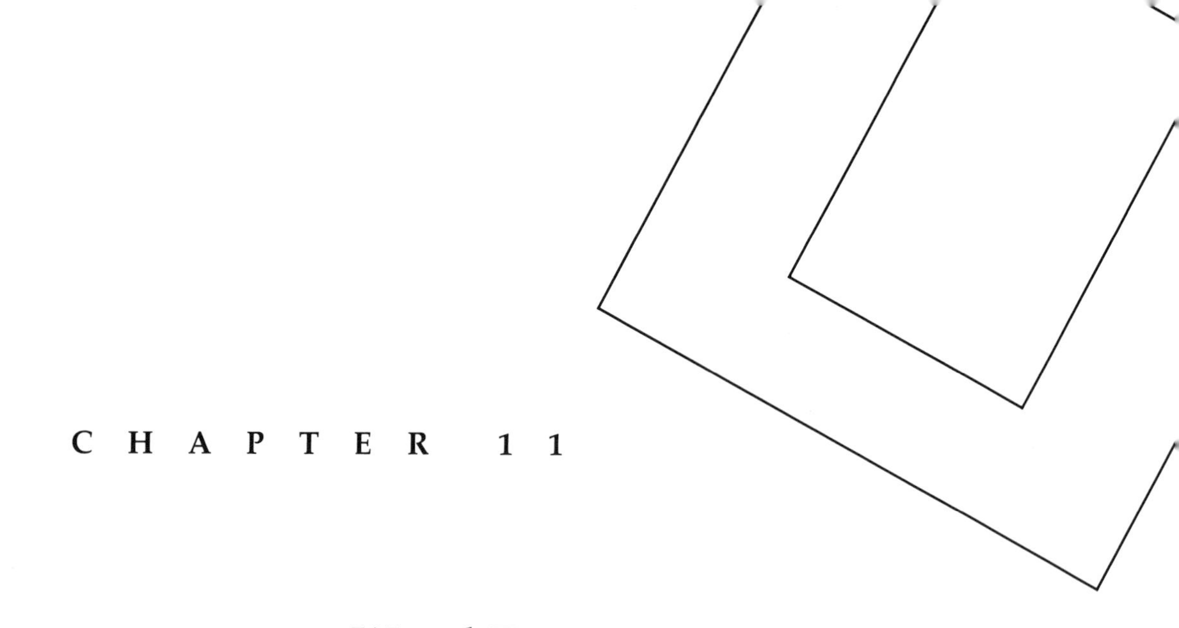

CHAPTER 11

Wood Frames

Wood has a number of drawbacks that should be considered before choosing it as a framing material. Raw wood emits peroxides that can injure a print. Wood frames take up much space in storage, which may lead to careless handling. This tendency is accentuated because the print and mat cannot be so easily removed from them as from other types of frames. The manufacture of wood frames requires a large work area and many tools and supplies like glue, nails, and finishing oils. And then there is that air of finality about a wood frame: will there be someone around in the future to open it up and take a look inside to see that all is well?

The choice of a wood frame can be justified only on aesthetic grounds, not conservation ones. Be that as it may, the visual appeal of a finely finished wood frame is so great that wood frames will be around for a long time, so it is important that they be used correctly.

TERMINOLOGY: PARTS AND FUNCTIONS

Molding

This comprises the frame itself. Manufacturers make molding in long sticks that can be bought finished or unfinished. To make a frame they are then cut or *mitered* at a 45° angle and joined at the corners.

Molding has different parts. The *face* is the surface toward the viewer. The indented area behind the face is called the *rabbet*, and the *lip* is the part of the rabbet that retains the glass. The terms *side* and *back* should be self-explanatory. (See Figure 11.1.)

Painted or enameled wood frames should never be used until the paint has thoroughly cured, because dry but uncured paint emits vapors known to be particularly harmful to photographic emulsions. Gesso and gilt finishes do not present the same kind of problem.

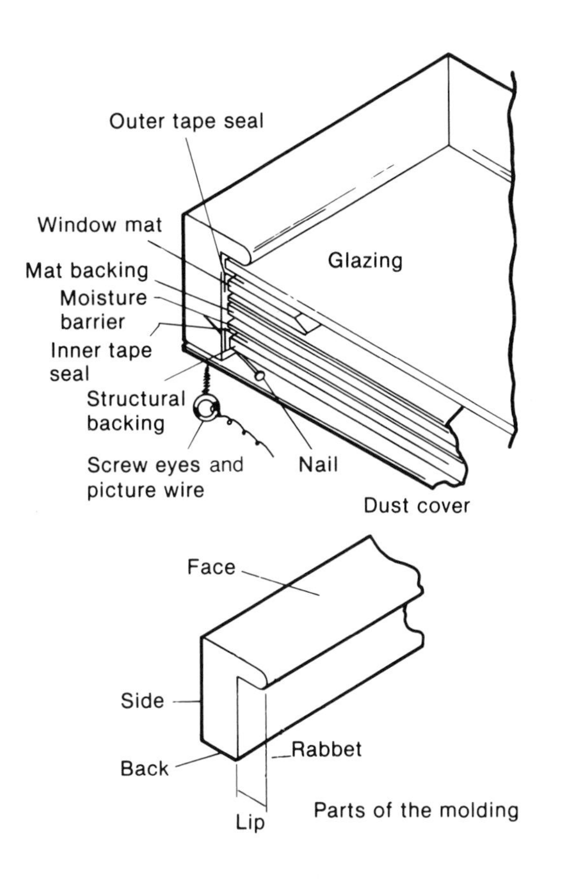

Figure 11.1
Parts of a frame.

Avoid softwood frames. By *softwood* we mean the wood from any evergreen tree like pine or cedar. (Hardwood comes from deciduous trees, those that shed their leaves annually.) The resins of softwood remain volatile for years and will have an effect on prints. Because of its low resin content, basswood is considered the best wood for conservation-grade frames.

The rabbet of a wood frame needs to be sealed with polyurethane varnish as a conservation measure. Even hardwoods contain some resins, which the varnish prevents from reaching the print.

An important factor in choosing molding should be frame strength. Molding that is too thin will not have strong enough corners to support the weight of mat, glass, and frame on the wall. The result can be a pile of broken glass and wood on the floor.

In the demonstration photographs in the section on joining, only one type of molding is shown. This is for the sake of clarity and not because one particular style of molding is preferred for conservation framing.

Glazing

This consists of either glass or ultraviolet-filtering plastic. The chief function of glazing is to provide the surface of the print with physical protection while allowing it to be viewed with minimal distraction.

Outer Tape Seal
This seal keeps insects and airborne pollutants from entering through the front of the frame. It cushions the glass during handling and transit, prevents glass chips from getting inside the frame, and may stop corner breaks if the frame gets dropped. Suitable tapes are 3M's Scotch 810 Magic Transparent or Polyester Tape 8411.

As an alternative, you might also consider plastic molding liner available through some frame shops.

Window Mats
Window mats or hidden separating strips keep the print from contacting the glass. Temperature changes often cause condensation on the inside of the glass. If the print touches the glass in the presence of water, it can adhere in spots and be ruined. The gelatin emulsions of photographs are particularly susceptible to this kind of damage, known as local ferrotyping. Separation of print and glass is a significant conservation measure in framing.

Print
The print is attached to the mat backing board.

Mat Backing Board
This is hinged to the back of the window mat. This piece is never attached directly to the frame. It should always be possible to take the entire mat and print unit out of the frame without damage.

Moisture Barrier
A moisture barrier can be made of either Mylar sheeting or aluminum foil. It stops the migration of chemical contaminants from the back of the frame. If sealed around the edges, it also gives temporary protection in case an overhead water leak causes water to run down the wall (a surprisingly common cause of print damage). A moisture barrier is an important conservation measure; it is absolutely essential if an acidic material like chipboard is used for the structural backing.

Inner Tape Seal
This seal is made of the same material used for the outer one. It holds the moisture barrier to the rabbet and protects against bugs, dust, and contaminants.

A ventilating gap can be left in one of the upper corners of the frame to allow trapped moisture to escape through the tape seal. If the decision is made to completely seal the back of the moisture barrier, allow the print and mat to acclimate themselves in a cool room with a relative humidity under 50% for at least two weeks. If this is not done, high temperatures can later drive absorbed moisture out of the mat board. This moisture gets trapped in the frame and raises humidity to unacceptably high levels.

Structural Backing

This backing holds the print and mat package in the frame. Nails or brads driven through it at a shallow angle secure it to the inner face of the rabbet. The structural backing protects the rear of the print from punctures and cuts.

The best material for backing in wood frames is archival multiuse board, a nonacidic gray corrugated cardboard that holds nails and brads well. Versions of it are available under brand names of Process Materials Corporation, Light Impressions, and Conservation Resources. (For other kinds of frames, a good choice is a neutral pH, extruded polystyrene foam core board covered with ABS cap sheets. This foam core board does not have the strength to hold nails and brads as well as the PMC paper board, so it will be better used in places like metal frames where this is not a consideration. Monsanto's Fome-Core® board is the leading type of this material.)

Most commercial frame shops content themselves with chipboard, a stiff gray paperboard about ⅛ inch thick. It gives great strength for the weight, but because of the chemical impurities in it, a moisture barrier of Mylar or aluminum foil should be used between it and the mat.

Never use regular corrugated cardboard or wood paneling for the structural backing. Even a moisture barrier cannot provide adequate protection to safeguard the print thoroughly from corrugated or wood backings.

Nails

Nails hold the structural backing in place. Glazing points, staples, or brass escutcheons can be substituted. To prevent ferrous fasteners from rusting, spray them with a clear lacquer artist's fixative.

Dust Cover

The dust cover of kraft paper gives the back of the frame a finished and professional look, and it is a good place to record data like date, owner, framer, materials used, medium, catalog number, restoration methods used or attempted, and, of course, the artist. Also it helps keep out dust.

It is easy to replace a dust cover, so do not hesitate to take one off an old frame to look inside.

Screw Eyes and Picture Wire

This is the most common way to hang pictures. Experience leads us to think that a sizable amount of damage to prints in private hands comes about due to their falling off the wall. Even curators and gallery owners need to keep in mind that a drop of several feet onto a hard floor is likely to cause serious problems, so an adequately strong hanging system is an important part of conservation framing.

Wall Spacers

These do more than just protect the wall behind the frame. They prevent condensation on the frame back by allowing air to circulate. In case of an overhead leak, wall spacers will keep out some water flowing down the wall. They also help in getting the piece to hang vertically. 3M makes soft rubber Bumpons with an adhesive back, but cork squares or small pieces of balsa wood can be substituted.

JOINING

The technical term for assembling a frame is *joining*. Long sticks of molding are *chopped*, in the framing vernacular, into appropriate lengths for the frame. Each length has a 45° angle at both ends. The joining operation itself comprises all steps from nailing and gluing the frame corners together to final finishing. We discuss chopping in some detail under "Molding."

Most steps in joinery do not have a direct bearing on print conservation per se, but the process is so basic to framing that it is pointless to talk about other conservation methods without covering this first. A high-quality frame can be joined with a minimal investment in tools and equipment. Ornate frames may require specialized finishing techniques, but since most photographic and other print media are not usually framed this way, these methods are not covered here. Expensive power tools serve only to speed up production in what remains essentially a handcraft.

Construction of frames has to be carried out in an area entirely separate from the work sites where prints are handled. The tools, chemicals, and debris of a frame-making operation all pose serious threats to print safety. (See Figure 11.20.) Either establish a segregated working area or purchase frames made by an outside concern. The following sections describe tools and supplies needed for joining.

Vises and Clamps

Some way to hold the molding firmly while nailing and gluing is essential to joining. The bench vise shown in the demonstration photographs is a versatile and efficient tool for doing this. A good one costs about as much as a mediocre double-miter corner vise. Do not confuse the bench vise with a carpenter's vise, which has a large wood face and several draw screws. The carpenter's vise has only limited application in frame joining.

Bench vises have to be adapted by covering both gripping faces with chipboard to avoid leaving clinch marks on the molding (Figure 11.2). Use double-sided or adhesive transfer tape to attach the clipboard.

Double-miter corner vises are built especially for framing. They hold two pieces of molding at right angles, with the corners ex-

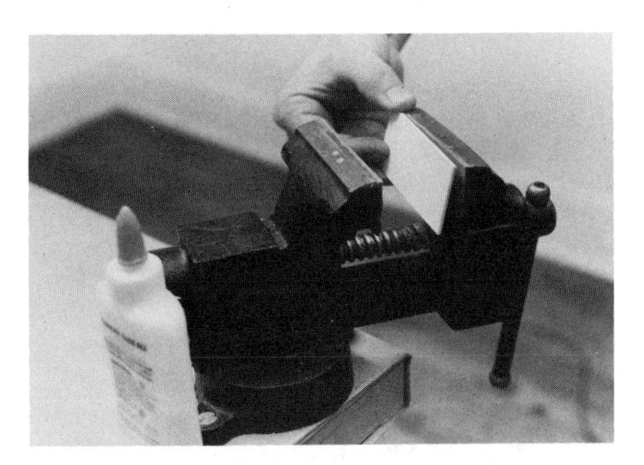

Figure 11.2
Cover the jaws of a bench vise with paper board so that it doesn't make clinch marks on the molding.

posed for nailing. A good production model like the Stanley has fast, delicate release grips and costs around $190. The cheap ones are hardly worth the trouble.

Four corner clamps apply equal pressure to all corners of the frame while the glue dries. You can choose from two basic kinds. One has metal tie rods with aluminum corner blocks, while the other consists of a strap drawn around the frame perimeter like a belt. (We have found that the entire frame can leap out if the strap is drawn too tightly, so be careful!) Corner clamps make no provision for nailing, so they should be bought as a supplement to a vise, not as a substitute. Glue alone is not strong enough to hold any but the lightest frames.

Hand Tools

The basic kit consists of

- ☐ A hand drill with brad bit
- ☐ Needlenose pliers with a wire cutting section
- ☐ Tack hammer with flat head *or*
- ☐ Resin-head hammer
- ☐ Nail punch
- ☐ Sanding block
- ☐ Half-inch brush (utility-grade nylon bristles are adequate)

Power tools are not required for joining most frames.

Supplies

Purchase supplies in quantity before starting, and keep an adequate stock handy. Basics include the following:

- ☐ A selection of wire brads, sizes ½ inch to 1½ inch
- ☐ Fine and very fine grades sandpaper
- ☐ Polyurethane varnish

☐ White glue (Elmer's or Titebond)

☐ 00 or 000 grade steel wool

☐ Wood-finishing oil (boiled linseed oil, Watco, or Val-Oil)

☐ Nail hole filler (S&W putty or wall paneling crayons)

Molding

An operation that plans to use more than several hundred feet of wood molding a year should look into chopping its own molding. The best tool for the job is the Lion Miter-Trimmer, which uses a lever-operated guillotine blade. With measuring option it sells for around $250. Although the Lion looks a bit archaic, it gives as precise a cut as is humanly possible.

Power miter saws are an expensive alternative to the Lion. They make economic sense when one plans to cut large quantities of wood and metal molding. Most have electric motors for the saw blades and compressor-driven hydraulic lines to force the head down for the cut. Consult the ads in framing magazines like *Decor* for suppliers of the Morse and Pistorius miter cutters and Power Miter Box Saws.

The poor person's method of chopping remains the miter box. In addition to a good miter box, you need several C-clamps with chipboard-padded gripping faces and a backsaw with 12 or 16 points to the inch. Use a square piece of wood clamped in the rabbet to avoid crushing the lip during cutting.

You might also consider buying molding prechopped. Many frame shops use nationally advertised chop services that ship the same day an order is received. Individuals or small institutions that do not do enough business to warrant an account with a chop service can buy through the mail from custom cutting services or from the local framer.

Will it pay to do your own chopping? When calculating the costs, keep these factors in mind: storage, inventory of molding, waste and surplus, and maintenance of the work area. Remember that the supplier pays for any cutting mistakes when he does the cutting.

Cost and supply will not be the only considerations when you buy molding. You also want to know whether the molding affects the prints. Start with the general rule that most hardwood molding is preferred. Newly painted wood, as mentioned above, gives off vapors, so avoid it until the paint cures thoroughly. Softwoods should also be avoided, permanently, because of the resin problem. This means pine, cedar, and all the other evergreens, including redwood.

Old wood presents a special problem. If it seems wormeaten, throw it out or fumigate it. In any case, clean its surfaces thoroughly with methyl alcohol to sterilize it. Refinished old wood can also present problems. Oxalic acid, used for bleaching, and the chemicals used in dipping and stripping, seep into the grain of the wood. Their persistence and high level of activity make them suspect.

FIGURING FRAME SIZES

The length of the rabbet determines frame size. Neither the outside dimensions nor the distance along the inside edge of the lip have anything to do with the size of the mat the frame will hold, so you can safely ignore them. The only thing that counts is the size of the box made in the back of the frame by the rabbet when the corners are joined (Figure 11.3). The 45° angle of the corner miter means that the outside dimensions will be larger, and the lip edge shorter, than the desired distance. Needless to say, this principle holds true for all kinds of frames, whether wood, metal, or plastic.

We mentioned earlier that allowance has to be made for changes in print size caused by humidity fluctuations. The same holds true for mats. Mats that are fitted tight and flush against all four sides of the frame will buckle and take on a wavy appearance over time, with gaps appearing between the mat and the print. If left unremedied, the mat sets up in this configuration.

How much tolerance should you allow? A good rule of thumb calls for leaving a tolerance of slightly less than the width of one rabbet. Technically speaking, the width of the rabbet is equal to the size of the back of the frame's lip. On conventional moldings this varies from ⅛ inch to as much as ½ inch. If the width of the rabbet on a frame is ⅜ inch, you would leave a tolerance of about 5⁄16 inch — *slightly* less than the width of the rabbet. Figure this tolerance along both the height and the width of the mat, whether cutting the mat to fit the frame or vice versa.

Check with your molding supplier about this allowance if you plan to have the supplier do your chopping. Either have the frame cut with an allowance to your specification, or plan to cut your mats smaller than the nominal size. Practice varies from supplier to supplier.

Do not use a matted print to check frame sizes. A ruler has much more resistance to nicks and stains than does a valuable print in the environment of a woodworking area.

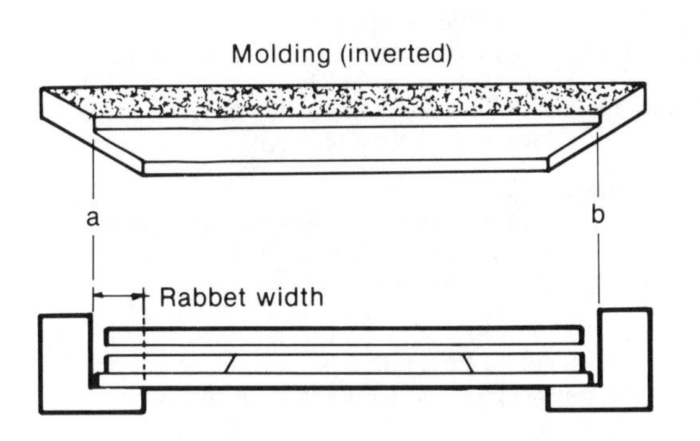

Figure 11.3
Rabbet length. The nominal size of the frame is from point (a) to point (b).

ASSEMBLY

After gathering your tools, lay out the four pieces of molding. Examine each for straightness and surface defects. Replace unsatisfactory pieces, and sand out passable ones with minor blemishes. A word about sanding: use a sanding block at all times to avoid rounding off the crisp contours that give a frame its distinctive character (Figure 11.4).

Eight brads will hold all but the largest frames together if used in conjunction with glue. Choose the right length of brad by holding it on the face of the frame at a slight angle pointing inward. The right length will not protrude into the center of the frame. Brads can be tacked into the top and bottom moldings only. This means that they will not be seen from the sides, which are the only edges of a frame commonly viewed (see Figure 11.11).

Before setting the brads, make holes in the corner with a hand drill. Cut the head off a brad with the needlenose pliers to make a drill bit (see Figures 11.5 and 11.6). These homemade bits get dull after a while, so make new ones as needed. They are cheap.

Use a small piece of cardboard to spread a thin layer of glue over both miter faces of the corner being joined. Do not overglue, because any excess only squeezes out. Do only one corner at a time (see Figures 11.7–11.16 for step-by-step illustrations for assembling a frame).

Clamp the same molding — the piece without predrilled holes — in the bench vise. Grip only the rabbet part, with the lip

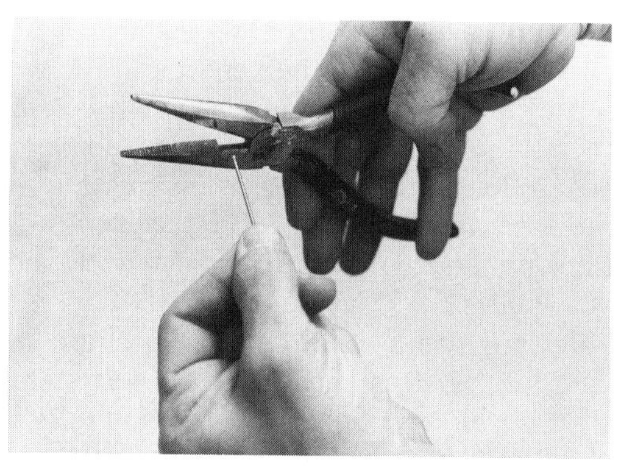

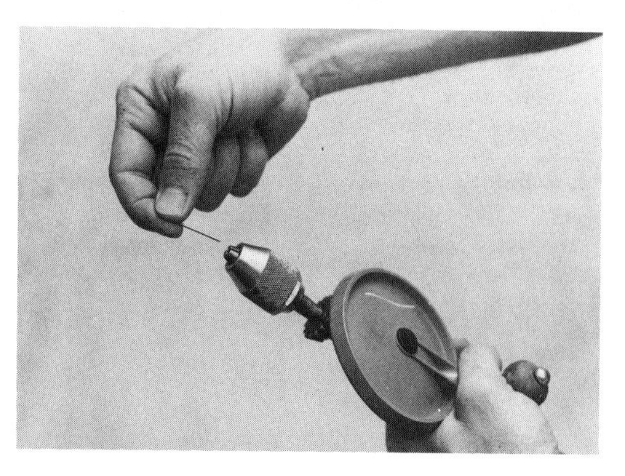

Figure 11.4 (above left)
Preassembly sanding of the molding makes the final finish work go more quickly. A sanding block and a vise are needed to keep the sharp edges and smooth lines from getting blurred.

Figure 11.5 (above right)
Cut the head off a brad one size smaller than the ones used to secure the corners. This makes a cheap drill bit of the right size.

Figure 11.6 (left)
Put the drill bit brad into a hand drill. Power drills are a useless piece of overtooling for this job.

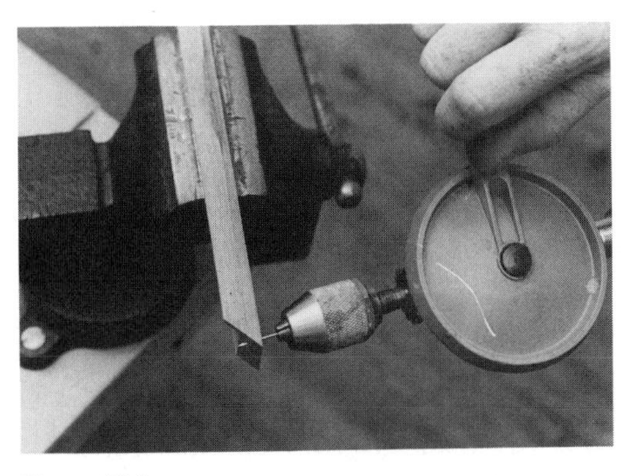

Figure 11.7
Clamp the top (or bottom) section of molding in the vise. Drill a starting hole all the way through to the mitered edge. The hole should angle slightly out to the side. (A little experience will give you an eye for the right angle.)

Figure 11.8
Inspect and lay out the molding sections before assembly.

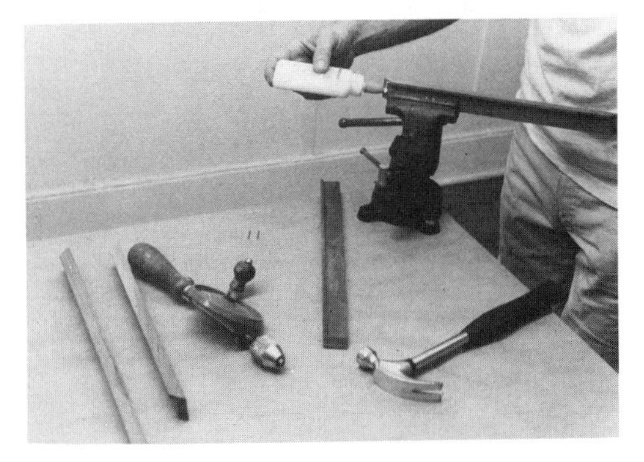

Figure 11.9
Take out the drilled section and replace it with the side section that will receive the tip of the brad. Put on a thin line of glue on the miter . . .

Figure 11.10
. . . and smear it to a thin layer. Take care that none of the glue gets onto the outer surface of the frame. If it does, wipe immediately with a wet cloth.

Figure 11.11
Choose the right length brad by putting together a corner and angling the brad slightly toward the outside of the frame.

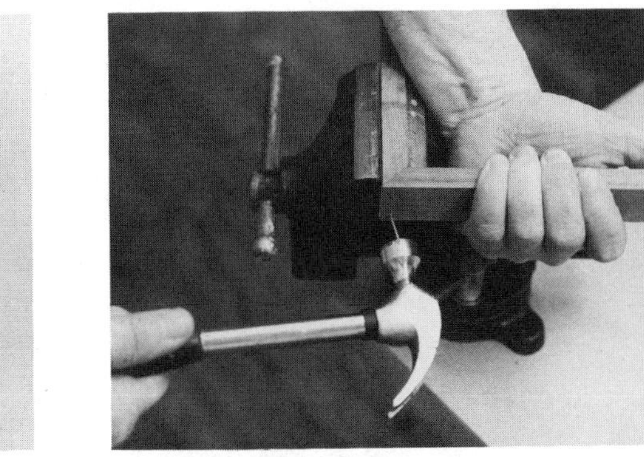

Figure 11.12
Drive both brads into the corner one after the other. A firm grip is essential; we find it easiest to pull the hand-held section toward the body. Use the *face* of the frame, not sides or back, to line up the sections.

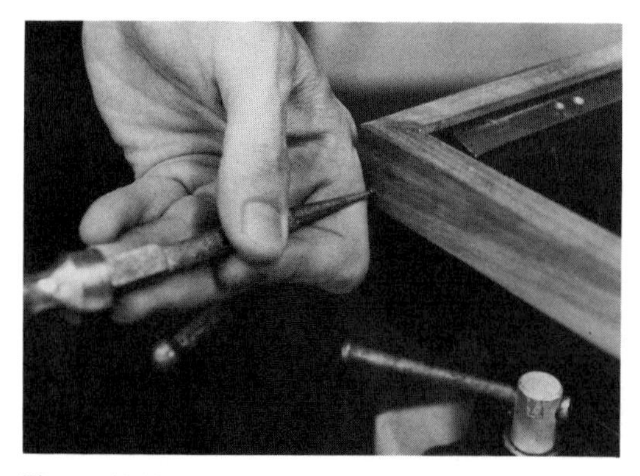

Figure 11.13
Instead of driving the brads all the way in, set them with a nail punch, about ⅛ inch under the surface of the wood. This avoids crescent-shaped dents and makes it possible to hide the nailheads.

Figure 11.14
If you make two identical sections like this . . .

Figure 11.15
. . . the result will be something like this. Each corner should be the mirror image of the other, rather than identical.

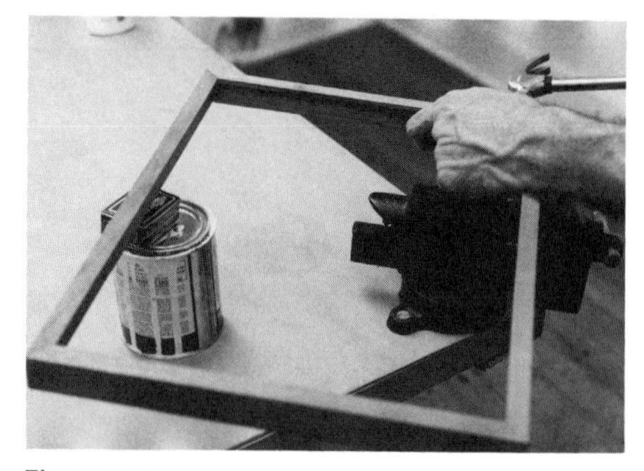

Figure 11.16
Support the far corner when nailing the two corner sections.

extending over the top; the inner edge of the miter should be almost flush with the edge of the vise.

Hold the other, predrilled piece of molding against the piece in the vise. Put one nail into one of the drilled holes with your free hand. Line up the faces of the two pieces exactly. Sides and back have to line up too, but the face is most critical.

It is important at this point to have a very firm grip on the piece of free molding. Do it whatever way you find most convenient. One way that works well is to reach across the molding in the vise, so that you are hammering toward yourself.

Tap the first brad in, and then set and drive in the other brad. Both should angle slightly toward the frame center (Figure 11.12).

Do not try to hammer the brads flush with the molding surface. This will leave crescent marks from the hammer edge that cannot be sanded out later. Instead, use a punch to set both brads. Drive them about ⅛ inch below the surface. Do this right away to avoid hammering on the joint after the glue dries (Figure 11.13).

Take the corner out of the vise and inspect your work. If it is wrong, take it apart and do it over. To take a corner apart, put the molding back in the vise, put a square in the rabbet of the crosspiece, and hit it sharply. This pulls the brads right through the drilled holes. Pull them out with the pliers and start again. Any extra holes can be filled later. Do not try to use a claw hammer to take out the brads.

Once you finish the first corner, pick up the other two pieces of molding and make another L-shaped corner.

Stop before you go on to the next step. We have noticed, in training framers, an understandable tendency to make two identical L-shaped corners. This leaves you with two half-frames. The second corner has to be the *mirror image* of the first. If in doubt, lay the frame out on the table to see that it goes together right, *before* nailing (Figures 11.14 and 11.15).

You now have two corners finished. Set up to nail the third corner. Put the side of one corner in the vise and hold the other corner against it. Prop up the far corner with some boxes and cans so that you do not put too much stress on the freshly nailed joints (Figure 11.16). (If you are doing many frames, use a piece of 2 × 4 lumber cut to length. The correct height will not vary much from length. The correct height will not vary much from frame to frame.)

Repeat the process for the fourth corner. Make certain that no glue has squeezed out onto the visible surface of the frame. If any has done so, wipe it off with a damp cloth before it dries. Wipe thoroughly.

SEALING THE RABBET

In a work area removed from the vicinity of the print, paint the rabbet with polyurethane varnish or artist's matte medium. Apply the sealant only to the rabbet, and wipe off any excess that gets onto the visible faces of the frame (Figure 11.17). Any varnish or medium that dries there will prevent the wood from absorbing finishing oil, and will show up as a light spot. Sealing the rabbet prevents vapor from the wood from reaching the print. This is an especially important conservation measure when using a narrow or concealed mat.

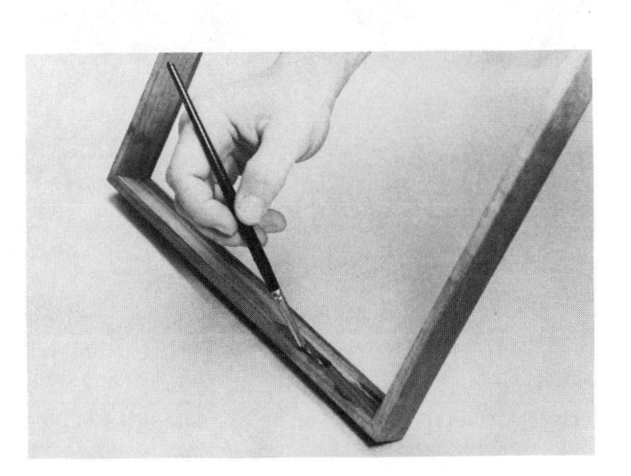

Figure 11.17
Seal the rabbet with polyurethane varnish or artist's matte medium as part of the finishing process.

FINISHING

Wood frames can be finished in a hundred different ways, from gold leafing to veneering. Here is a simple way to put on a natural finish; for other techniques we suggest any of the numerous fine books on woodworking.

To dress the corners, put the frame back in the vise and use a sanding block. Use fine-grade paper. These measures are important to avoid erasing the sharp, rectangular look of the frame.

The miters probably overlap on the sides. Erase this overlap by sliding the sanding block along the side with the projection, and around the corner so that the projecting piece gets pushed in toward the frame center. Bring the sanding block sharply around the corner, repeatedly, until the overlap disappears.

This usually fills any little gap in the miter. If a crack remains, fill it with plastic wood from the hardware store, or with a mixture of white glue and frame sawdust (Figure 11.18). The plastic wood has to be mixed with oil before it is used to fill the crack. Let the filler dry, and sand the corner again.

Sand the faces of the corner flat. These were lined up during nailing, so there should be little to sand.

Go quickly and lightly over the entire frame with very fine sandpaper. Wipe thoroughly with a tack rag or a lightly moistened paper towel to pick off all the sawdust.

The frame is now ready for oiling. Suitable oils are Watco Natural Danish Oil Finish, Val-Oil, or boiled linseed oil.* The Watco has a hardening resin that reacts with air to make a durable finish that does not dull with time, and it hardens the wood surface against scratching.

Soak some grade 0 (fine) or 00 (very fine) steel wool with oil and flood the visible frame surfaces with oil. Let stand about 15 minutes, wipe off the excess with a rag, and apply another light coat (Figure 11.19). Burnish the second coat with the steel wool as it is put on, wipe again, and set aside to dry. A third coat can be applied the next day.

Figure 11.18
Fill nail holes at the corners with woodworker's putty or dark, stained plastic wood.

* Do not confuse boiled with raw linseed oil. "Boiled" refers to a complex distilling process, so you cannot just boil some of the raw.

Figure 11.19
Apply oil to the frame's visible surfaces with steel wool; let the first coat stand for fifteen minutes; wipe off and reapply. Buff the second coat again with very fine steel wool.

A word about fire safety. Volatile finishes need to be handled with care — first, because of the safety of the human beings involved, and second, because you are probably working where valuable prints are housed. Do not allow smoking or open flames (this includes pilot lights!) in the area where volatiles are used. Guard against spontaneous combustion by soaking all oily rags and steel wool with water. Store them in tightly capped metal containers until disposal. Provide adequate ventilation.

Now back to the frame. If you desire a glossier finish, wait a week for the oil to dry completely and buff the frame with a coat or two of butcher's paste wax or similar hard paste wax. This produces a high-gloss finish.*

Nail holes have to be filled right after oiling. You can use the white glue and sawdust mixture, but it tends to remain visible. A better choice is the filling putty sold by framing suppliers expressly for the purpose. Or you can substitute the crayons made to hide nailheads on plywood wall paneling. A color darker than the wood filled will not show up as readily as a lighter color.

FITTING THE MAT

Fitting refers to mounting the glass and mat into the frame. At this juncture in framing, you can take conservation measures that help preserve the print.

Whether or not the print and mat should be hermetically sealed into the frame remains a point of debate. We suggest that some provision for ventilation be made. The reason is simple. A sudden increase in heat — caused, for example, by having the frame hang several hours in direct sunlight — can force moisture trapped in the paper to form vapor. When the frame cools, this moisture condenses inside the frame in liquid form instead of dissolving back into the mat board and print. Water stains result if the moisture cannot escape.

* An excellent medium-gloss finish can be achieved without oiling by using Renaissance Micro-Crystalline Wax. This wax, which comes with directions for application, creates a durable finish that resists both moisture and alcohol.

Figure 11.20
We realized after finishing the demonstration photos on frame joinery that we had a perfect visual argument for keeping prints out of the area.

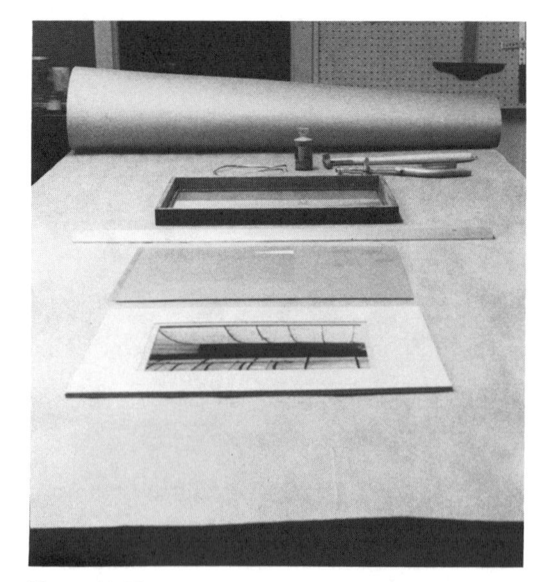

Figure 11.21
If you must use the same area to fit the print into the frame as you did for the joining, make certain it looks like this before you start. Note the fresh table cover of kraft paper and the absence of spills.

Figure 11.22
And now for the fitting proper. After cleaning the glass, apply Scotch Magic Transparent Tape around all four sides. Only a thin border should actually adhere to the glass.

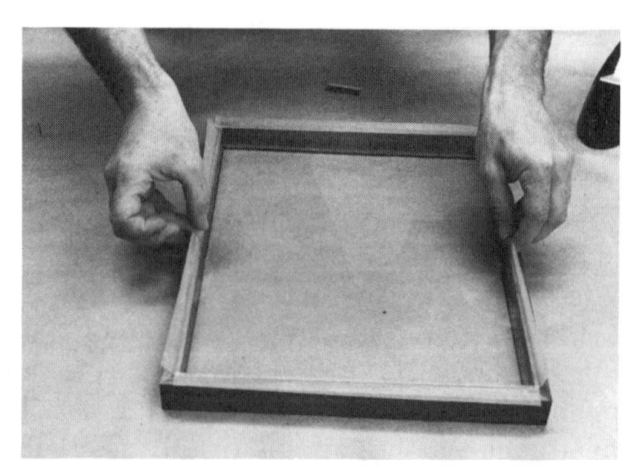

Figure 11.23
Slide one end of the glass into the frame, and then push down the opposite end.

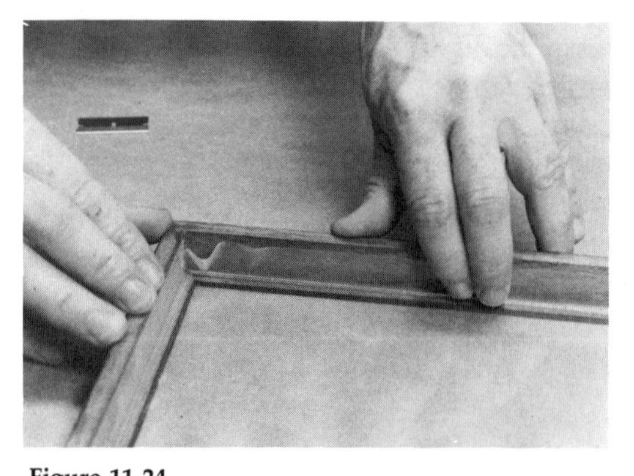

Figure 11.24
Seal the tape to the rabbet by running a fingertip along it. The razor blade in the background is for trimming the overlap at the corners.

Figure 11.25
The three components remaining to be fitted into the frame: the matted print (white), the moisture barrier of clear plastic (just barely visible), and the mechanical backing (gray).

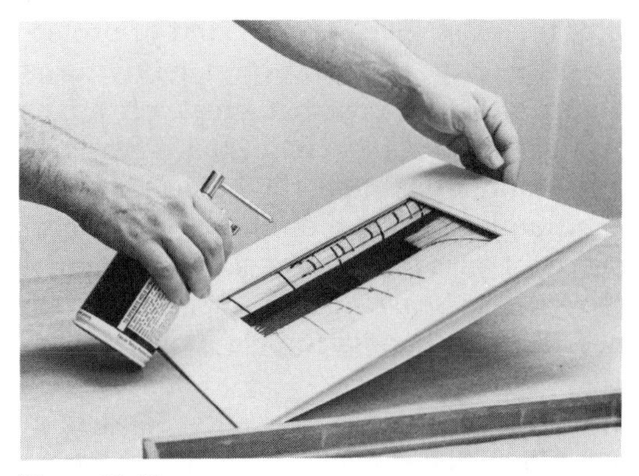

Figure 11.26
Dust sealed in the frame will cause gnashing of the teeth. A clean drafting brush can be used, or a can of compressed gas with a spray nozzle like Dust-off or Omit.

Figure 11.27
Seal the moisture barrier around all four sides after putting the mat in place. A small venting gap is left at one corner.

Figure 11.28
Secure the structural backing with brads every 2 inches. If using a tack hammer, the frame has to be butted against a padded bar or table lip to give something to strike against.

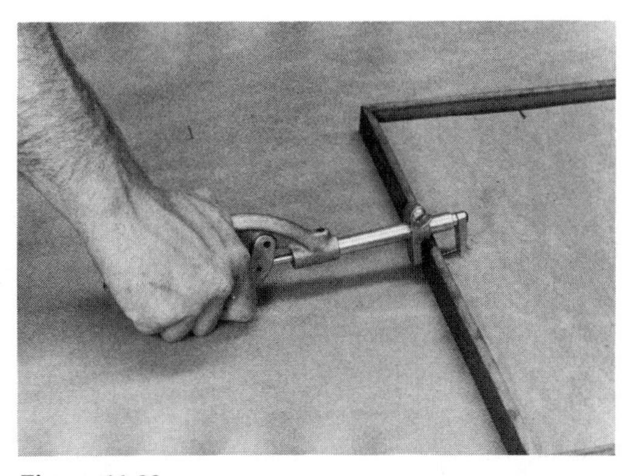

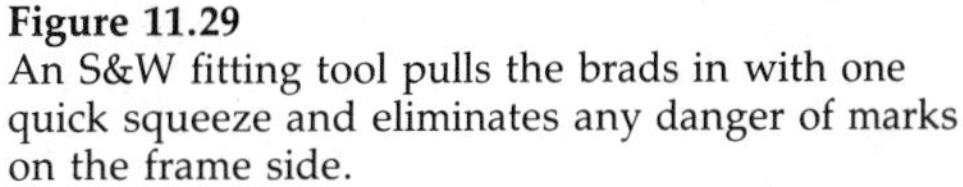

Figure 11.29
An S&W fitting tool pulls the brads in with one quick squeeze and eliminates any danger of marks on the frame side.

The alternative to venting would be to allow the mat and print to dry for several weeks in an atmosphere of low relative humidity and low temperature — say, 30% to 50% relative humidity and 50°F. Framing would then have to be done under these conditions as well to prevent condensation on the chilled materials from canceling the effects of the prolonged drying.

Prints do need to be more or less sealed off from the surrounding environment. The inner and outer tape seals and the moisture barrier perform this function. These seals protect against common dangers: overzealous housecleaners who spray glass cleaner directly onto the front of the frame,* or gaseous pollutants such as sulfur dioxide, dust, insect intruders, and wet walls.

---

* The right way to clean glass in a frame is to spray some glass cleaner onto a rag or paper toweling, and then to wipe the glass. Glass cleaner sprayed directly onto the glass often seeps around the bottom edge and gets drawn up inside the frame by capillary action.

Tape seals on the edge of the glass give another kind of protection as well. Inspection of many traveling shows frequently turns up little slivers of glass and glass crumbs that break off edges because of repeated stressing. Tape eases some of this stress and contains the particles so they cannot penetrate the frame's interior.

Start fitting by laying the frame face down on the work surface. A thoroughly cleaned piece of glass or other glazing material should be already cut to size. Glass is to be cut in a space separate from the fitting area. Once cleaned, handle the glass only by the edges to avoid fingerprints.

Set the glass gently into the rabbet, one edge first, and then angle the far edge down until it fits snugly. Hold up the frame and glass, and look down its length toward a light source to spot any specks of dust. Remove dust specks with a brush.

Apply 3M's Scotch Magic Transparent No. 810 Tape along the edges of the glass to make the outer tape seal. Put the tape on in four strips, one along each side. Overlap each length at the corners to get a completed seal.

After the tape has been joined to the glass (a narrow strip will suffice), join it to the sides of the rabbet. Trim any excess at the corners with a sharp knife.

Look at the frame from the front to see that no tape is visible. Check the glass again for dirt and dust. Check the front of the mat for smudges and dust. Use a kneaded rubber eraser to lift any smudges from the mat front; the kneaded rubber does not leave particles trapped inside the frame.

If the print itself has a delicate or friable surface, use a can of Dust-Off or similar type of canned air to blow off dust from the print surface. Prints with a more durable surface can be dusted with a drafting brush.

Lay the matted print face down in the frame on top of the glass. Pushing on the back of the mat with your fingers, lift up the entire unit, turn it over, and look for dust. Then set it down again, face down. If no dust is visible, proceed to the next step. If dust is present, dust again.

Now set the moisture barrier on the back of the mat. Type S Mylar* or a similar inert polyester film is the easiest to work with because of its relative rigidity. Aluminum foil can be obtained more readily and is more impermeable than polyester, but you may find it more difficult to cut exactly to size.

Tape around the edges of the moisture barrier the same way you did around the glass — except this time leave a small gap at one of the upper corners. The gap allows excess moisture to vent out the frame. It goes in an upper corner because water runs downhill and is less likely to get trapped in the top of the frame.

On top of the moisture barrier, insert the structural backing. As

* Some kinds of Mylar come with invisible coatings applied to them. Types D and S are plain, without any coating.

mentioned earlier, this backing can be PMC archival multiuse board or chipboard.

Nail the structural backing to the sides of the frame. The brads are angled slightly toward the front so they penetrate only the structural backing element and then go into the wood. If you use a tacking hammer to set the brads, the edge of the frame has to be seated firmly against a padded stop. A raised lip bolted to the end of the work table is the best kind of stop. Space the brads about 2 inches apart for optimum strength.

Occasionally a brad will poke through the side of the frame. Leave it in place and use its length as a guide in driving the rest of the brads. After finishing, pull it out, set a new brad next to its space, and fill the hole on the frame edge.

There are a couple of tools that will speed up the nailing process. One is the fitting tool made by S&W Framing Supplies. It has a squeeze grip and padded jaws that protect the frame edge and that can be adjusted for different-sized brads. One squeeze on the pliers-type grip and the brad is set to the correct depth. The Fletcher Model No. 3 Driver uses glazer's points instead of brads. It is a gun device that looks like a heavy-duty stapler.

DUST COVERS

The dust cover for the frame back can be made in the conventional way from kraft paper or from polyester film like Mylar. The application methods are different for each one (see Figures 11.30–11.33).

Thin, flexible kraft paper comes in rolls of several hundred feet. A small investment in a roll holder makes it possible to tear off the desired piece with just a flick of the wrist. This is a great convenience for large framing operations where kraft paper is used for other purposes such as covering work surfaces. Everybody working with it should be made aware that kraft paper should not stay in direct contact with artwork for any length of time.

Individuals doing a limited amount of framing can consider making kraft paper from grocery bags. Generally this paper is a bit thicker and has creases and folds. These can be reduced by lightly moistening and flattening the paper in a dry mount press or with an electric iron.

Other supplies needed are white glue, sponge and water, and a sanding block or razor.

Run a thin bead of white glue along the back of the frame, which is lying face down. Spread the glue with a small piece of cardboard so that it forms a thin layer that gets tacky quickly. Do not apply too much glue.

Lay an oversized piece of kraft paper on the back of the frame. Run your finger all the way around the edge to seal it. Make sure that the paper is pulled slightly taut, though all creases and wrinkles do not have to be pulled out.

Trim along the edges to remove excess paper. A razorblade can be slid along the edge to do this. Another method is to crease the

Figure 11.30
To put on the dust cover, first run a thin line of white glue along the back.

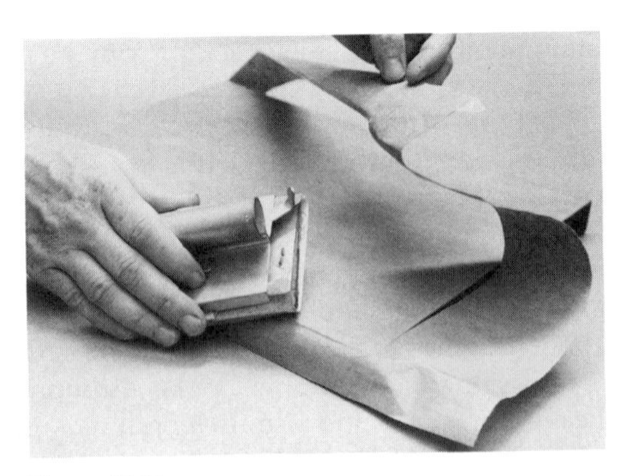

Figure 11.31
Lay a piece of kraft paper over the back, pull tight, and burnish down with a fingertip. A perfect trim can be made by running a sanding block along all four sides.

Figure 11.32
To tighten the paper drum taut, lightly moisten with a sponge and allow to dry.

Figure 11.33
The back of a completed frame has dust cover, wall spacers, screw eyes and wire strap in place.

paper around the edge of the frame and to run a sanding block along the fold to separate the two pieces. With this method you do not risk nicking the frame, and you get a nicely feathered edge that follows the frame contours exactly.

Wet the sponge and lightly moisten the entire dust cover. This tightens it up to produce a taut, snappy-looking piece. (Moistening will not work as well with the thicker kraft paper from grocery bags.)

Apply any information to the back with a pencil.

A polyester film dust cover lets you look inside the frame without removing anything, and it can create a more stable internal environment when extreme humidity might be expected. To monitor and control internal humidity, a small sachet of silica gel with an indicator dye can be put inside the polyester dust cover. A glance will indicate the presence of excess moisture. When silica gel is used, the frame contents should be predried in a cool, low-relative-humidity environment for a week or more.

Cut a sheet of polyester to a size 1/32 inch smaller than the frame back. Round the corners with a fingernail clipper or a graphic arts corner rounder, so that the corners will not snag and pull off.

Run Scotch Adhesive Transfer Tape along the back of the frame molding. This tape comes in 1/2-inch and 1/4-inch widths; use the widest possible size for strength.

Set the polyester sheet in place and burnish along the edges.

HANGING ATTACHMENTS

The hanging attachments have the sole purpose of keeping the frame on the wall. Let this fail just once, and all your careful work comes to naught. The most common type of hanging hardware consists of the traditional screw eyes and a hanging strap between them (Figure 11.33).

The best material for the hanging strap is braided picture wire. Even the thickest solid wire snaps if it gets a kink in it, but the braided wire can be twisted and bent repeatedly without failing. A good all-purpose picture wire is the number 3 weight; for extra strength it can be doubled back on itself. Useful weight limits for the different sizes are:

No. 1	34 lb
No. 3	68 lb
No. 5	102 lb
No. 8	153 lb

The only time to use solid wire is when hanging a frame from the picture molding around the top edge of a room. This architectural convention is not common in modern houses, but it is often found in museums and galleries. Two pieces of wire run down from hooks slung over the molding to the sides of the frame, where they are attached. The solid wire can be painted the same color as the wall to make it less obtrusive. Any length with a kink should be cut out and discarded to avoid a sudden failure.

Screw Eyes

These come in sizes from numbers 219 to 200, the lower the number the larger the size. Standard sizes, depending on the length of the threaded shank, are further divided into "full" and "half" sizes, while the dimensions of the eyes stay the same (see Figure 11.34). Usual sizes for framing are numbers 212, 214, 215, and 216 (very light frames). Keep plenty in stock, because a particular size can be hard to find at the hardware store.

Sink the screw eye into the thickest part of the molding to exert the greatest possible purchase. Check the shank length carefully to avoid screwing through the front.

Screw eyes, full- and half-size

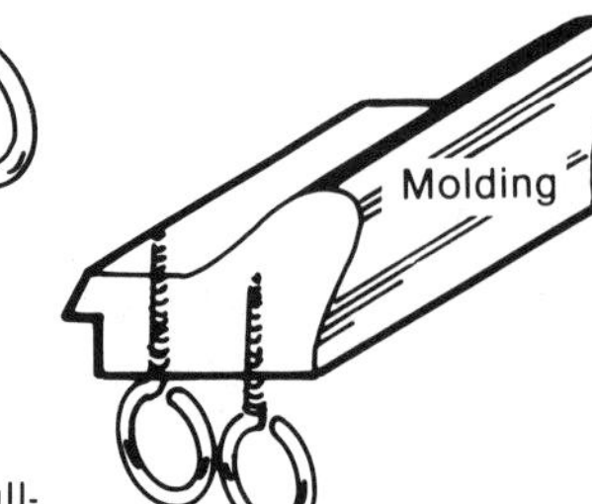

Be careful where you sink them!

Mirror hanger, showing wire strand properly double wrapped to prevent slipping

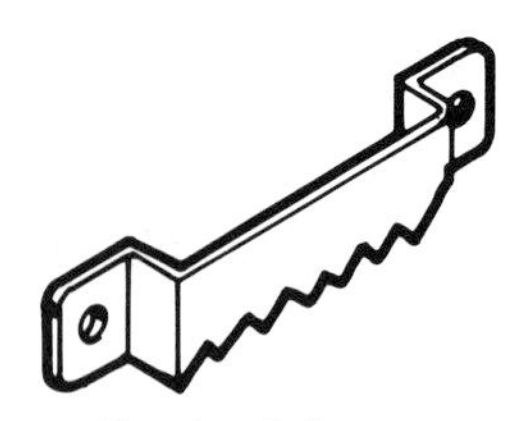

Sawtooth hanger

Figure 11.34
Hanging hardware.

Mirror Hangers
These give greater strength than screw eyes, and hold heavy frames with a great deal of safety. They consist of a flat plate with one or two screws through it, and a metal loop secured by wrapping one edge of the plate around it.

Sawtooth Hangers
The sawtooth hanger does not need any picture wire. It is fine for light frames, quick and convenient to install, but for anything heavier than a few pounds it must be considered something of a risk. Sawtooths are nailed to the top of the molding back right at the center. Then they are hung from a nail with a big head that is driven into the wall.

Wrapping the Wire
Techniques for wrapping the wire apply equally to wood, metal, and plastic frames. To run a single strand between two screw eyes or mirror hangers, feed a short section of wire through one loop. Wrap it around and pass it through the loop again. Pull tight. Wrap the short leader around the base of the longer section right where it enters the loop. Again, pull tight. Continue wrapping out toward the center of the frame, but do it more loosely with each turn.

Now feed the other end of the wire through the other loop. Pull to adjust the play in the wire, and repeat. As a rule of thumb, the wire should be just slack enough to pull up about an inch at the center from the axis of the two eyes.

Two factors influence how close to plumb a frame hangs. The higher the eyes on the back of the frame, and the tighter the wire, the closer the frame will hang to a true vertical. You want the frame vertical, but do not let the wire stick out above the top molding.

Heavy frames need more support than a couple of screw eyes and a strand of wire. The first option is to put a couple of extra screw eyes in the bottom length of molding, each of them several inches in from the corner. Start the wire at one of the bottom screw eyes, run it up and through the upper eye on the same side, across the back into its opposite number, and down to the last eye. Loop and tighten it there to adjust the slack. This kind of wiring gives support to the bottom of the frame and uses the weight of the entire assembly to hold the sides together.

Still heavier frames should hang from two separate picture hooks. Choose the heaviest possible screw eyes — or better yet, two-screw mirror hangers. Wrap a double loop of wire around each one. Adjust for level by tightening up the wrap at the base of the loop on the low side of the picture. Frames this heavy should generally be crosswired as well. Put screw eyes in the center of each side, run wire across the back to the ones opposite, and twist until taut (Figure 11.35). This reduces the chances of the sides popping apart, bowing, or sagging.

As a general guide to frame weights, consult this table:

Frame Weight (lb)	*Hanging Configuration*
0–10	Sawtooth hanger(s), *or* 2 screw eyes (215, 216), no. 2 wire
10–30	2 screw eyes (214, 215), no. 3 wire
30–50	4 screw eyes (212, 214), no. 4/5 wire *or* 2 screw eyes (212, 214), two wire loops of no. 3/4 wire, cross wire
50–up	2 mirror hangers, two wire loops of no. 5 wire, cross wire

As a final touch, add corner spacers on the bottom two corners. As we mentioned before, these can be made from bits of balsa wood or cork stuck on with adhesive transfer tape, or you can purchase rubber Bumpons.

Figure 11.35 Hanging configurations for different-weight frames.

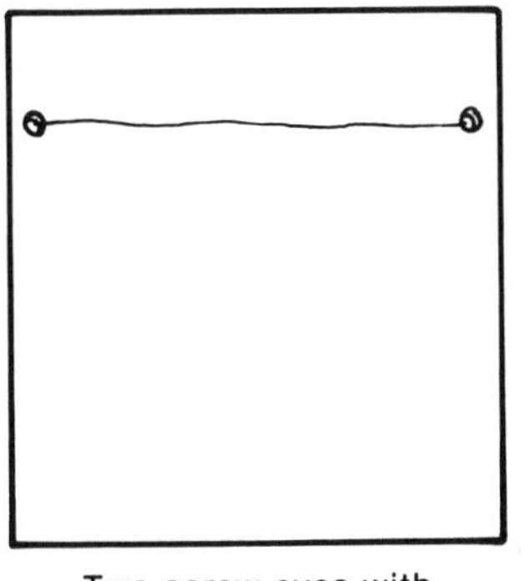
Two screw eyes with wire strand

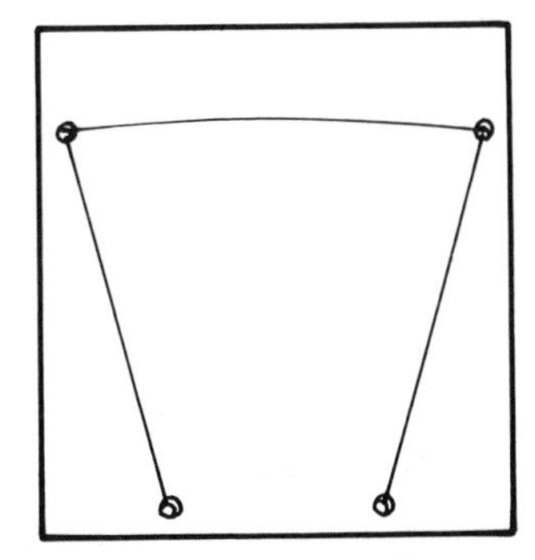
Four screw eyes with wire self-reinforcement

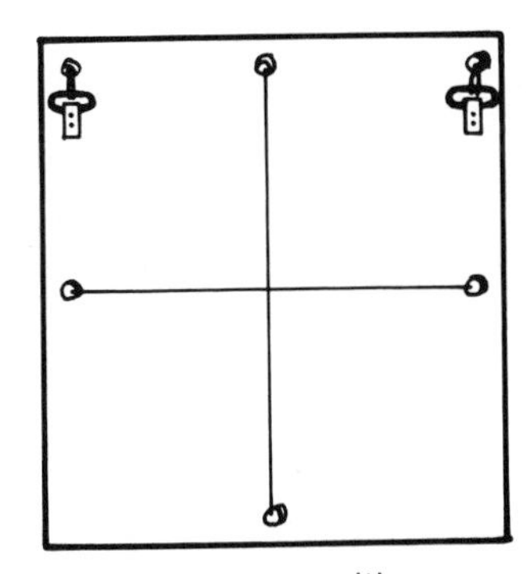
Mirror hangers with separate wire loops and frame crosswired

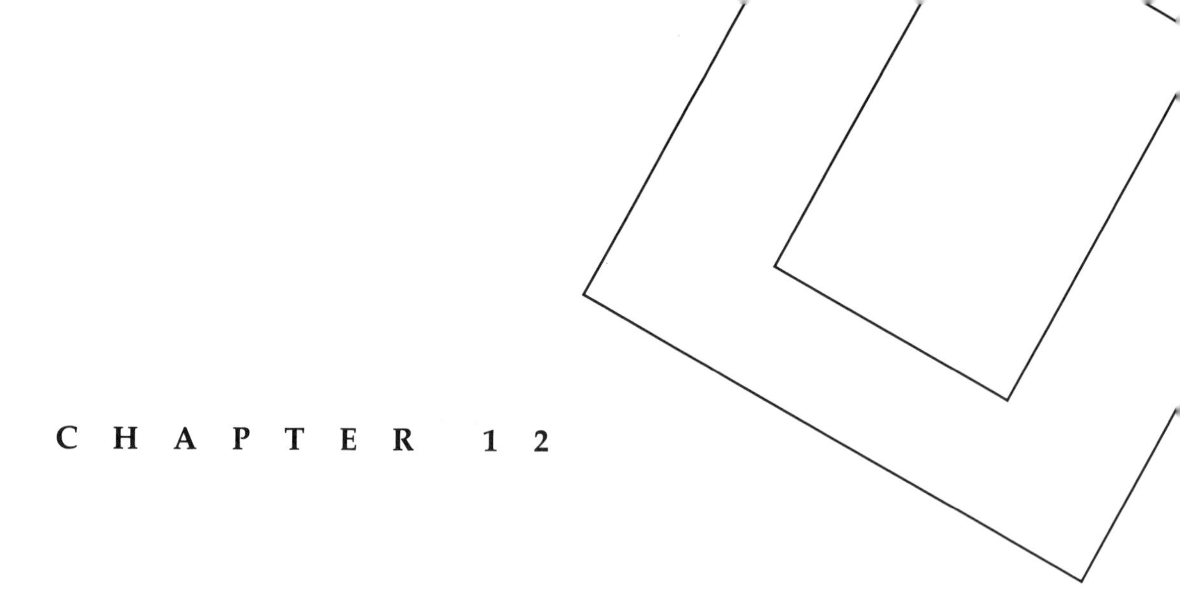

CHAPTER 12

Choice, Cutting, and Handling of Glazing Materials

The glazing material in a frame is usually glass, but because of weight, breakage, and ultraviolet light, other materials can be substituted. A working knowledge of how to use glazing materials can be an important asset in the conservation and framing of prints. Because of the simple tools and techniques used, it is often feasible to do your own glazing.

GLASS

Glass is the most traditional glazing material used in picture frames. It comes in a variety of thicknesses and qualities. It is sold in *lights* (single sheets) that do not come in stock sizes to match picture frame stock sizes, so usually it has to be cut to fit.

Exotic glasses come in many different thicknesses, but three standard weights of ordinary glass are used in most framing applications.

- ☐ *Picture weight* is approximately 1⁄16 inch thick. It can be used in frames up to 24 × 30 without too much danger of breaking. Because it is thin, it is very transparent, and it does not add excess poundage to the frame.
- ☐ *Single weight* is 3⁄32 inch thick, and will do quite well in frame sizes up to 30 × 40. It is easier to find in most hardware stores than picture weight.
- ☐ *Double weight* is double the thickness of picture weight — that is, 1⁄8 inch thick. It is used for very large frames, but because of the extra weight involved, it requires adequately strong pro-

visions for hanging so that it will not tear the frame from the wall. Internal bracing of the frame should also be used.

A note of caution about quality is in order. Glass manufacturers used to sell their wares pregraded as A-quality (presorted for defects) and B-quality (unsorted direct from the factory). They no longer do this for ordinary glass, and everything is B-quality. The most common defects will be bubbles and surface flaws. If you are buying glass cut to your specifications, avoid unseemly hassles with your supplier by insisting upon inspection before the glass is cut.

Back in the good old days, it used to be considered a defect if the glass had a slight greenish tint. Now it seems that all glass sold has this greenish coloration. To see what we mean, hold a piece of pure white mat board behind a piece of glass, and compare the covered part with the uncovered part. In buying quantities of glass, it might be a good idea to run this test and reject any lights that have a pronounced tinge, although this will of course be a subjective test.

Glass is the only glazing substance that has a long, proven history of properties that makes it safe to use with prints. It is chemically inert, dimensionally stable, static free, and uniform in composition from manufacturer to manufacturer. If you buy glass, you know what you are getting.

The drawbacks to glass lie mostly in practical application. Glass can break, sometimes with what seems like little cause; it is heavy when compared to plastic materials; and its thermal conductivity can cause condensation.

When an entire show has been framed with glass and will be traveling for exhibition, or when a private collection is moved, you can protect the prints from the effects of breakage by stripping lengths of masking tape crosswise in both directions on the front of the glass. Never do this with any of the plastic glazing materials, because the adhesive of the tape can bond to them permanently. In case the glass breaks in transit, the tape holds it in place and helps prevent movement of the sharp edges against the print surface. If the show has mixed glazing materials, note on the back of each frame with glass that it should be protected in this fashion.

NONREFLECTIVE MATERIALS FOR GLAZING

Many people object to glass in frames because reflections can hide the picture. When you need to conserve the piece in the frame, the so-called "nonglare" glasses do not solve this problem. The conventional type of nonglare glass is made by abrading the surface to give it a matte texture. The frosted surface does reduce reflections, but it dulls the color and sharpness of the print, and unless the glass touches the front of the print directly, the optical effect of the frosting is largely to obscure the image. Either you can't see the print, or you risk ruining it by having it adhere to the glass. Nonglare glass is not acceptable for conservation framing.

An acceptable alternative is sold under the trade name of Denglas by Denton Vacuum in Cherry Hill, New Jersey. Denglas is made to reflect very little light while having excellent optical characteristics in every other respect. Where regular glass reflects approximately 8% of the light that strikes it, Denglas reflects less than 1%. This effect is achieved by a multilayer optical coating on both sides of the glass. The manufacturer asserts that this coating is chemically neutral and cannot create changes in the print.

Denglas can be cut easily in the same way as conventional glass (see the instructions in this chapter). It does cost more, but has certain corresponding advantages in terms of ultraviolet transmission. While it does not eliminate all ultraviolet light, it will substantially decrease the amount that reaches the print surface, especially at the lower end of the spectrum where damage is greatest. Denglas does not provide *complete* ultraviolet protection, nor does the manufacturer claim this benefit. But when used in conjunction with other methods to reduce ultraviolet radiation, such as filtering the light sources and keeping illumination levels to a minimum, it should provide a helpful additional degree of protection.

Denglas has other features to take into account when you consider a glazing material: it does not generate static that can pull material off the surface of the print; it can be placed at any distance from the surface of the print, which makes it useful for framing pieces with thick mats and for shadowbox type frames; and it resists scratching better than plastics, and can be cleaned with all nonabrasive cleaners.

POLYMETHYL METHACRYLATE SHEETING

This substance is usually known by some trade name such as Plexiglas. It will adequately substitute for glass when excess weight or breakage must be prevented. Do not confuse it with inexpensive acrylic or other kinds of plastic sheets that have little of the excellent working quality of Plexiglas and that can be chemically less stable.

The big advantage of Plexiglas and related material for picture glazing is its low weight and its resistance to breakage. And even if broken, Plexiglas is less likely to damage the print than is glass. At this writing, Plexiglas is believed to be chemically inert enough so that it can safely be used in close proximity with most prints.*

Plexiglas and similar types of sheeting are not cut the same way as glass. They either have to be scored deeply with a knife and snapped against a sharp edge, or cut with a power saw. Snapping tends to be a little more problematic than glass-cutting, and the

* The one exception may be prints treated with thymol as a fungicide, discussed later in more detail.

use of a table saw of course requires a considerable investment in equipment. See "The Simple Way to Cut Plexiglas" for an easier method.

This kind of glazing material comes with a protective paper covering on both sides. Do not remove the paper before cutting, because the cutting operation will leave scratches on the surface.

Many glass houses supply polymethyl methacrylate cut to customer specifications. Some manufacturers put their trademarks in a corner of the sheet. This little advertisement will not be acceptable.

Plexiglas is very versatile. It can be cut, sawed, filed, sanded, and bent into exotic shapes. The mail-order art-supply house of Dick Blick sells special heating tools for bending it.

The broken or sawed edges are not very attractive when left unfinished. If you plan to use Plexiglas with clip frames or in other applications where the edges are exposed to view, the edges need to be flame-finished.

After cutting, place the Plexiglas, with protective paper still on it, into a vise. Remove burrs and cut marks with a rough file, working lengthwise down the piece until the edge is quite flat. Take off the paper and put the sheet back in the vise after padding the jaws. To make the milky-white edge translucent like the rest of the sheet, use a propane torch with the flame set low. Run the flame lightly up and down the edge until the edge turns clear and smooth.

Plexiglas can also be drilled and glued. It comes in clear and opaque colors as well as clear, so a complete frame can be made from it. This is an area we leave for creative craftspeople to explore on their own.

In large frames all types of plastic sheeting tend to flex because they lack the inherent rigidity of glass. In this situation, we suggest that a double mat of two 4-ply boards will help to prevent contact with the print.

Unlike glass, Plexiglas will scratch. As a matter of fact, it scratches quite easily. Naturally, this has a marginal effect on its safety from a print conservation standpoint, but it is a practical problem. The only solution is careful handling.

Plastic materials like Plexiglas generate static electricity when rubbed. For most photographic prints, this does not create difficulties. However, you should remember this property when framing materials that have a friable, crumbly surface — like charcoal or pastel drawings. When working with some contemporary pieces that combine delicate media with photographic imaging, this becomes important because static electricity can actually pull part of the picture off the paper. It can also be a problem with very old and deteriorated prints, or prints made on light and fragile paper.

An antistatic polish is recommended by Rohm and Haas, the makers of Plexiglas. It is designated AR-8, and is sold by Rogers Anti-Static Chemicals, Inc. For maximum effectiveness, apply it to both sides of the sheet before framing, and use it thereafter when cleaning the front of the frame. It leaves a thin detergent film that is nonvolatile and acid free.

Plexiglas reacts with many solvents. Since no print should be kept in an environment where there are many solvent fumes anyway, ordinary precautions should suffice. However, we have read an interesting article in the *Journal of the American Institute for Conservation* about the effects of the fungicide thymol on Plexiglas. According to a report by a paper conservator at Yale University, a print was framed with Plexiglas glazing and backed by a sheet of thymol-impregnated paper to protect against fungus infestation. The frame was stored wrapped. Two months later, when examined, the Plexiglas had discolored to a deep yellow, and the print and mat had fused to it. Apparently the wrapping had confined thymol vapors inside the frame. The collector who owned the print had used the same procedure years earlier without damage, so it is speculated that the damage occurred as a result of changes in the formulation of Plexiglas.

One can learn a few things from this sad tale. One is the danger of overly aggressive conservation measures. The better method of protection against fungus would have been storage in a cool, dry space, instead of packing a thymol-impregnated sheet in the frame. Another is the importance of recording with the print all conservation and restoration measures used — for example, the common one of vapor-treating a print with thymol.

Treating photographic prints with thymol is not, in any case, a good method of treating for fungus, and you will want to avoid framing any kind of print treated this way with Plexiglas.*

THE SIMPLE WAY TO CUT PLEXIGLAS

We mentioned earlier that Plexiglas-type sheeting can be either snapped along a score line or cut with a power saw. Wall-mounted glass cutters with heads especially designed for the purpose work well enough for snapping, but the cost makes them prohibitive for occasional work. We have found that scoring and snapping by hand is a tricky procedure at best. The money and space, combined with a certain amount of danger, makes a table saw less than an optimum solution for most purposes.

As an alternative, we suggest getting an inexpensive hand-held saber saw, the kind used by home handypersons (see Figure 12.1). Usually they cost under $20 if you get the single-speed model with no fancy attachments. A plastic-cutting blade can be picked up at the place of purchase. Also needed are a straight strip of wood, like a 1 × 3 furring strip, and two C-clamps with padded jaws. The jaws can simply be padded with a piece of rag.

Draw the outlines of the piece to be cut on one side of the

* There seems to be little cause to worry about the minuscule amounts of thymol sometimes mixed into the paste for Japanese tissue hinges.

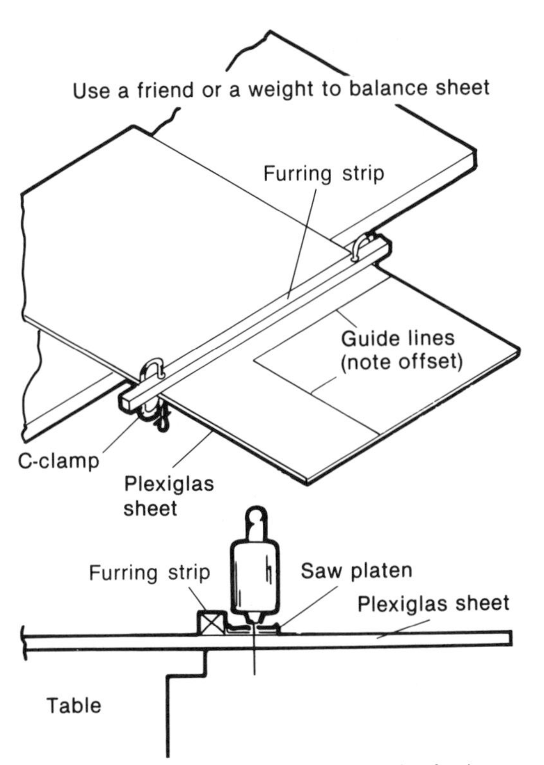

Figure 12.1
A simple jig for cutting Plexiglas.

protective paper. The paper is not removed before cutting. You will find that if the straight edges of the sheet are used, you need only make two cuts.

Measure the distance from the saw blade to the edge of the saw platen — the little shoe that rides on the surface being cut. You are going to use the furring strip as a guide bar along which to slide the platen, so it is necessary to know exactly how much the guide bar will have to be offset.

Position the Plexiglas so that part of it sticks off the edge of the work surface. The cut line has to be off the surface so that the saw blade can go through.

Clamp the furring strip to the Plexiglas with the C-clamps, padding the bottom part of the jaw so it will not mar the Plexiglas. Of course, the guide strip needs to be offset by the amount measured before. You may find it helpful to draw the offset lines on the paper as well, but take care not to confuse them with the cutting lines.

Put the leading edge of the platen on the Plexiglas sheet and start the saw. Push the saw slowly but firmly along the cut line with the side of the platen against the furring strip. Exert some slight pressure sideways to keep the saw true to the guide strip. After cutting all the way down the guide strip, turn the work, reposition the guide strip, and repeat for the other cut(s).

When the cutting is done, put the piece in a vise and file it to remove burrs and roughness. Remove the protective paper. If the exposed edges will be exposed to view, flame-finish them as described in the section on applications.

ULTRAVIOLET FILTERS

High-energy radiation found at the ultraviolet end of the spectrum will damage any print on paper. Photographs made on resin-coated papers have been found to be even more susceptible to damage from ultraviolet radiation and they exhibit a characteristic pattern of emulsion lifting when exposed to high doses over a short period of time. The dyes in color photographs fade rapidly in ultraviolet radiation. A solution to the ultraviolet problem is to use glazing that absorbs the dangerous part of the spectrum, but that allows the visible part of the spectrum to reach the print.

Rohm and Haas sells a type of Plexiglas that they designate as UF-3; it filters out a wide band of ultraviolet rays. Because it also absorbs part of the visible spectrum, UF-3 has a light yellow tinge, but it is not pronounced enough to cause objections in most cases. Rohm and Haas's other formulation, UF-4, gives less protection but does not have the yellow tinge. A similar ultraviolet filter for glazing can sometimes be found in glass houses under the trade designation of OP-1.

A good time to consider the extra expense of ultraviolet-filtering glazing would be when a show will be sent traveling on exhibition and there is no way to foresee what kind of illumination it will get. The two most common sources of heavy ultraviolet radiation are sunlight and fluorescent bulbs. Cloud cover and window glass do not remove an appreciable amount of ultraviolet from sunlight.

ABCITE

As mentioned above, Plexiglas-type glazing materials have a strong tendency to scratch easily. Abcite is a special type of Plexiglas with a very hard coating of clear plastic that resists scratching. It has a very high price tag, however, and can be considered only for extremely valuable pieces. It does have the added advantage of being less electrostatic.

CUTTING AND HANDLING GLASS

Safety should be paramount in the mind of anyone who works with glass. Glass is easy to work with, and after a time you can forget that it can be dangerous, both to your person and to the print. Sometimes the two dangers can be combined — when you get blood on a print!

Adherence to some simple rules will reduce the dangers to a minimum.

☐ Never cut glass in the vicinity of a print.

☐ Never clean glass in the presence of a print.

☐ Follow established procedures when cutting glass. For example, do not try to knock off a protruding chip from a piece that has been improperly broken. Have patience.

☐ Carry one sheet of glass at a time, suspended vertically from the center of the long side. Carried flat, glass can break of its own weight.

☐ Glass edges are razor sharp. Treat them that way.

☐ Heavy tools will break glass when dropped on it. Hammers are used around glass only when necessary, and must be handled with care.

☐ When bringing a large sheet to the work table, hold it vertically so that it comes against the edge of the table about two-thirds up the side. Gently swing the top down onto the work area while supporting from the bottom, and slide the sheet the rest of the way onto the counter. *Do not* carry it vertically, switch abruptly to horizontal, and drop it.

The term *glass cutting* is actually a misnomer, because glass is broken, not cut. The separation is started by running a hard metal wheel along a line with enough pressure to produce a very slight fissure. Stress is then applied about the axis of the fissure to generate a break.

The tools you need are few and cheap, though you can spend as much as you like on more elaborate versions. The basic kit consists of a glass cutter, a wood yardstick, an oiling bottle, and, for close cuts, a pair of grozing pliers (Figure 12.2).

A draftsman's dusting brush can be used to sweep glass chips off the work surface, but a clean dustpan brush will be less costly and more convenient. Do not use the same brush you use to dust prints; the glass-dusting brush may contain small pieces of glass.

The work surface has to be flat. To convert a regular tabletop, cover it with a large sheet of chipboard. Some glass shops pad their tables with carpet remnants to reduce breakage and to speed freehand cuts.*

What kind of glass cutter should you use? An inexpensive hardware store type does just fine for window glass. These usually have steel wheels honed to an angle of 120°. More expensive versions have carbide wheels that last longer and that have more sharply honed edges. Either way, get one with a ball end handle to start difficult cuts.

Any of these cutters costs only a few dollars. The jump in price up to a wall-mounted unit is quite substantial; the latter is currently

* When cutting freehand, an experienced glass worker can draw the desired shape on the glass with the tool (often tracing over a paper cartoon). The scribed glass is then put onto the carpet padding, a sharp downward push right on the line starts running the cut, and all that is left to be done is to pick up the pieces. It is harder than it sounds.

Figure 12.2
Basics for cutting glass: some oil, a prescription bottle with a wad of cotton in the bottom, and a glass cutter. You also need a wooden straightedge, and maybe some grozing pliers.

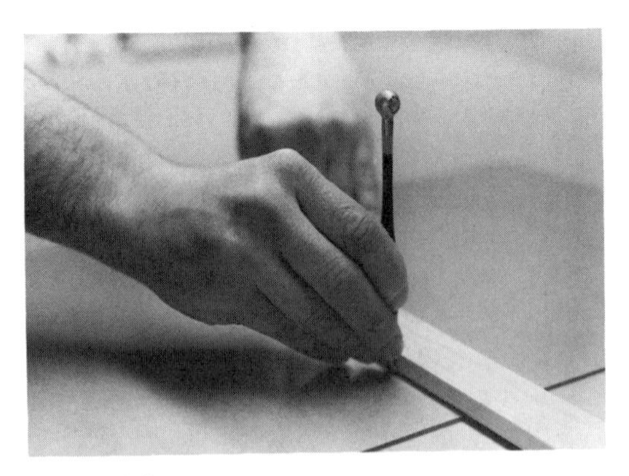

Figure 12.3
The correct way to hold the glass cutter is perfectly erect, with the tip cocked slightly toward you. If the cutter tips to the side, the wheel can roll on its beveled edge and skip over parts of the score.

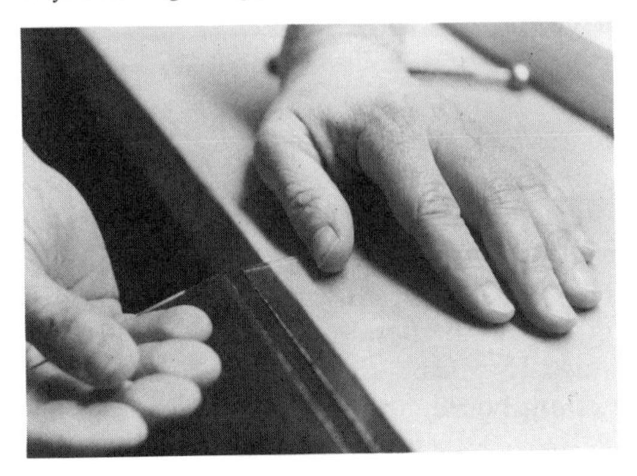

Figure 12.4
Parting the glass at the table edge. Push down and pull away from the score while holding on to the extended piece of glass. The part on the table gets held down firmly with the other hand. Remember to start at the point where the score ended.

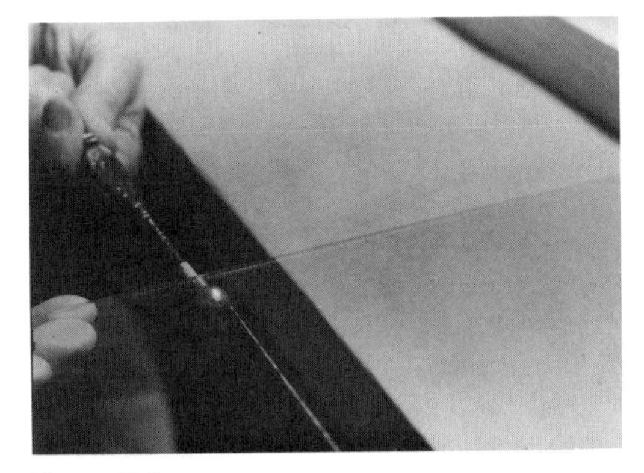

Figure 12.5
It didn't break like the photo showed? Try tapping under the end of the score until a fissure starts all the way through the glass, then attempt the break again.

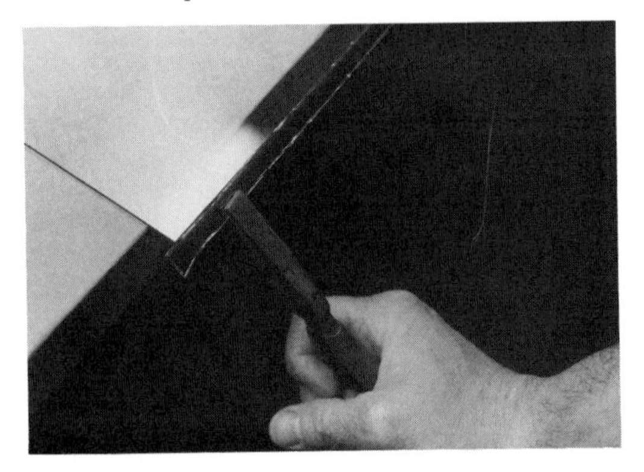

Figure 12.6
Really narrow strips of glass shouldn't be broken off with bare hands. This photo shows how grozing pliers can be used to grip the narrow strip during the break.

close to $900. It is like going from a pogo stick to a Mercedes-Benz. Wall-mounted units offer some advantage for the price. For example, the Fletcher Model 7554 cuts either glass or plastic up to ¼ inch thick. It has measuring stops to consistently produce identically sized sheets, and wall mounting can save expensive work space.

The reason to use a wood yardstick, as opposed to a metal one, is that this cuts down breakage. Lumber yards often give these items away as promotional items.

You can make an oiling bottle from a prescription pill bottle by stuffing some cotton batting in the bottom and pouring over it some light oil like 3-in-1. Heavier oils can be cut 1:1 with kerosene.

Dip the head of the cutter into the oil-soaked cotton every second or third stroke. This keeps the cutting wheel lubricated, so it turns freely during the score (more about this later).

Grozing (also called nibbling) pliers have flat jaws that are about ⅜ inch wide (Figure 12.7). Stained glass workers use them primarily for nibbling away at the edge of an irregular piece of glass so that it will fit exactly into a picture. You will use them mostly to hold narrow strips of glass when breaking them away from the main piece. Sometimes, but not often, a bad cut can be salvaged

Figure 12.7
Glass-putting pliers.

Grozing, or nibbling, pliers have wide jaws for holding narrow strips of glass when making a break.

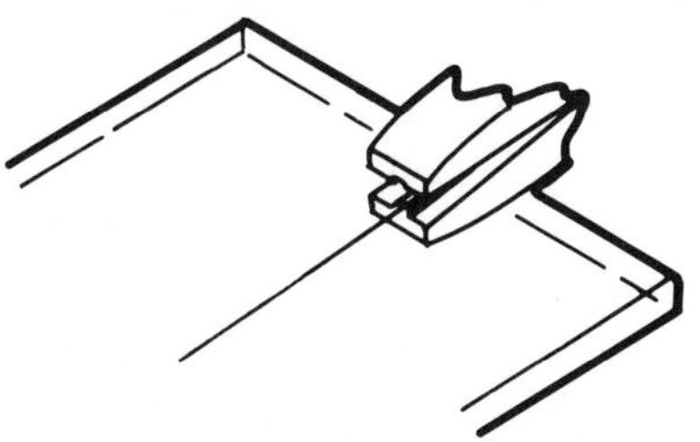

Cut-running pliers have a raised point in the center of the jaws. When the point is positioned at the end of the fissure and the jaws are squeezed together, the break starts automatically. These are a luxury.

by using grozing pliers to crumble away little chips of glass from the edge to get an exact fit.

You might find an additional type of pliers called *cut-running pliers*, useful if you do much glass cutting. These have a small raised knob in the center of one of the jaws. After a score has been made with the glass cutter, the knob is put directly under one end of the score and the jaws are closed together to break the glass at this point. This starts a break running the length of the cut and speeds the work.

Pliers for working glass come in both metal and plastic models. Some even have inserts for the jaws that can be replaced when wear takes its toll. Plastic pliers have the advantage of being less likely to break glass accidentally when dropped, and they reduce fatigue a little.

Cleaning

Cleaning is an important first step in cutting glass. By its nature, glass cutting creates a myriad of small particles that accumulate in the work area, and these must be regularly removed by sweeping off the surface. Other types of grit and dirt can also cause problems if left on the surface of the glass, because if they lie in the path of the glass cutter they will make the wheel ride up momentarily and lose contact with the glass. This causes a small discontinuity in the fissure, and when the break reaches this point it can branch off in any direction (usually not the one you desire).

So, rule one is to clean the glass thoroughly. Also check the work surface for anything that sticks up above the surface before setting the glass down. Any small hard object will act as a fulcrum, and the weight of the glass on either side can cause a break. Common household window cleaners like Windex work satisfactorily and leave little or no residue. Some of the more expensive foam cleaners in pressurized cans that are sold for graphic arts use seem to work more efficiently, but since manufacturers do not list ingredients as a rule, we are hesitant to recommend any particular brand.

Ordinary paper toweling can be used for glass cleaning, but in the long run you will find it preferable to use one of the inexpensive "lintless" cloths sold commonly in grocery stores as dish towels. These leave behind many fewer particles that can get trapped in the frame.

Grease and oils can be difficult to remove completely from glass. The only sure cure is to clean it several times, lifting and turning it in the light to catch any remaining spots. It helps to wash your hands thoroughly before starting.

In framing, glass has to be cleaned twice: once before cutting, and a second time before installation in the frame. Do not neglect the second time, because the cutting operation will leave behind oil from the cutter, finger smudges, and small glass particles from the scoring.

How Glass Is Cut

Remember that glass is not actually cut. Rather, what we are doing is making a thin fault line or fissure on the surface by passing a cutter wheel along it. The wheel does not dig out material by scratching or gouging. The sharp edge fractures the glass part way into the interior by pressure exerted downward in a very narrow line. Making a clean fissure with no gaps or gouges is the only secret to getting a clean break.

The fissure, commonly called a *score,* is made with a *single pass* of the wheel. A second pass ruins the fissure and crumbles the glass into a rough trench. This results in an erratic break, and ruins the cutter wheel as it plows through the glass chips left from the first pass.

The score line should be almost invisible. Too much pressure on the wheel digs out a ragged furrow and leaves white grit and glass flakes on the surface.

The cutter must be moved rapidly and uniformly, at approximately a foot per second or faster. A slow pass does not make a deep enough fissure, and a pass with erratic speed develops a fissure of varied depth.

Practiced glass cutters use their ears to tell if the cutter is traveling at the right speed. If you are going at about the right speed, the glass will "sing," emitting a thin clear note during the entire length of the stroke.

Most faulty breaks occur because of a discontinuous score line. If a fissure does not run smoothly from one side of the glass to the other, the break starts to propagate irregularly when it reaches a gap in the line. It will snake spontaneously into just the areas it should not, in compliance with the infamous Murphy's law.

The causes of a discontinuous score are several. The most common are:

- ☐ *Dirt* on the glass causes the wheel to lose contact with the surface. Prevent by thorough cleaning.
- ☐ A *flat cutter wheel* slides on the flat section part of the time. Prevent by regular examination of the wheel and by not pushing the wheel back and forth in the fissure.
- ☐ *Binding* of the wheel happens because of loose debris in the wheel slot or because of lack of lubrication on the wheel axle. A flat wheel can also result. Prevent by frequent use of the oil bottle and by looking for foreign matter in the wheel slot.
- ☐ *Improper angling* of the cutter during scoring occurs if you do not hold the cutter vertical enough, so that only the cutting edge touches the glass; the angled sides of the wheel can then drag and make it skid. Hold the cutter exactly vertical over the score, cocked only slightly (about 5° or 10°) into the direction of travel.

Running the Score

Once you have made the score correctly, breakout is done by bending the glass around the axis of the line formed by the score. A very small amount of bending is all that is needed.

Running the score always starts on the side where the scoring ended. You must run the glass cutter right off the edge in order to get a point from which the break will propagate.

There are several techniques you can use in making the break. The simplest is to hold the glass in both hands with the score side up. Push down on the sides and up in the center under the score. Concentrate pressure near the end point of the score.

Another method is to rest the glass on the work surface, partly hanging over the edge. Position the score directly over the edge. Hold the glass flat against the supporting surface and push down on the protruding part. Use either your hand to push down, or, if the part coming off is very narrow, grasp it with grozing pliers.

A small-diameter metal rod can be used as an anvil. Lay it on the work surface, put the score directly above the rod, and push down on both sides.

Cut-running pliers, if you have them, can also be used. Put the knob of the lower jaws directly under the end point of the score and squeeze.

If the glass does not break easily, look carefully at the score to see that it goes right up to the edge. If it looks good, a break can usually be started all the way through from top to bottom by trapping from the underside with the ball end of the cutter. Tap lightly until you can see that a crack runs all the way through, and then try again.

It is always a good idea to wipe down the score line with a rag before starting the break. A well-lubricated cutter lays down a thin line of oil that gives the illusion of a continuous score, when in fact the wheel may have skidded. Wiping the glass reveals these bare spots, and sometimes they can be scored again before trying the break.

Step-by-Step Glass Cutting

☐ Clean the work surface. Clean the glass.

☐ Put a wood yardstick on the glass as a guide for the cutter. If working on a kraft paper surface, you can draw a guideline on the paper to show exactly where the cut goes. Make allowance for the distance between the cutter wheel and the edge of the yardstick when positioning the yardstick on the glass.

☐ Place the cutter firmly on the surface at the far edge of the sheet of glass. Draw it rapidly and firmly toward you, and off the near edge.

☐ Wipe and check the score for continuity.

☐ Move the sheet of glass toward you so that several inches hang over the edge of the work surface.

☐ With the fingers of both hands underneath and thumbs on top, lift the glass. Bend it simultaneously up under the score and out toward the sides. It should separate easily.

☐ If this does not work, set the glass down and try tapping the underside with the ball end of the cutter until the fissure goes all the way through at the end. Try breaking again.

☐ When a narrow piece is being cut off one side — narrower, say, than 4 inches — it would be an excellent idea to use grozing pliers to grasp the narrow piece. In this case, put the fissure over the edge of the work surface, grip the glass near the end of the score, and push down. The leverage of the plier handles and the narrowness of their jaws means that you can safely exert much greater pressure than with the bare hands alone.

☐ Cut glass is extremely sharp. For safety, we suggest that the edges be lightly sanded with some garnet paper wrapped around a wood block. Pass it lightly back and forth to dull the sharpness.

☐ A little-known fact about glass is that it can partly heal itself. Do not delay breaking a long time after making a score, because the edges of the fissure can spontaneously anneal to one another and make the break more difficult.

Cutting Circles and Ovals

It is a common practice with many framers to cut circles and ovals by hand without the aid of any expensive machinery. The only additional material you need add to the basic glass-cutting kit is a thick piece of felt or some outdoor carpeting.

If you plan to do production quantities of circle and oval cutting, there is equipment that can be purchased for the purpose. Several manufacturers make cutters that work on the same principle as a beam compass, with a glass-cutting head in place of the pencil. Such a model is Fletcher's No. 42 "Gold Tip" Circle Cutter, which, with attachments, will cut circles from 3 to 48 inches in diameter. Machines for cutting oval mats are often designed for double duty as glass cutters to make oval cuts; cost of these machines runs from $500 up.

Hand-cutting glass circles and ovals for framing is possible because of the leeway offered by the frame's rabbet, which will hide any minor deviations from a perfect cut, as long as they are not too severe. So even if the cut looks a bit off, try it in the frame before discarding it.

The rabbet should serve as your template. Lay the frame face up on a piece of kraft paper. With a pencil reach under the face

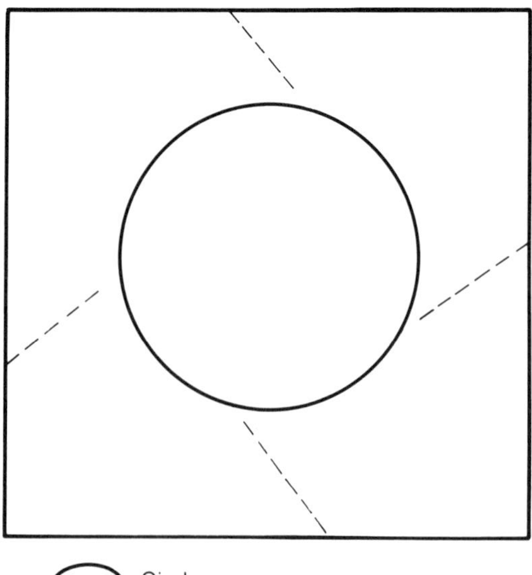

Figure 12.8
Relief scores when cutting glass circles and ovals.

Note that even when the score for the circle has been run, the center piece does not immediately drop out because the fissure does not run through the heart of the glass in a perfectly straight line.

lip and trace the outline of the rabbet onto the paper. Go around several times to make a thick, solid line. Remove the frame and lay a sheet of glass over the tracing. To make things easier, put a couple of masking tape loops in the center of the circle so that glass and paper can be turned together while cutting.

Trace the circular mark around the entire perimeter with a glass cutter. Remember to keep the cutter vertical and only slightly cocked in the direction of travel. Unless you have a swivel joint in your wrist, it will probably be necessary to make at least two passes, one for each side of the oval. Turn the glass over after completing the score and take off the kraft paper.

The next step is to run the score. Before you get ahead of the instructions, let us point out that this does not mean that you immediately get a circle of glass at this point. First lay the glass, with the score side down, on the felt or carpeting. Push down lightly all the way around the perimeter of the circle. The circle will not just drop out of the center of the sheet quite yet. Rather, the fissure should now appear on both sides of the glass, going through all the way around.

Look all the way around the circle to see that the fissure is continuous. At any points where it has skipped, apply extra pressure. If this does not work, try tapping with the ball end of the cutter.

When the fissure is complete, relief scores have to be cut to take off the outside sheet. Start the relief scores a fraction of an inch (about 1/16 inch) from the edge of the circle and take them out to the edge of the glass sheet. The likelihood of the relief scores propagating into the center of the circle decreases if the scores run almost at a tangent to the circle, and at an angle of 45° to the edge of the sheet of glass. Four relief scores, one to a side, will do the job.

Break off the extra glass along these scores, tapping as necessary. The most efficient way to do this is to remove adjoining pieces in sequence.

PART FOUR

Display and Copying

CHAPTER 13

Planning an Exhibition of Framed Prints

If you create, conserve, or own even a modest collection of prints, sooner or later you face the prospect of putting these prints before a public. This means you must confront the conflict between the safety of your prints and the inevitable dangers that attend upon exposing them in public.

This section has been written for those who, whether they are archivists for a local historical society, librarians, gallery owners, workshop directors, private collectors, or artist-photographers, are responsible for their own showing. We lay a firm foundation by beginning at the beginning.

The emphasis here is on planning for one particular show, and on how to ensure in advance that the prints on display do not end up the worse for it. Using this knowledge, all exhibitors can be as creative and innovative as they please. There is always a need for fresh approaches to exhibition design, and those responsible for installing a show should not feel trammeled by aesthetic rules about topics such as the proper color of walls or good taste in decor. Such details we cheerfully leave to your discretion.

EXHIBITION AREA SURVEY

Methodical inspection of the area where an exhibit will hang provides a surprisingly large amount of useful information, even if the site has already housed many previous shows. If properly conducted, such a survey can be an important conservation technique in itself. This is a time when all factors in print care come into play — and often, major dangers can be averted with minimal effort.

Physical Dimensions
Measure the wall space to determine the number of running linear feet available for the exhibit. Make certain to exclude obviously unsuitable parts of the room or rooms, like windows, areas next to free-standing pipes and above hot air registers, and exits and entrances.

This initial measurement provides information about whether partitions and free-standing screens need to be constructed to hold some of the work. Also, planning for traffic flow and security can begin in advance of hanging the exhibit. In drawing up a floor plan, you want to ensure that viewers can see the entire show in an orderly sequence, and that any partitions that are planned do not create blind spots for vandals and thieves.

Wall Condition
When a floor plan has been drawn up, the time has come to decide whether painting will be required, either for decorative or practical purposes. Oil paint must be allowed to cure completely, not just to dry to the touch, before you put valuable prints on the wall. The reason is that, as oil paint cures, it releases vapors with peroxides in them. Peroxides affect the stability of photographic emulsions, especially on black and white prints [1]. Even with adequate ventilation, the curing process continues for long after the paint has become dry to the touch.

Paint manufacturers do not make it a practice to specify curing time, but a useful clue can often be found in directions on the can label for the earliest washing date. If the directions say that the paint should not be washed for a week after application, try to paint at least a week in advance of the actual installation.

Paints that use zinc white or titanium oxide as a pigment absorb ultraviolet light and should be preferred when choosing a wall covering. Talk it over with a salesperson at a reputable paint store to find out the composition of various paints, and if the store does not have the required information, ask the salesperson to consult with the manufacturer's representative. Most discount stores, obviously, do not provide this kind of service.

Lighting
Satisfactory lighting illuminates the work on the wall so that it can be seen clearly, without threatening it with ultraviolet radiation or heat. Start this part of your survey by listing all sources of illumination, which will usually include fluorescent lights, incandescent lights, and window light.

Ceiling tracks with movable spots are a favored method of lighting exhibitions, both because of their versatility in highlighting individual frames and because incandescent bulbs emit little ultraviolet radiation. Check the wattage of the bulbs and the location of individual spots to ensure that prints do not heat up from being too close to a light source. An easy way to test for overheating is to tape a sheet of black paper in place of a frame and leave it illuminated for half an hour. If it is warm to the touch, that area is getting too much light.

The inverse square law of light propagation states that an object twice as far from a light source as another gets only one-quarter as much radiation. Thus, moving prints and lights only slightly farther apart greatly increases protection from overheating.

Since track lighting is usually a fixed distance from the wall, you have two options for cutting back on radiant heat: either reduce bulb wattage, or crosslight, using spots to illuminate areas to the right or left instead of directly ahead on the wall.

Besides being cool, fluorescent lighting has the advantage over incandescent of providing more nearly "neutral" illumination instead of the warm, orange colors of incandescent. Not all fluorescent lights, however, are truly neutral; in fact, there are eight standard colors of fluorescent tubes.* At this point, you should think about whether the tubes in the fluorescent fixtures might be replaced to bring about a better color balance.

Fluorescent bulbs give off large amounts of ultraviolet radiation, second in intensity only to daylight. If these will be the major source of light, consider shielding either the lights themselves with ultraviolet filter sleeves† or protecting the framed prints with ultraviolet-filtering glazing material. Also, keep in mind while doing the survey that fluorescent tubes cannot highlight a particular frame like a spotlight, and so additional fixtures may be required.

Window light can be very pleasant and natural, but it is sadly rich in both ultraviolet and infrared radiation. Areas of wall that get direct sunlight should not be used for hanging prints. Window light necessitates protective measures, particularly for older, fragile prints that can be very sensitive to yellowing or fading. Some suggested solutions would be yellow plastic screens like those used in shop windows to protect fabrics, or drapes or blinds, or — once again — framing with ultraviolet-filtering glazing materials. Some combination would be best in rooms brightly illuminated from outdoors.

A generally accepted standard for the proper level of illumination is 5 footcandles (50 lux) at the picture plane. The Gossen Panlux Meter does a good job of measuring environmental illumination to find out precisely how many footcandles of ambient light are falling on your display.

By contemporary architectural standards, 5 footcandles is an extremely low level of lighting. You may encounter resistance from colleagues and visitors to safe levels of illumination — unless you take steps to make the light on the pictures seem brighter than it

---

* They are: cool white, warm white, deluxe warm white, white, deluxe cool white, daylight white, soft white/natural, and plant growth tubes. Deluxe cool white approximates skylight, but blends poorly with incandescent; the best choice for neutral illumination in conjunction with incandescent spots would be white.

† For a source of supply, contact the mail order houses listed in the back under Suppliers.

is. A simple way to do this is to condition the viewers' eyes by dimmer lighting in adjacent areas. The brighter light falling on the frames pleases the eye and makes the light seem more intense. This is more aesthetically pleasing than having each of the prints huddled in its own murky little puddle of darkness.

Humidity and Temperature Fluctuations

The next step in the survey will be to find out whether the display site undergoes extreme changes in temperature and humidity during the average 24-hour period. You will also want to look for possible sources of water damage, and find out whether air comes directly into the rooms from outdoors during times of public access.

Canadian standards for temperature and relative humidity in a public viewing area call for a stable temperature of 70°F (21°C), varying not more than 3°F in the course of a day; also, relative humidity should be maintained in the area of 47% to 53%, fluctuating daily no more than ± 2% [2]. No analogous standard has been set in the United States, but most U.S. archivists would agree with the Canadian standard.

Many exhibitions, of course, have to be held in areas that cannot even approximate these conditions. So bear in mind that paper objects do not have any consideration for our convenience or budgets, and they deteriorate in a humid climate. Use these figures to determine how seriously the environment deviates from the ideal, and set a goal for correcting the situation as soon as possible. (These standards do not apply to storage areas, which can be kept colder and drier without regard to human comfort.)

An inexpensive way to measure temperature extremes is to use a dial-type thermometer with resettable pointers for highs and lows. You can order such a thermometer from a scientific supply house like Edmunds, and log the readings daily for a week or so to determine whether you have problems in this area.

Look to see whether the building has an automatic 24-hour thermostat. In cold northern climates the temperature may drop abruptly during the night, when the furnace is turned down. There will be an attendant rise in relative humidity, which can cause condensation and buckling of the prints, in addition to other problems. If the thermostat cannot be reset, another solution might be to remove large and valuable pieces to a more stable environment each night.

Use a sling hygrometer, also available from scientific supply houses, to monitor the relative humidity. Take a number of readings during both the day and night, if possible, and record the average humidity in addition to any extremes that occur. For the short-term hanging of a show, great disparities between high and low humidity levels are more to be avoided than a somewhat high overall humidity level. Best of all is a stable environment with low humidity.

If you have determined that you will have a stable environment, it is a good idea to frame the show in similar conditions. This gives the paper a chance to acclimatize to the future surroundings that it will face. If this option does not exist, and great extremes cannot be controlled, a least leave some extra room in the frames between the rabbets and the sides of the mats to allow for expansion and contraction.

Air-conditioning units need a thorough checking out. Do they work on an automatic 24-hour cycle, and if so, can they be disconnected from it to prevent cycling? Are they well drained, to prevent leaks? Are they in good condition, and large enough to control the atmosphere when many people are present? We each sweat out between one and two quarts of water a day, and a large group of people can turn a room into a steam bath very quickly.

If the area is not air conditioned, or not sufficiently so, plan to ventilate the space by other means, particularly during receptions. If fans are used, locate them so they blow air out of the area, exhausting heat and humidity. "No Smoking" signs should be posted, if not already in place, and all smoking prohibited. Placing large ashtrays near the entrance will remind smokers of the rule and provide a place for them to extinguish and leave their cigarettes.

Look for ways to route visitors through at least one antechamber. If possible, no doors should open directly from the outdoors into the exhibit area itself. This will help maintain a more or less stable climate. Major museums have systems to both filter and wash air coming in from outdoors; however, research into their effectiveness indicates that many environmental contaminants like sulfur dioxide still ride into display areas in the fibers of visitors' clothing. This problem is particularly severe during wet weather.

Liquid water can be an unexpected and unwelcome intruder. Survey the site for possible causes of sudden inundations. Inspect overhead pipes for signs of recent unrepaired leaks. If there is a suspended ceiling, get up on a stepladder, remove several panels, and look for concealed pipes. If there is an automatic sprinkler system, it should have been inspected recently; these sprinklers have a tendency to go off accidentally as they age.

Find out the location of restrooms. In hotels, especially, these are a cause of sudden flooding, particularly late at night, though the problem is endemic to any public building. If the site is a basement room, are there floor drains? Do they work? Test them by pouring a couple of cups of water down them.

Examine the walls for signs of wetness caused by condensation or leakage (if you are in the basement, look doubly hard). A good indication of whether walls cause humidity to condense is their temperature; if they are markedly colder than the ambient temperature, count on condensation to be a potential difficulty.

One solution to wall moisture problems is to put blocks behind the frames to keep them from collecting water migrating down the wall. Seal the picture backs with polyester sheeting and tape. Obviously, the best solution would be to avoid hanging anything at all on these walls.

Atmospheric Contaminants

Although common sense tells us that brief exposures to polluted air are unlikely to damage framed prints, the quality of air in the display area can have a significant effect on pieces displayed for prolonged periods. If you plan to hang valuable works for a long time, consider the quality of the atmospheric environment to which you are exposing them. Recent advances in gas monitoring technology mean that for an investment of less than $300 for instruments and supplies, you can now check the area for levels of the oxides of sulfur, nitrogen, and carbon, and for hydrogen sulfide. The less expensive monitoring devices, such as the ones from the Draeger company, work by drawing air samples into a clear cylinder of chemicals that change color according to the amount of the target gas present in the sample. By matching these against a reference color, you can determine the levels of corrosive contaminants.

You may find that these materials lie outside your budget, but even so there are steps you can take. The federal government in the United States issues daily reports on the Air Quality Index. Among other things, this rating is based upon the level of ozone, and because ozone tends to disperse in a fairly uniform manner through the atmosphere, the report indicates how much of it is probably present in the immediate environment. To reduce ozone contamination, check for sources in the immediate environment, such as photocopying machines, bactericidal machines, and electronic bug killers. These generate ozone and should not be operating in display galleries or storage areas.

Air conditioning and climate control reduce the presence of pollutants, especially particulate matter. During your survey of the display area, examine the filters in whatever air conditioning you find. Dirty filters, besides wasting energy, do not remove airborne particles.

During winter in wet climates, an attendant in a cloak room can reduce the amount of atmospheric pollutants by encouraging visitors to check outer garments and umbrellas. The gases that you want to keep out of the display area often make their entrance dissolved in water droplets and trapped in the folds of coats.

Security

The topic of security embraces an unhappily broad range of problems caused by humans in an exhibition area. Begin the security part of the survey with a step-by-step examination of each possible entrance, including windows and fire doors. Look for locks on each entrance, and find out who has keys into the area itself. Are the locks in working condition, and of the type to stop at least an amateur burglar? Remember that crime is a venerable arena for the American spirit of free enterprise, where even amateurs have a surprising degree of sophistication. Many police departments have a community relations office that specializes in doing security surveys; you might ask for the assistance of such an office in planning needed changes.

Find out whether the building has an alarm system that gets turned on in the evening and during weekends. If your exhibit is hung in a building open to the public day and night, security guards will probably be on duty at least during the evening; meet with the chief of the security detail and make known your schedule and the names of personnel who need access during off hours.

If the show is going up in an institution that does not have its own guards, and if valuable pieces are part of the show, consider hiring temporary security personnel. If insurance is being taken out for just this one show, this may reduce the premium.

Another imporant security measure is to limit access to a single entrance during open hours. Municipal fire codes require that at least two exits be available for emergencies. When it is feasible, keep all but these two locked from the inside as well as the outside.

Fire Prevention

Fire prevention and control represent another area of security that the planning survey should not neglect. In addition to an automatic sprinkler system, a well-protected building should have some kind of fire detector. If the site does not have a centralized system, think about buying and installing the kind of smoke detectors sold for home use. They are cheap and provide at least a minimum of warning.

Take a careful look at the fire extinguishers. Are there enough, and are they visibly marked and readily accessible? Equally important for the curator and print conservator, what kind are they? The antique soda-acid type rated for class-A (wood) fires is probably the most dangerous to prints. They shoot a high-pressure stream of dirty, chemical-laden water at the fire — and at anything else in the vicinity, including the prints. In addition, they do not work too well. If these are present, try to replace them with more modern fixtures rated A-B-C for wood, electrical, and chemical fires. These use either carbon dioxide or foam to smother the fire; besides being safer around prints, they are safer for people because they put out fires more effectively.

Next in this section of the survey, check out the telephone. There should be one in the immediate vicinity. Is the number of the fire department posted conspicuously near it?

Quite a few fires are electrical in origin, and consequently the wiring deserves attention. The entire science of electrical wiring and code compliance is far too complicated to cover briefly. If you are dealing with an older setup that seems to have serious defects, call an electrician.

Before going to the expense of hiring a professional, even an inexperienced layperson can decide if basic problems exist that warrant immediate attention. Start your examination at the point where electricity comes into the building. This point, called the *service,* will be either a circuit-breaker panel or a fuse box, and is most often located in the basement.

Circuit-breaker panels, which have an array of small switches, will generally be of fairly recent construction and can be considered relatively safe. However, if a circuit breaker has to be reset frequently this indicates that the circuit it services has been overloaded. Some of the heavy appliances on that circuit should be disconnected.

Fuse boxes cause many problems, which is why they have generally been superseded by circuit breakers. Stand on a piece of dry board to minimize chances of shock, and unscrew each fuse to make sure that its socket has not been wrapped in metal foil to bypass a frequently blown fuse. Look for other illicit shortcuts like pennies stuck in the socket, and examine the amperage rating of each fuse. Circuits built for lighting and other small loads require only a 15 ampere fuse, and a bigger fuse indicates that the circuit may be overloaded. Track lighting in particular tends to overload circuits as additional spots get put on, so check the rating of the track system itself if a high ampere fuse is found on that circuit. Heavy-duty appliances like air conditioners require their own separate circuits.

If the electrical service seems satisfactory, continue the inspection on the site. You will need a screwdriver and flashlight. First, shut off the power at the main. (That is why you need the flashlight.) Go around to each outlet and wall switch and unscrew the face plate. Look inside to make sure that the wiring is in good condition, that connections are tight, and that the system is grounded. If you cannot tell whether the wiring is grounded, get someone knowledgeable about such matters to help you. An ungrounded system indicates that the entire plant should be rewired, and this may be taken into account when doing long-range planning.

Open any junction boxes you find, and look inside. If the boxes seem to be a hopelessly overpacked jumble of wires loosely wrapped with electrical tape, they require the attention of an electrician. If there is a suspended ceiling, lift out a few panels and look for wires installed above. Check everywhere for loose connections when wires run into junction boxes or outlets, and make certain that all cables are securely attached to the wall rather than flopping loose. By going slowly, even an amateur can spot immediate fire hazards like loose or bare wires and flaking insulation.

Chemical fires often start with improperly stored petroleum products. Paints, varnish, shellac, lacquers, and thinners should all be stored in metal cabinets away from stairwells and in places where there is otherwise little fire danger. Look also for oily rags, which can heat up by themselves and ignite spontaneously. These should be stored in airtight metal containers after being soaked in water.

We may seem to have gone on overlong about the dangers of fire. However, a major part of the entire corpus of nineteenth-century photography has already perished in a series of major and minor fires, and anyone who thinks that fire prevention does not belong in a book about the care of prints should read the account in William Welling's book [3] of major collections lost to the flames; nineteenth-century underwriters considered photographic galleries a "special hazard." Each such fire diminishes our heritage.

THE EXHIBITION LOG

An exhibition log helps maintain control over all kinds of shows, whether work by a group, a historic display, or a one-person show. It helps in planning both layout and framing and in locating individual pieces, and will serve as a reference tool for writing up material to be published. In addition, it can later be produced as evidence about the condition of particular items in case insurance claims are filed.

The easiest way to start a log is with a three-ring binder and some prepunched paper. If you want to get elaborate, make up a form with spaces for the appropriate information. Type out the various headings, and then have it copied onto punched paper.

Each print should have a separate page. Enter the following information about each print, or make a note when it is not available: artist-photographer; title, if any, or subject matter; printmaker, if different from artist; return address; media used; date of print; image size; sheet size; any visible damage; and a description of the mounting as received. Log each piece as soon as it comes in or is added to the show.

Make a note each time a print is matted, framed, or sent anywhere for other work, so that you know at a glance where it is currently located. After the show is over, enter the date when the work was returned to storage or to the original owner, and record the method of shipment. The log will serve as your permanent reference on the contents and condition of an exhibition.

STANDARDIZED FRAMING

We have already mentioned how ordering a quantity of the same size frames reduces labor costs for a chop service. In addition, your matting will go much faster. If reusable frames are stored in the stockroom, their cost can be written off over the course of several exhibits.

Standardizing the framing does not mean that all frames in the show have to be the same size. A judicious mixture of 11 × 14, 14 × 18, and 16 × 20 frames, for example, breaks up the monotony induced by a large vista of identical formats. On the other hand, a certain uniformity of style gives an exhibit the visual coherence and continuity that enables viewers to compare different prints without being distracted by clashing styles.

Begin to plan the framing operation by consulting the exhibition log. Make a list of prints by their sheet sizes. Framing to conservation standards means that no print will be trimmed to fit a frame.

Rather, the frame must accommodate the size of the paper. Next to the sheet size, jot down the size of the mat window. Sometimes this will differ from the image size, such as when some information on the border has to remain visible.

Putting down both sizes side by side helps you decide whether sheet or window size will determine the size of the frame. For example, a print whose image closely approaches the borders of the paper will probably need a larger frame than a smaller image with the same size sheet. That is because the larger image will need wider mat borders to look good.

Now go down the list of prints and assign a standard size frame to each one. In this way all you have to do is add up the number of each size, and either order them cut or pull them out of stock.

At this stage, you might also assign each print an arbitrary lot number by going through the log and numbering each page from front to back. Use these numbers to match prints with mats and frames as these get done. Do not go overboard, though. If you have only ten prints or so, there is no point in getting this complicated.

The list of frame sizes can also be used to start planning the layout of the show. You can tell, at least, whether the running linear footage of the frames exceeds your wall space.

EXHIBITION INFORMATION

Physical protection of print integrity rightfully commands the primary attention of those responsible for valuable photographs. However, disseminating information about your prints should also be a significant part of print conservation. Pieces of paper locked away in dead storage have little or no social value unless some part of the public has access to them, and so an exhibition of prints can actually enhance their worth. Providing information about the prints should be considered a meaningful custodial task.

The kind of reference material that accompanies an exhibit will vary greatly depending upon the nature of the work being shown. It goes without saying that a retrospective survey of the life work of a major photographer like Frederick Sommers would merit a more comprehensive publication program than would be fitting for a show of first-year photography students. Yet in each case both the viewer and the artist will benefit equally from a well-done introduction.

Well-endowed institutions can call upon the expertise of a host of personnel in preparing information for the public. As is the case with many enterprises, the variety of means available expands in direct proportion to the money spent. For the less fortunate, attractive means do exist for inexpensively informing the public, and we suggest that they be used to the fullest.

Labels

Labels should convey the following data: title or subject identification; artist, if more than one person is being shown; date; medium; donor or collection name, if from an institutional collection; and, when appropriate, historical or critical comment. Works for sale usually have the price shown.

Labels can be made in a number of ways, as outlined below.

Method 1 Set the information on a personal computer such as a MacIntosh and print out the labels. Leave plenty of space between the blocks of copy for each label. You get the sharpest, most attractive type by using a laser printer. If your personal computer does not have one, you can often rent time on one at a copy center near you. While there, have the labels copied onto a sheet of clear plastic film of the type used for overhead projectors. Cut the labels apart with a razor knife and a straightedge. Stick the labels to the wall with 3M's Positionable Mounting Adhesive, 3M's Scotch Brand Double Sided No. 415 Tape, or photographic spray mount adhesive from a can. Because the plastic is clear, the wall color shows through and makes the film less noticeable.

You will still have the original copy, and this can be used to make up a checklist for the show. Type out a cover sheet with the name, location, and dates of the show. Have it photocopied for a handout at the door, and it will serve to publicize your exhibition. To make it more attractive run it on a nicely colored paper like ivory or gray.

Method 2 Number keys are a less obtrusive alternative. The numbers on the wall can be keyed to a list handed out at the door. Numbatabs, sold at most stationery stores, come in rolls of numbered, peel-off transfer paper circles that can be stuck down low next to each frame. Transfer lettering sheets, the kind used by graphic artists, have numbers you can burnish directly onto the wall.

Method 3 Typed labels can also identify the work. Get a sheet of peel-off labels from any stationery store. Type your information onto them and stick them on the wall. A carbon ribbon typewriter gives the sharpest results. A variant on this idea is to paint several sheets of index card stock or thick paper at the same time your gallery is painted. You can type the labels directly on this painted sheet and then glue or tape them to the wall.

INFORMATION SERVICES AND PUBLICITY

Planning a show gives you the chance to use a wide range of media for publicity; we mention only a few here.

Press releases should go to all local media well in advance of the show. Editors are most interested in facts, and the more concise the better. Start the release with vital information like subject, dates and hours, and location, and be sure to give a contact person and phone number where that person can be reached to arrange further coverage. Then go into explaining things like the idea of the show. Make it pithy.

Any handout should also promote the organization mounting

the show. Give people who are sympathetic to your aims something to take with them to show their friends and acquaintances. Include a summary of what the exhibition is about, what the purpose of your institution is, where the event is located, and the dates and hours it is open to the public.

Checklists we have already covered. They have additional uses. They can be invaluable tools for collectors and scholars at a later date; if kept in the files, they will serve as a permanent record of the activities of your collection.

A simple checklist in booklet form can be made on sheets of 8½ × 11 paper folded lengthwise, or on sheets of 8½ × 14 paper folded in half. These can be stapled in the centerfold if more than one sheet is used. Printing on this size paper means that a cheap offset process can be used, or even a plain-paper copier. If you routinely make 35mm or 2¼-inch-square copies of prints in your collection, consider illustrating the checklist with contact prints from the copy negatives. (Not every print need be shown.) Simply paste each contact print onto the checklist next to its caption and have copies made from that. The quality will be mediocre compared to conventional halftone processes, but these rather stark images still identify subject matter and composition.

Posters have been used so often to publicize exhibits of fine art and photography that they have developed into an art form of their own. A high-quality poster, however, eats up even a large budget quickly. We have known many small galleries that produced beautiful posters but that went broke in short order. In our opinion, expenditure for posters should be tightly budgeted, and the decision to produce one should come after all other forms of publicity have been taken care of. Do not count on the sale of posters to turn a profit.

EXHIBITION DESIGN

The aesthetics of display design lie outside the realm of print conservation. However, there is no way to put on a show unless you have control of the mechanical details.

LAYOUT

The simplest approach to placing the framed prints on the walls is to start with framed prints and an empty gallery. Lay the prints face up with tops to the wall, going around the room in the sequence you want the show seen. This allows for juggling until everything fits, and gives you the chance to see what the show will look like before the first hanger gets nailed in place.

You can also use the log to make a list by number of the prints as you want them seen. Then take the amount of wall space and subtract from it the width of all the frames. Divide the remainder by the number of frames to get the number of inches between each two frames. Hang the frames in sequence with a uniform amount of space separating them.

If the results are too mechanical, try another method. Make a rough sketch of each wall and indicate height and width. Diagram where each frame will be hung. It is useful to purchase some graph paper and assign an arbitrary scale size to each block. You will need some such visual method of presentation when using staff personnel to hang a show.

EXTRA WALL SPACE

After the measurements and calculations have been made, what happens if you come up with a space between frames that is about minus 2 inches each? Short of overlapping some of the frames, you have two choices: edit the show or make more wall space.

Short of a complete renovation, the best way to increase the usable wall surface lies in putting together some panels that can be utilized for many different purposes.

PANEL CONSTRUCTION

Panels for most display applications can be built with a minimum of tools and time. You need a hammer, a saw, some finishing and drywall nails, a carpenter's square, and paint or enough fabric to cover the panel. A helper would be helpful, but one person can do it all.

Construction material consists of 1 × 2 and 2 × 4 lumber and ¾-inch-thick Homosote™ from the lumberyard (Figure 13.1). Homosote is sold in 4 × 8-foot sheets and larger. It cuts like butter

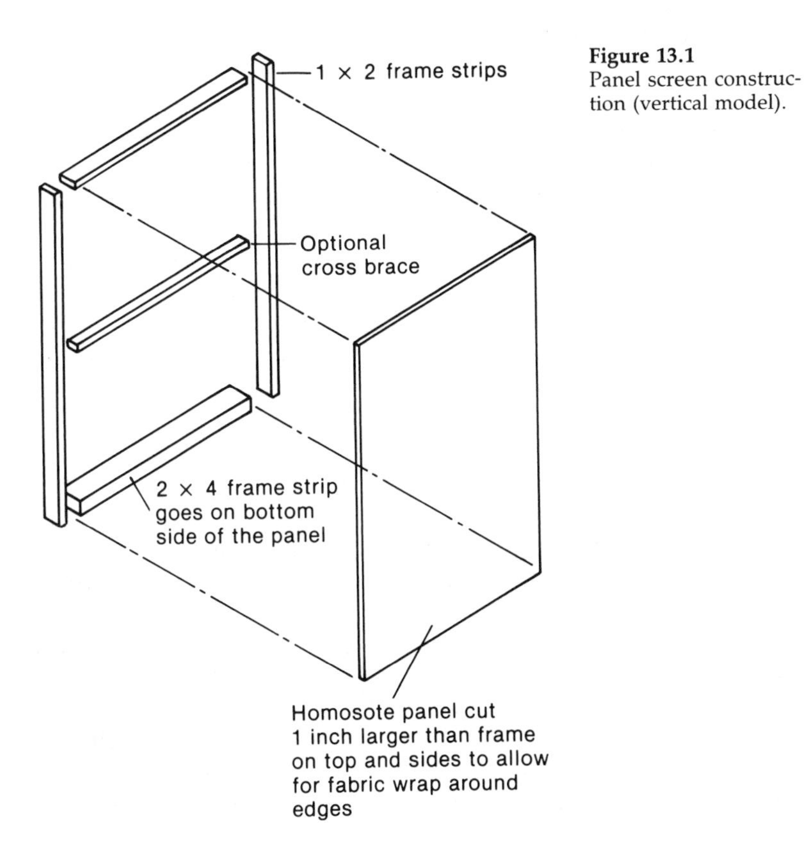

Figure 13.1
Panel screen construction (vertical model).

with any kind of saw, and can also be broken along a scribed line, though the edge that results is somewhat rough.

Screen panels require a frame made of 1 × 2s. For strength the bottom side can be a 2 × 4 stood on its side. Cut the lumber to size and nail it together with the finishing nails. (We are not specifying measurements, because different applications require different sizes, but for a good size room 4 × 8-foot panels work well.)

When you plan to cover the panels with fabric, make the frame about 1 inch smaller all around to allow the cloth to wrap around the edges of the panel. If the frame seems loose and floppy, do not worry. The Homosote braces it.

After the Homosote panel has been cut to necessary size, nail it to the frame using drywall nails. If you are building a two-sided screen, do the same on the other side. If you plan to paint the surface, sink the nailheads with a final blow of the hammer and then cover the dent with some spackling compound. Burlap is the best material to cover panels with, because nails and pushpins can be shoved through it repeatedly without having to repair the Homosote underneath.

If you are thinking about substituting plywood for Homosote, you have not priced plywood lately — at least, not in the thickness that would be needed to give comparable purchase for frame hangers. If you decide to cover the panels with fabric, you can fasten the fabric around the edges any way you like, but we find that staples work well. Use the heavy-duty construction kind put in with a big gun like the Arrow stapler.

When panels have been finished, they can be either hinged or joined with angle braces to support one another, or used on the walls. You will have to work out the details for your particular application, but make the joining technique removable so that the panels can come apart for storage or rearrangement.

Variations on this basic scheme will be needed for particular applications. For free-standing stanchion partitions (details in a moment), substitute 2 × 4s on the sides for the 1 × 2s. The 2 × 4s are secured to ceiling and floor before putting on the Homosote. Another note: do not add interior braces on wall-covering panels; they interfere with the hanging strip. (Patience! we'll explain.)

PANEL STRUCTURES

Hinged screens can be made for open areas of the gallery, greatly increasing the available wall surface as well as creating an interesting variety of spaces for people. Screens can be of three types: angled, 90° offset, or T-shaped (Figure 13.2). Hinge these or brace them with release pin devices so they can come apart for moving and storage.

Island wall spaces in the center of the gallery resemble a small room dropped down in the middle of the site. These you can either make open for entry by adding a small doorframe, or you can close them and hang frames on the outside only. Use four or

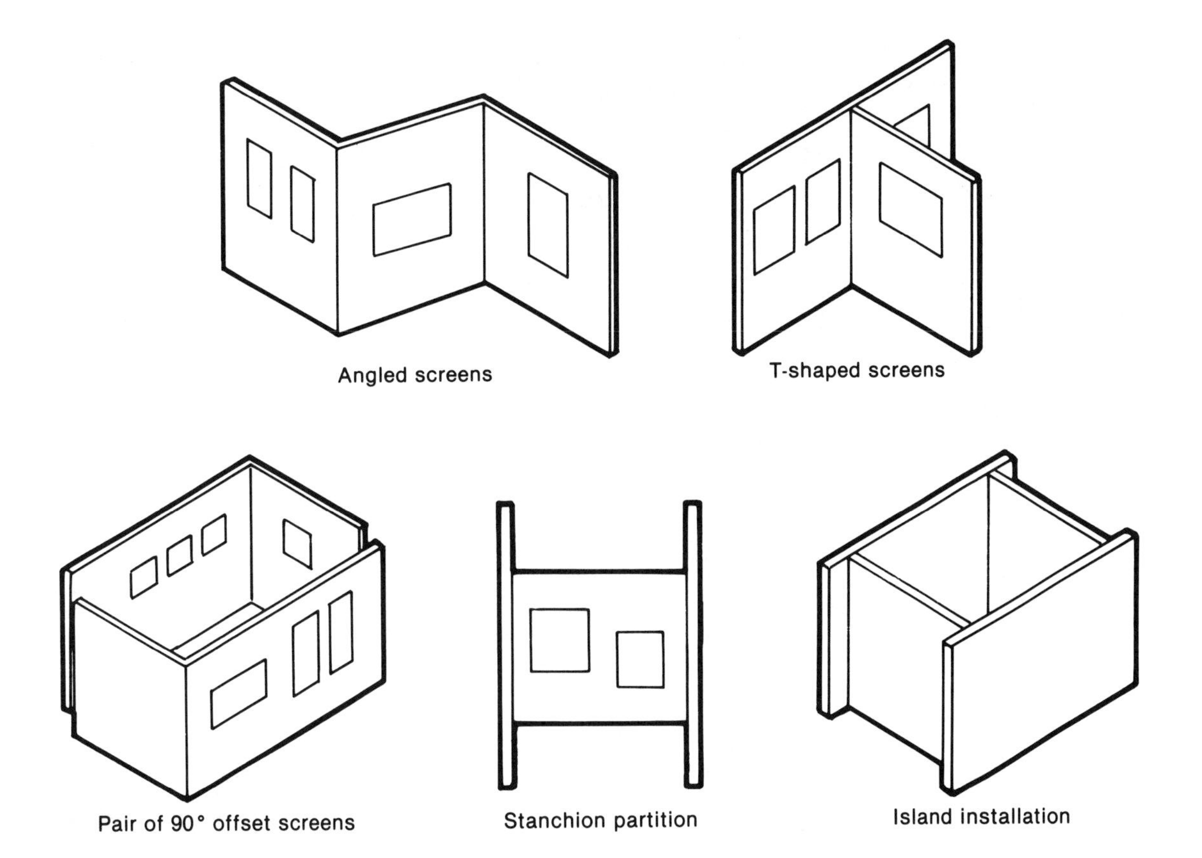

Figure 13.2
Wall partitions.

more of the panel screens. Gallery goers seem to enjoy the kind of intimate corridors that you can create with these.

Stanchion partitions are a semipermanent addition to the gallery space. Attach two load-bearing supporters to the ceiling and the floor, and support a panel between them. The open space at top and bottom increases the visual area of the gallery, lets light into dark corners, and gives the staff personnel a chance to keep track of people's location.

RECLAIMING EXISTING WALL SPACE

Some walls seem at first glance to be unsuitable for hanging because of their material or because of built-in obstacles. Problem walls made of substances like brick or cinderblock and walls with windows or with pipes running down their faces can be reclaimed for viewing surfaces by hanging panels in front of the obstructions.

Panels are hung so that they can be taken off to restore the wall to its original state. The secret of putting them up in the first place is to start by installing a hanger strip of 2 × 2 lumber on the wall about 7½ feet above ground level (Figure 13.3).

On brick or cinderblock walls, use a level to mark a guideline for the strip. Drill a series of holes along the line with a power drill equipped with a masonry bit. Pound screw-anchoring plugs into the holes, and draw guide arrows just beneath them on the

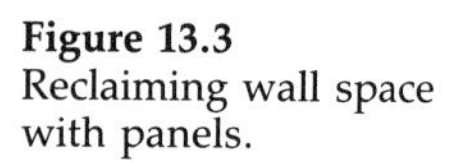

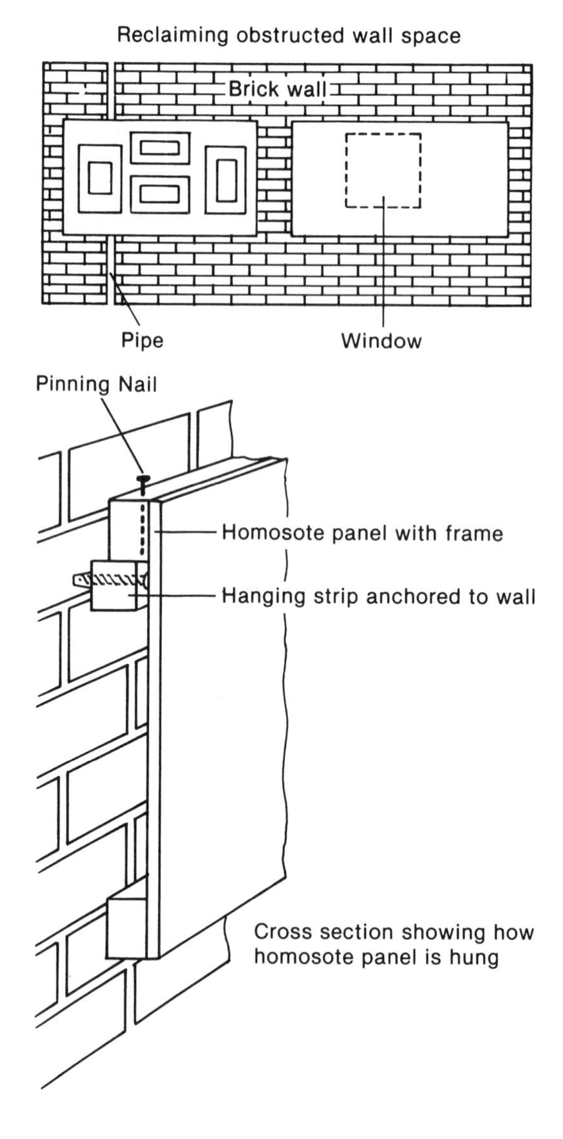

Figure 13.3
Reclaiming wall space with panels.

wall. Place the strip on top of the holes, and drill small guide holes for the screws at the sites indicated by the arrows. Now screw the hanging strip to the wall, and paint to match the wall surface. For brick walls, a dark brown will be least noticeable. When drywall construction is encountered, use molly toggles in place of the screw anchors.

A Homosote panel is then hung on the strip. This panel should be framed on four sides for rigidity; it will be the hanging surface for framed prints. To attach it to the hanger strip, drill several small holes from the top through the uppermost horizontal frame, and drop a long nail with a large head into each hole. Three or four nails per panel will hold it firmly in location, and they can be pulled out to remove the panel for storage.

Of the commercially available partitions we investigated, none combined the versatility of Homosote panels with their economy. The free-standing panels sold for convention hall booths usually cost hundreds of dollars apiece, and since they are built to minimize transportation costs they are very lightweight, with a foam-core interior. This makes them useless for hanging glass-framed prints. There are systems of movable museum walls; if you have the budget to get these, consult an architect before laying out the money.

SUGGESTIONS ABOUT INSTALLING THE SHOW

The actual mechanics of hanging the show should be fairly clear by now, but some points still need to be mentioned. First, about the number of hangers used per frame. Depending on its rating, a single hanger can hold up to a 50-pound frame. This may be fine in a private home, but you might want to consider using two hangers per frame. This gives the frame greater stability, so that even if it gets brushed by a careless viewer it will continue to hang straight. It also gives you a backup in case one hanger works loose.

Then, about cleaning the glass in a frame. The glass always seems to get fingerprints on it during hanging. Don't squirt glass cleaner directly onto it. To apply cleaner, always spray it onto paper toweling and then put the towel onto the glass. Under pressure, cleaner can leak into the frame and get onto the print.

A Tool to Speed Hanging

A friend of ours on the staff of the International Museum of Photography at George Eastman House recently showed us this simple way to speed the process of hanging a show. We pass it along as a suggestion.

The problem: how to hang a group of frames at a common height for viewing. The usual way is to decide on this height, bang a couple of nails in the wall at that distance from the floor, and run a string between the nails. To determine where the top should be, as you hang each picture you measure up from the string to a distance equal to one-half the height of the frame. Then you have to measure down from the top to determine where the hanger should be installed. In theory it sounds complicated; in practice it eats up time.

Enter: the show-hanging stick (Figure 13.4). Take a 1 × 2 piece of lumber (usually eight feet will do). At the height where you want the common center of your frames, draw a thick line on the stick. This is usually around 54 inches to 60 inches from the floor.

Next, mark on the stick the heights of all your frames. If you use standard sizes, this is easy. For a 10-inch-high frame, make a mark 5 inches above the line and 5 inches below and label it. Do the same for the 14-inch and so forth. Make sure you put labels next to each one. If you have any odd-sized frames in your show, put them on as well.

Now here's how you use the stick to hang your show quickly

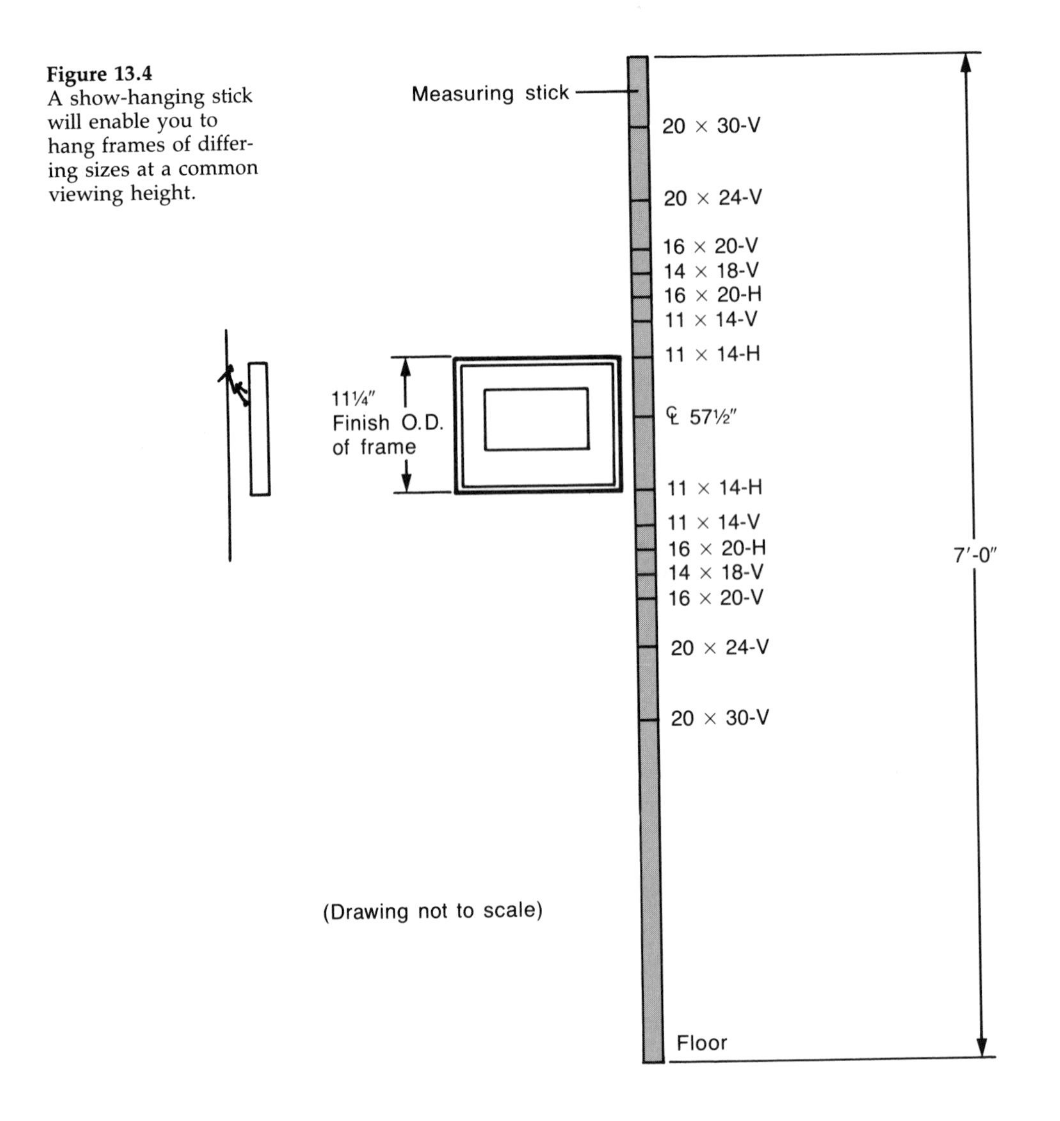

Figure 13.4
A show-hanging stick will enable you to hang frames of differing sizes at a common viewing height.

and without wasting any effort. We assume you have all your framed prints, an adequate supply of hangers, a hammer, bubble level, and measuring tape. Starting in the first corner, measure out to the centerline of where you want to hang the first picture. Put the stick against the wall with its bottom on the floor. Note where the top of the frame should go by the mark on the stick, either with a faint pencil mark or by holding the bubble level against the stick. Measure down from there with your tape to where the hanger goes.

A quick word here about where to put this hanger. You can tell how far down from the top it has to be for a particular frame by pulling up on the wire so that it makes a bow and then measuring the distance from the high point to the top side of the frame. If you are using metal section frames, the devices on the side that hold the wire slide up and down so you can adjust where the high point of the wire will be. To speed hanging you could standardize this distance at around 4 inches.

Nail up the hanger, hang the frame on it, and level it by putting your bubble level on the top edge. You will want to use a standard distance between pictures, so take that distance and add it to half the width of the next frame. This gives you the distance to your next centerline. For example, if your distance between pictures is 6 inches and the next frame is a vertical 16 × 20, add 6 inches to 8 inches; measure 14 inches to the centerline of the next frame. Put your stick in place and repeat the process. Follow these steps exactly as you go around the room. Even when you have frames of many different sizes, you will hang them all at a common viewing height without problems.

The same stick can be used to position one print on top of the other. On the back of it, mark the viewing height again. Mark a standard separation distance, say 6 inches, and then mark and label tops and bottoms for the different-sized frames in your show. For the lower frames, only the bottom edges change while the top level remains the same. For the frames above, vice versa. It sounds harder than it really is. Try it and things will get clear right away.

REFERENCES

1. Larry Feldman. "Discoloration of Black and White Photographic Prints." *Journal of Applied Photographic Engineering*. February 1981:1–9.
2. Raymond H. Lafontaine. *Environmental Norms for Canadian Museums, Art Galleries and Archives*. Technical Bulletin no. 5. Ottawa: Canadian Conservation Institute, p. 2.
3. William Welling. *Collector's Guide to Nineteenth-Century Photographs*. New York: Collier Books, 1976, pp. xiii–xvi.

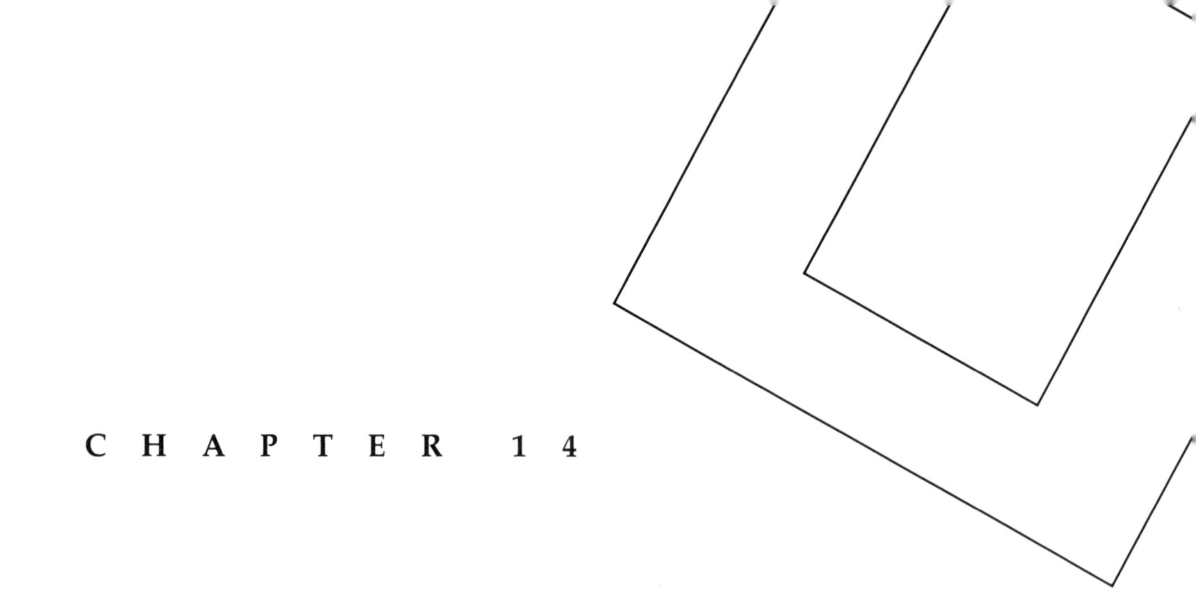

C H A P T E R 1 4

Copying, Duplicating, and Video as Tools for Photographic Conservation

Making copy prints, duplicating film, and copying images onto video tape or disk can all be useful in caring for your photographs. Here we want to discuss each process in turn, to show what choices you have in each area and how best to use available tools and techniques.

COPYING

Copying comprises any way of making a print or other photographic reproduction that approximates the appearance of the original positive. Institutions need to do this for a variety of reasons.

Preservation To protect images that are deteriorating, or for which no original negative exists in the files.

Restoration To restore the original appearance of a photograph by filtering out stains or increasing contrast to compensate for fading.

Reference To provide prints for reference, study, and access by researchers using the collection, and to reduce the handling of originals.

Public Information To service outreach and exchange programs, and to prepare images for audio-visual presentations, newsletters, exhibitions, and displays.

At a minimum, a basic copy service should be able to provide 5 × 7 and 8 × 10 black and white copy prints, 4 × 5 copy negatives, and 35mm color transparencies.

Priorities for Copying

Limitations of time and money make it necessary to establish your priorities for copying. High-priority candidates for copying include:

- ☐ Prints (and negatives) that exhibit staining and fading due to the presence of residual processing chemicals
- ☐ Photos adhered to brittle or flaking mount boards
- ☐ Photos that are frequently accessed
- ☐ Photos for which exhibition-quality copies are required
- ☐ Photos in scrapbooks or adhered to unsuitable housings
- ☐ Negatives for which no positive print exists
- ☐ Photos for which no original negative exists

Negatives in the collection for which no positive print exists should be contact printed as part of the copying program. For research purposes, 8 × 10 enlargements are useful and can often be made at the same time.

Copy Stands

Copying does not need to be difficult or mysterious. One way to simplify the process is to acquire the right tools for the job. A basic instrument is the vertical copy stand (Figure 14.1). A camera mounted on a head slides up and down on a post above a hori-

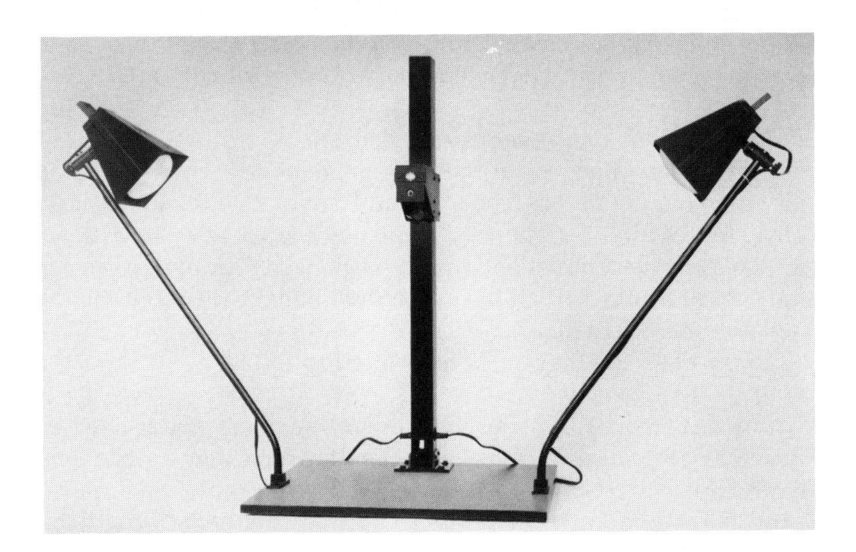

Figure 14.1
A basic copy stand like this Kinex model simplifies the copying process.

zontal copyboard. Lights located on either side illuminate the copy on the board from an angle of 45° from the axis of the lens. The copyboard may have an easel or a sheet of plate glass for holding the copy flat, although this is not absolutely necessary.

Some copy stands also have a lightbox built into the copyboard so that film originals can be backlit for copying.

The cost of a basic copy stand is between $200 and $350. As with most things, you can spend more. The additional dollars will get you the ability to use a larger camera (4 × 5 or 8 × 10 format), more elaborate capabilities for positioning the original, timers and motor-driven film advancing for automatic exposure, and sophisticated focusing systems.

Most stands can also mount video cameras for copying still photographs.

We have used the new, low-priced stands made by the Kinex Corporation and found them to be solidly constructed with a large number of useful features. Kinex has made stands for large-format copying for twenty years, and is willing to provide customized units at moderate prices. Copy stands are also made by Bencher, Bogen, and Spiratone.

Optics

When buying equipment for copying, get the best optics you can afford. You need a lens that has the flattest field possible and the best edge-to-edge definition. This means that it is absolutely sharp at the same distance from the center and the edge of the lens.

For 35mm copying we suggest a normal length (not wide angle) macro lens. The 55mm Nikkor Macro makes an excellent choice.

For large formats, the magic word is *apochromatic*. These lenses are built primarily for reproduction, they have absolutely flat fields, and they reproduce all three colors to exactly the same size. Most apochromatic lenses are f/9, and they are rather dark-focusing. This is the price you pay for the kind of optical excellence in this type of lens. You can buy them from Fuji, Schneider, and Rodenstock; all the manufacturers of lenses for large-format cameras make an apochromatic lens as part of their line.

Polarizing Filters and Controlling Reflections

For the general purpose of reducing glare, you can use polarizing filters on the lights of the copy stand and on your camera lens. This kind of filtering lets you completely control stray reflections. It increases the contrast and improves the color saturation, and in the case of a print with a heavily sulfided image it restores all the detail of the original.

If you have to choose between filtering the lights or the lens, filter the lens.

Spiratone makes 10-inch reflectors for the photo floods used on their copy stand, and they have clips for attaching a polarizing filter. The Kinex copier has arms that screw onto their polarizing filters and keep the filters away from the heat of the light,

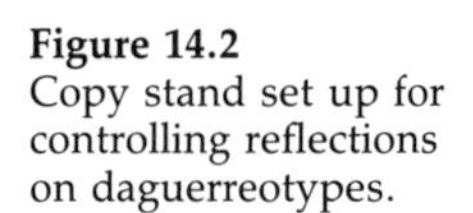

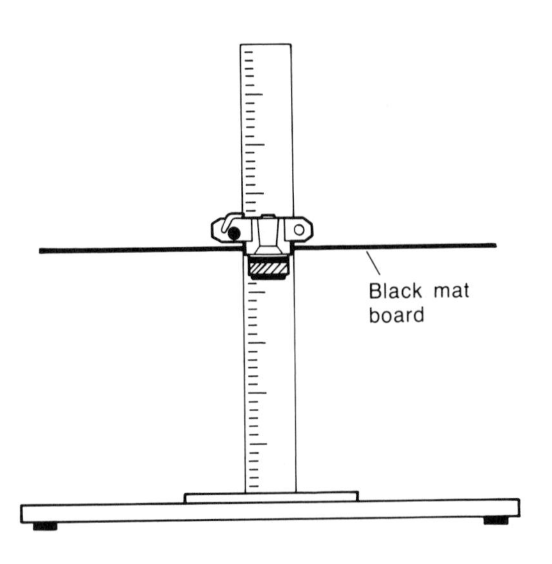

Figure 14.2
Copy stand set up for controlling reflections on daguerreotypes.

but keep them in a direct line with the axis of the bulb when its position is shifted.

When you copy cased images like daguerreotypes that are mounted behind glass, you might also try an alternative technique for controlling reflections (Figure 14.2). This should give you better color saturation and truer tonality. Take a large piece of black mat board and cut a small hole in the center for the copy lens to poke through. Mount it parallel to the copyboard at the height of the camera and photograph through it. The black board eliminates any possibility of reflection but does not remove the specular highlights of the image or the metal mat.

Film Formats for Copying

The standard size format for most copying is 4 × 5 film. Wherever there is a choice, the largest possible format gives the best result; in other words, 2¼ format negatives are to be preferred over 35mm, and so forth.

Use of 35mm negatives gives acceptable quality when you need copy prints of medium quality, such as 5 × 7 prints to be used in halftone reproduction. When making exhibition-quality copy prints or large display prints, 4 × 5 negatives give the required results.

When 35mm slides are needed for audiovisual presentations, color film is frequently chosen even for black and white originals because it can be quickly processed in a commercial color lab.

Copy ("second generation") slides can be made on a Honeywell Repronar unit specifically designed to accept most SLR camera bodies. Clearly mark copy slides as copies to avoid confusing them with originals in the collection; their quality, while suitable for viewing, makes them unsuitable for printing.

Copying Black and White Prints

Copy negatives can be made directly on the copy stand from clean black and white positives.

An excellent film to use for this purpose is Kodak T-Max 100 Professional Film. It has a long straight-line portion on its densitometric curve that makes it superior to Plus-X, and is readily available in sizes from 35mm through 8 × 10. It takes conventional processing, and the fine grain structure makes it possible to make nongrainy enlargements. If you are accustomed to using other films, be sure to run test strips and make sample prints when switching to T-Max, because the film base is clearer and may give a misleading impression at first.

T-Max gives results generally superior to the films that Kodak has recommended in the past for copying: Kodak Professional Copy Film 4125 and Kodak Ortho Copy Film 5125, but because it is panchromatic it does require working without the convenience of a safelight in the darkroom.

Removing Stains and Discoloration

Rather than submit an original print to the dangers of chemical treatment, it makes sense to remove objectionable stains and discoloration by copying with filtration.

The causes of stains can be many, from chemical reactions caused by mounting adhesives to material spilled on the print. In any case, the effect can often be minimized or completely reversed by copying onto panchromatic film, through a filter that is the same color as the stain. For some reason, many stains are yellow or brown; most such markings can be taken out with Wratten 12 or 25 filters. Polycontrast filters, the kind for enlarging, can also be used.

Again, T-Max makes a good choice for a copying film. An alternative — especially when you have to use a small format like 35mm — is Kodak Technical Pan Film 2415, which can also produce increased contrast and which has higher resolution.

Increasing Contrast

When a monochrome original like an albumen print or a silver gelatin print has faded to a light yellow-brown, you can often restore the image to its original contrast range with little loss of detail.

Technical Pan Film used with a Wratten 47B Blue Filter is an excellent starting point when you face this problem (Figure 14.3). The inherent contrast of the film and the blue filtration work together to restore lost contrast.

Copying Color Prints

Modern color materials show improved resistance to fading and color changes during dark storage. With the right light source, filtration, and exposure time, you can use them to correct for color shifts and fading in your prints, and to preserve the image close to its original form.

Figure 14.3
Restoring lost contrast with Technical Pan Film and Wratten 473B Blue Filter.

Where the original negative can be used, it is preferable to make a new print from that rather than copying. While there may also be color shifts in the negative that correspond to the changes in the print, these can be more readily corrected for by filtration during printing than during copying, with its necessary increase in contrast and loss in detail.

You have a choice of film types for color copying. When you intend to make a color print, a color negative film is the best material. In 35mm and 2¼ formats, Kodacolor VR Gold 100 produces acceptable results for most purposes. Kodacolor does increase the contrast in the copy print, but it does not require you to use color densitometry to match camera filtration with the chemistry at a particular lab.

Kodak also makes Vericolor Internegative Films specially for color copying: 4112 Film in sheets, 6011 Film in rolls, and 4114 Film for copying transparencies. These must be used with color densitometry to get the full benefit of their capability; follow the manufacturer's instructions for exposure, filtration, and testing.

When you are copying color prints for reproduction in a publication, the printer will prefer a transparency. Kodachrome and Ektachrome slides and transparencies can serve the purpose equally well. Follow the recommendations for exposure and processing. Because of its better dark-keeping qualities, use Kodachrome when you expect the copies to find their way into your files.

It is sometimes suitable to use color to copy monochrome originals. The image may have distinctive colors that are an inherent part of its visual appeal: an albumen print has distinctive purple and red shades, while a daguerreotype or an ambrotype may have bits of color painted onto it.

Handling Materials for Copying

When you handle materials during copying or duplicating, you need to take the same precautions with the photographs as you do elsewhere. In the case of an institutional collection, originals should never be simply handed over to an outside laboratory under the assumption that its personnel will handle them with proper care and understanding. If you do use an outside vendor, at a minimum you want to review with them the standards you require from them in terms of handling and security.

Copying also exposes a photograph to a number of special dangers. Prints mounted on curved or distorted mounts should not be pressed flat with a cover glass, unless it is absolutely clear that the mount board has the resilience to accept the pressure. Unmounted prints should not be dry mounted to increase their flatness. Cased images such as daguerreotypes and ambrotypes may require support from underneath to avoid the danger of breaking their hinges. Old glass negatives, which were often made on glass that was not entirely flat, should not be put into contact frames where the pressure can crack them.

Many commercial photo studios rephotograph and do extensive hand work on deteriorated family photographs to remove tears and rips, blemishes, and other signs of damage. These "restorations" have little relation to copying for conservation, where the intent is to show the original without modification.

Running a Copy Service

If you establish a copy service for your institution, it should be housed in a secured room with its own locks. This prevents equipment and supplies from disappearing and protects the prints from browsing and unauthorized handling. Only your own personnel should be allowed to do copying. While you may not need a great deal of space, the copy room should meet the same standards for cleanliness and atmospheric conditions that you maintain in other parts of the institution where prints are handled, and it must be possible to control the lighting in the room.

DUPLICATING FILM

The purpose of film duplicating is to replicate as exactly as possible the tonal range of one negative onto another piece of film. Like so many things, it's simple to do and difficult to do well.

Purpose of Duplicating

You may have a need to make duplicates of film originals for a variety of reasons. Some duplicate negatives will be for archival purposes and should be processed to the highest standards for permanence. Others, which are working copies, may be considered expendable and need not be made to the same rigid specifications.

- ☐ Replace unstable originals (cellulose nitrate and diacetate films, color, films with residual processing chemicals); process for permanence.
- ☐ Improve ease of handling and reduce chance of breaking original negatives (e.g., glass plates); process for permanence.
- ☐ Make negatives with a more suitable contrast range for modern papers; process for permanence.
- ☐ Reduce handling of original negatives where there is a high demand for prints.
- ☐ Supply other people and institutions with negatives for making their own prints.

Densitometric Control of Film Duplicates

The ability of film to stop light depends on three factors: how much of the light striking it is reflected backwards, how much is absorbed by the silver grains, and how much is scattered and dispersed. For practical purposes, this relation can be simplified by measuring the difference between the light that strikes the film on one side and the light that comes through on the other side.

The common way to find the effect of exposure and processing on a piece of film is to use an instrument called a densitometer to measure its ability to stop light. Collected in a systematic fashion, this data gives you the ability to predict the way a negative will print and what the final image will look like.

The latest generation of lightweight, inexpensive densitometers are easy to use and give accurate results. They let you control both your exposure and processing to get accurate results. We suggest to anyone who needs to do much duping that they start with some basic instruction in densitometry. Most photographic texts* explain

* The authoritative text on photo science, with an extensive discussion of densitometry, is *Photographic Materials and Processes* by Leslie Stroebel, et al. Focal Press: Stoneham, MA, 1986.

it in detail, courses in photo science all include a section on it, and trained photographers know it sufficiently well to explain the basics to somebody who needs to use it in a lab.

Direct Duplication Method

For duping black and white film you can choose between two methods: direct duplication and making a film interpositive.

The film for direct duplication is Kodak Professional B/W Duplicating Film 4168 (formerly SO-015). One exposure and one development of this film are all that is required to produce a duplicate. Because this film is relatively slow, most duplicates are made by contact printing emulsion to emulsion, which produces a laterally reversed image. It can be handled under safelights and processed in Dektol, DK-50, or a film processor. Previous questions about its stability have been resolved.

There is, however, another problem with 4168. It is intrinsically contrasty, and its narrow latitude means that it does not make good duplicates of old negative materials that have a great range in contrast. When you attempt to compensate for this limitation, the result is likely to be loss of detail in the shadows or "plugged" (pure white) highlights.

When cellulose nitrate and diacetate negatives start to deteriorate, they often buckle and change shape and can only be printed by projection (rather than contact printed). Since the presence of such negatives is often the reason for starting a film duplication program, you might want to consider the two-step interpositive method.

Interpositive Method

The two-step interpositive method gives you optimal control over the quality of the duplicate film, and it can even increase the printability of a negative. You can do the same things in making an interpositive — dodging, burning, and contrast control — that would ordinarily be done in making a print. When the negative is made by contact printing from the interpositive, it can then be used to make straight prints. These controls, it must be said, are best used with discretion when dealing with archival materials, lest you distort the appearance and meaning of your original negative — which it is, after all, your intent to preserve.

For the greatest uniformity of results, controls are best applied when making the interpositive. You can then use standardized exposure and processing times for printing all your negatives.

Kodak T-Max 100 works well as a general-purpose film for making interpositives. Its inherent speed means that it can be projection printed in an enlarger, and it can also be contact printed. Other alternatives include Kodak Commercial Film 4127 and Kodak Super-XX Pan Film, which is suitable especially when it is desired to reduce the contrast range of the original.

When it is necessary to duplicate a very large number of negatives, as is the case with a collection of nitrate film, budget constraints can limit the time and materials available for the project.

In this case, it is possible to make only the interpositives, processed to archival standards, and to contact print negatives as needed. The interpositives are stored as the archival record.

Quality Control of Duplicates

The only way to ensure that the duplicates produced by any process meet the required standards is to use a densitometer. It enables you to match the contrast range of your negatives to your printing paper. A densitometer enables you to determine whether the correct exposure is being given to materials like nitrate films, where yellowing may change the required exposure.

When you replenish standing chemical baths, process-control film strips should be run through the chemistry on a regular basis to check its level of activity. These strips are pre-exposed and sold by Kodak.

In addition, you must control the quality of the duplicates by visual inspection at each stage of the processing. The laboratory technician should examine each negative for stray artifacts as it dries, and the film should be checked again when it is being sleeved.

Color Duplication

For duplicating color negatives, the recommended method used to be contact printing them onto Kodak Ektaflex PCT Reversal Film, but we understand this film is no longer available.

Starting in 1977, the Preservation Services Division of the National Archives began an investigation of duplication methods for their color negative materials. Their research covered all known methods of color duping — separation negatives; three-color holography; direct duplication with silver-dye bleach material; and CRI, Kodak Motion Picture Stock 5249. While the CRI produced good copies, it did not meet their standards for keeping 100 years with a 20% maximum single dye fade.

The method for duping color negatives on which they finally settled after extensive tests was to use Kodak Ektachrome 5071 transparency film with the E-6 process. "The simplest approach is to use a common degree of filtration as orange mask compensation plus the standard recommended Ektachrome duplicating filters. . . . This can be accomplished by the addition of 30–50 cc's of cyan."*

The National Archives personnel report that they can make very satisfactory negatives for enlarging ratios of 2.5 to 3x. This method lets you use one type of film for copying negatives and positives, and the film can be processed at a commercial lab. Incidentally, when using an outside lab, it's sound practice to use the same one

* Richard R. Youso, "Innovative Color Preservation Methods," *Second International Symposium: The Stability and Preservation of Photographic Images*. Society of Photographic Scientists & Engineers: Springfield, VA, 1985, pp. 238–250.

all the time in order to achieve consistent results. This is true for any film work. Each lab sets its processing machinery to slightly different standards, and while the variations may not appear when you do your receiving inspection, the end result of constant shopping can be a wildly variegated collection.

Disposal of Originals

We have indicated in the chapter on film storage that it is preferable to keep original cellulose nitrate and diacetate negatives in storage — housed separately from the rest of the collection, in buffered envelopes, with proper ventilation, and so forth — so long as they do not show signs of deterioration.

A duplication program is not a sufficient reason to dispose of these materials. There may be pressure to destroy them. Resist it. We have heard of cases where, due to inexperience with duplicating procedures, a series of very poor quality dupe negatives was made and the originals consigned to the flames. The result of haste can be the loss of viable originals and their replacement with inferior substitutes.

Positive Identification of Duplicate Films

When you start to create a series of duplicate negatives, especially in an institutional setting, you create the danger of confusing duplicates with originals.

There are ways to solve this problem, but first let us illustrate the difficulty. Take the hypothetical example of a situation that might be encountered by a museum holding a large collection of works by a single photographer. To make the situation more interesting, let us assume that a philanthropist had bought the collection from the photographer's heirs before willing it to the museum. Each successive owner of the collection has done some work with it.

Now, when a cataloguer encounters a negative in the collection, he or she has to consider the possibility that it is:

☐ An original negative made by the original photographer

☐ A dupe negative made by the photographer

☐ A dupe negative made by the photographer's heirs

☐ A dupe negative made by the philanthropist

☐ A dupe negative made by museum staff

☐ A copy negative made by museum staff

Needless to say, a difficult situation for the cataloguer.

It would be possible to examine the film's notch codes and check the dates when certain materials were used to determine the actual provenance of a negative, but the process is not necessarily easy or infallible.

Figure 14.4
A dupe negative that includes the image, gray scale, and label. Example courtesy of Michael Haggar.

A good solution is to dupe the film onto the next larger film size. A 5 × 7 negative is duped onto 8 × 10 film (Figure 14.4). This gives you three working areas on the dupe negative:

1. The image
2. A contact print of a 21-step gray scale
3. A label (e.g., "Smith & Babbit nitrate duplication project, Upstate Photo Museum, 1988–89")

The original gray scale can then be sleeved in the same envelope with the dupe. On the outside, processing information is recorded: developer, exposure, time, and so on. The negative is processed to archival standards.

The next-larger-size method has added benefits. It gives the image a margin of protection, so that handling and atmospheric pollutants will be likely to affect only the blank edge areas. It protects against confusion with the original. Finally, future conservators should be able to scan the attached gray scale to correct for any shifts in tonality caused during duping or storage.

We propose that you adopt this as the standard method for making all of your duplicate negatives.*

* We are grateful to Michael Haggar, formerly negative archivist at the International Museum of Photography at George Eastman House, for the large number of valuable technical and practical insights he has given us on the subject of copying and duplicating.

VIDEO

It is highly probable that electronic imaging will increase its role in the conservation and exhibition of photographic images. We are already at the stage where it is possible to electronically enhance a blurry, out-of-focus picture, and a time may come when this will be routine. The thought has become widespread that electronic media may prove to be the best way to preserve any photographic image.

We have not yet reached that happy day.

If you take your family photo album to the local camera store to have it copied onto a video cassette, or bring in your home movies so that you can watch them on TV, don't trash the originals. Video is an excellent way to access a collection of pictures, but it is not yet an archival medium. This holds true not only for individuals, but for institutional collections as well, where video is already being used to catalogue large numbers of images.

We do not yet see the kind of authoritative testing on the keeping qualities of video tape comparable to the extensive work done on film, paper, and photographic emulsions. What we do know, however, makes us skeptical about casual claims that a video cassette will last for hundreds of years.

Video tape consists of three layers:

Oxide layer:	0.2 mil emulsion with magnetic material dispersed in a binder to hold information; lubricates against friction produced by tape guides and head stacks in VCR
Base layer:	An 0.6 mil polyester film that provides mechanical support
Antistatic layer:	Lubricates and protects against dust buildup

The information that the VCR uses to reconstruct a picture from the tape is carried by tiny metal particles embedded in the oxide layer. (In the future, this will be supplanted by metal evaporated onto the base.) During recording, the camera assigns each particle a magnetic value. When the tape is viewed, it passes through a head that scans the value of these particles and translates them back into a picture signal. Anytime that the head fails to read a significant number of these values, you experience a "dropout," a condition that causes a white or black flash on the picture.

When playing back a video tape, the machinery operates at extremely critical tolerances. The working distance between the video head and the oxide layer is only 0.02 mils. To put this into perspective, that distance equals only 1/30 of the depth of the oil deposit that makes up an average human fingerprint (0.6 mils).

Anything that gets between the head and the tape distorts the signal and can clog the head. The severity of the problem becomes clear when you know that a piece of ordinary dust measures 1.5

mils in diameter, a human hair is 3.0 mils thick, and a cotton fiber is up to 5.0 mils thick. If speed and size were scaled up, it's said that the precision involved in moving a video head over the tape is equivalent to flying a B-52 bomber one inch off the ground.

Some video tape comes on open reels, but this is for mastering and professional use, and is rarely encountered outside of the studio. All other video tape comes in nicely sealed cassettes, which is fortunate considering how little room it allows for error.

There are four standard formats: VHS, the most common; Beta, with the same size tape but laced through the cassette in a different manner; VHS-C, which is VHS format but housed in a compact cassette; and the small 8mm cassettes.

You can get approximately four different grades of tape, depending on the manufacturer. Each has its own terms for grading tape, and each gradation sounds like the premium tape: high grade, extra-high grade, and so forth. Basically, there are tapes for:

- ☐ Time shifting — recording and one-time replaying of TV programs
- ☐ Camera use — making video movies
- ☐ Hi-fi tapes — deliver the best-quality music
- ☐ Professional or studio use — dubbing

Although the grade of tape shows up in the quality of the picture, we do not know any reason that it should affect storage quality.

Failure of the tape medium itself may not prove to be a major threat to the preservation of video images, but this is by no means certain. Contaminants in the manufacturers' formulations, for example, have been known to cause oxidation and powdering of older formulations. High temperatures can speed up the onset of this problem.

Handling and storage conditions affect the quality and longevity of the video image. Video tape prefers the same kind of environment that is most healthy for a human being: slightly cool (50°F to 75°F), moderately dry (30% to 55% RH), dust-free, and away from tobacco smoke.

Video tape also requires a number of specific measures to survive storage:

- ☐ Keep the camcorder and VCR cassette containers free from dust. It affects their mechanism as well as the tape.
- ☐ Remove the cassette from the recorder or camcorder for storage. Proximity to the recording head can leave a dead spot.
- ☐ Wind the tape all the way to the end to distribute tension on it uniformly.

- ☐ Store the cassettes upright — vertically or sideways — to keep the tape edges from being damaged.
- ☐ Play each cassette once a year. This keeps the magnetic particles on one layer from affecting the information on the adjoining tape. It also keeps the layers of tapes from sticking together or picking up a permanent curl.
- ☐ Never touch tape with bare hands; wear lintless gloves.
- ☐ Avoid frequent starts, stops, fast-forwarding, and rewinding. The tensions these create can cause jamming, tape drag, and damage to the tape edges.
- ☐ Before using a cassette, take up the tape's slack by turning one of the reel hubs until it is taut.

Fortunately, there are some things that you don't have to worry about. Magnets, for example. Most magnetic fields in the domestic environment will not harm cassettes. Metal detectors and X ray machines (in the United States) don't create enough radiation to damage video images, either, though in foreign airports the situation may be different.

The recommendations for storing video tape are not markedly different than those for film. It is important to provide a stable, cool environment; prevent smoking; not store or handle tapes with bare hands or in the presence of food and drinks. The one major difference: it is *critical* to play back the tape once a year to preserve it.

Machine-Readable Records for Archival Storage

We have wondered for some time whether some kind of video or digital medium would eventually solve all our problems with the archival storage of photographic images by encoding them all as digital bits.

An intriguing presentation by John C. Mallinson has changed our minds. Dr. Mallinson talked about a report to the Archivist of the United States, made by a committee he chaired. The assignment of the committee was to establish the criteria for using human-readable and machine-readable records in the National Archives. "Human-readable" means records that can be read with the eye in some form; for example, a book, microfilm, and the like. "Machine-readable" means that some electronic device is needed to retrieve the record; for example, a video cassette player.

After five years of research and reflection, the committee came up with one stunningly simple conclusion: all storage and retrieval strategies for the archive should be based on a human-readable medium.

The key reason for this is simple, and has nothing to do with the stability of the tape, disk, or whatever other base is used for encoding the data. The problem is the machinery. In the digital age, the rate of technological evolution means ongoing and inevitable obsolescence for electronic formats. And once a manufacturer stops supporting a machine, spare parts become impossible to acquire and the entire mass of records in that format becomes useless. And whether it is the head on a cassette player or any other component, the failure rate of playback machines means that in an institutional setting this failure *must* happen within a relatively few years.

It's an intriguing and disturbing thought. It has convinced us, more than ever, of the importance of preserving original artifacts. Of course, machine-readable records provide excellent educational and indexing tools, but the master records should always be made human readable. For a more extended exposition, we suggest reading Dr. Mallinson's rather witty and elegant article [1].

REFERENCE

1. John C. Mallinson, "Archiving Human and Machine Readable Records for the Millennia," *Second International Symposium: The Stability and Preservation of Photographic Images*. Society of Photographic Scientists & Engineers: Springfield, VA, 1985, pp. 388–397.

PART FIVE

Storage

CHAPTER 15

Storage of Prints

It need hardly be said here that proper storage can significantly extend the life of a print. This is especially true of collections, which are generally more vulnerable to disorganization and neglect than individual prints.

Most of us have come upon the stray snapshot of an unknown person from the distant past, tucked away in some odd spot like an antique dresser, and this snapshot has seemed none the worse for wear except for being a little dog-eared.

Few of us, however, will have found an intact, well-documented body of work that documents the building of a railroad or a family history. Such a discovery means a broader scope of action will be necessary. When neglected collections are found, at least a certain part of the whole will have suffered some damage. Prints will be waterstained or faded past recognition, glass plate negatives will have shattered, old nitrate film melted, and time erased the identity of the photographer.

Each person who takes responsibility for a group of prints has to decide what kind of handling the prints should get. What is involved here is setting priorities. Some collections consist of a relatively small number of valuable photographs prized for their aesthetic qualities; this situation requires treating each print separately as a precious object. Others contain a greater number of pictures, maintained perhaps for more utilitarian reasons such as research and historical documentation. These prints will, of economic and practical necessity, be treated in more standardized fashion. As examples on a grandiose scale, think of the Museum of Modern Art as opposed to the National Archives. One holds valuable manuscripts, first editions, original prints; the other has to retain almost every document turned out by the federal government. The approaches they take to conservation differ in kind as well as quantity, and you will have to decide which example offers the best model for the needs of your prints.

This brings us to another, and crucial, topic: organization. Before deciding what type of box to put the prints in, or whether to buy a cabinet system to store slides, your approach to storage must be organized. Start your storage efforts by doing an analysis of two areas.

First, who is your public and what needs must the collection satisfy? If the users are scholars looking through a large body of material for research purposes, rapid access to clearly identified prints will be your first goal. If they are members of the extended family who want to browse through memories and ancestors, protection during handling will have higher priority.

The second thing to examine closely is the question of resources. If you have an annual budget, that defines one parameter. If money is not readily available in large quantities, what about time? Can friends or volunteers contribute significantly? Are you personally able to make a long-term commitment to organizing and maintaining the collection, or will it be better off trusted to other hands? Questions like this must be answered if the collection is to benefit.

An informed choice of materials and equipment will obviously make a difference in how well the prints in a collection survive, but items billed as "archival" do not by themselves ensure the longevity of prints. As we emphasize in the section that follows, two things will be more important: a coherent organization of the collection, with identification and documentation; and the unremitting application of your intellect to its care.

Skepticism is a healthy trait of mind for a curator. Some makers of plastic negative enclosures, for example, claim that their product is "neutral pH." We hope so. All plastics should be free of loose hydrogen ions that could go into solution (which is the literal meaning of "neutral pH") — because plastic is insoluble in water. What you really want to know is whether the plastic enclosure will outgas plasticizers that tarnish the negative, whether it will melt to the negative when it gets warm, and other information like this.

By now, you should already have a good idea of the factors that affect print permanence. Good storage techniques give you a variety of methods for coping with dangers like high humidity and temperature, ultraviolet radiation, insects, mold, chemical attacks from the atmosphere and surrounding materials, and mechanical damage to the print.

You need to keep your photographs cool, dark, and dry. How best to do this requires a little elaboration.

STORAGE ENVIRONMENT

A complete standard for the proper environment for print storage has been published by the American National Standards Institute. Section 6 of the ANSI standard PH 1.48–1974, "Practice for Storage of Black-and-White Photographic Paper Prints," covers in exacting detail all the environmental conditions to be met in a room used for long-term storage of valuable prints. These conditions are summarized briefly here.

- ☐ *Humidity:* maintain between 30% to 50% relative humidity, never exceeding 60% and not cycling daily between extremes.
- ☐ *Temperature:* maintain between 50°F and 77°F, with daily cycling not to exceed 7°F. Temperature should not go over 86° for any prolonged period.
- ☐ *Air-entrained solids:* dust should be removed from air by noncombustible filters that remove 85% of the solid particles.
- ☐ *Gaseous impurities:* nitrogen oxides, sulfur dioxide, and hydrogen sulfide should be removed by air washers and activated charcoal filters. In addition, peroxides from bleached wood, glues, and varnishes in storage cabinets must be removed; these can be eliminated by all-metal furniture.
- ☐ *Light:* no light with high ultraviolet content should reach the print. This includes sunlight and fluorescent tubes with high ultraviolet levels.

In addition, the ANSI standard outlines certain specifications for construction of a room for the long-term storage of a large number of prints. Moisture barriers inside the insulation should be used to prevent interior condensation; prints should be protected against damage from fire or mechanical force and from water from floods, leaks, and sprinklers; and good housekeeping of the area should be performed. In addition, the ANSI standard recommends including an inspection area so that prints do not need to be removed to a different environment for examination.

These are stringent requirements that cannot easily be met by private individuals or small institutions. Keep in mind, however, that these standards are not set up because the people at ANSI happen to be purists with a rigid mindset who choose to make arbitrary demands on the rest of us. In fact, these standards are based on certain chemical and physical processes that determine precisely what happens to a print under less than optimum conditions.

Humidity control is the single most important factor in planning a storage area. High moisture content in the air increases the effect of residual chemicals in both paper and emulsion, and encourages the growth of mold. Too little moisture causes paper to lose its normal content of 6% to 7% water and to become brittle.

Fluctuations in temperature increase or decrease the relative humidity, and can quickly bring it to unacceptable levels. In addition, high temperatures increase the reactivity of chemical components of the print and its housing, speeding whatever disintegrative forces may be at work. As a rule of thumb, chemists calculate that a 10°C rise in temperature doubles the rate of a chemical reaction.

Different kinds of solids are regularly released into the atmosphere from sources like smokestacks, car exhausts, and construction sites. These reactive dusts can fade or stain an emulsion, and dust in general abrades or adheres to the surface of the print.

Nitrogen oxide, sulfur dioxide, and hydrogen sulfide are gases widely present in the urban environment, and they all interact vigorously with the silver in photographic emulsions to cause fading and staining. They also discolor and embrittle the paper support.

Ultraviolet radiation, found as a component of most forms of visible light, is a high-energy kind of wave that directly attacks paper fibers, causing them to break more easily; it also causes fading of dyes found in color photographs and printed image.

CREATING A PRINT STORAGE AREA

For the private collector or small institution, it is often possible to adapt a room so that some of the most important dangers to the print collection are controlled. Choice of a site can minimize the difficulties encountered. Start with the obvious, by staying away from attics and basements where extremes of temperature and humidity will normally be encountered. Choose an area that is large enough to house the collection and its projected growth, and to allow some space for print examination, but that is small enough so that problems of environmental control are kept to a minimum. The smaller the storage area, the easier it will be to regulate the atmosphere.

Site selection should also take into account possible hazards from overhead pipes, sprinkler systems, and fire. Where possible, the location of toilets on floors above should be avoided, and the electrical system should be inspected for code compliance. A smoke detector system should be installed.

Insulation and vapor barriers can help in stabilizing the environment. The greater the insulation, the more stable the temperature and humidity will be, even in the absence of other types of atmospheric controls. Walls, floor, and ceiling should all be insulated, even in interior rooms. Insulation facilitates the control of incoming air and increases the efficiency of air conditioning, dehumidifiers, and (where needed) humidifiers. It also enables you to localize the need for air filtering and purification systems.

In buildings where there is no central air conditioning, or where the air conditioning cannot be readily controlled for the benefit of the print collection, a small independent unit can be hooked up in the storage room. Its operation should be constantly monitored by a separate thermometer and readings taken regularly to determine the relative humidity in the room. If air conditioning cannot be installed, the minimum should be a small dehumidifier programmed by a humidistat.

Stagnant air facilitates the growth of mold and favors insect infestation, so the layout of the room should promote an even exchange of air throughout. Print boxes, for example, should not be pushed flush against walls at the backs of shelves, and the air conditioning unit or units should be placed so as to create a flow of air in all parts of the room.

Materials like fabric that retain humidity should not be used in the storage area. Carpeting and drapes absorb moisture during periods of high humidity and release it slowly into the air, thereby increasing the long-term relative humidity.

Protection from light damage covers three areas. Print boxes keep the prints dark during most of their time in storage. Outside windows should either be absent or have dark shades on them that are kept drawn. Illumination for print examination should be either low-level incandescent lighting not exceeding 5 footcandles in intensity, or ultraviolet-filtered, color-corrected fluorescent tubes.

Good general maintenance of the storage area helps control dust, insect infestation, and mold growth, and keeps equipment in top operating condition. Periodic checks should be made of any temperature, humidity, and filtration systems, and contents of the boxes must be examined at regular intervals not exceeding 2 years.

STANDARDS FOR ENCLOSURES AND THE PHOTOGRAPHIC ACTIVITY TEST

The American National Standards Institute has published a standard (IT9.2–1988) that governs filing enclosures and containers for storage. The entire document is highly specific, and many details in it are primarily of interest to manufacturers and the professional staffs of large institutions.

The most significant part of this standard, so far as the general public is concerned, is that for the first time we have a generally accepted photographic activity test for determining how storage enclosures will react over time with photo emulsions. Any material — with the exception of plastic — that passes this test is as safe to use in contact with a black and white photographic emulsion as current technology can make it.

The ANSI photographic activity test (PAT) indicates factors that cause fading (damage to the silver *image*) and staining (discoloration of the *gelatin* in the emulsion). The material used to detect fading is unprocessed colloidal silver, also called Carey Lea silver. The stain detector is a conventional (not resin-coated) black and white photo paper processed to D-min according to the manufacturer's instructions. These materials are wrapped into sandwiches according to the procedure specified in the standard, and are subjected to accelerated aging at 70°C, 86% relative humidity for a period of 15 days. A temperature/humidity chamber or a saturated solution of barium chloride is used to maintain the correct relative humidity. The standard specifies certain densitometric results that the detectors must pass for the material being tested to be declared safe.

If that seems complicated, it is. The test is not meant for general use by photographers. It was developed by the Image Permanence Institute, RIT City Center, 50 W. Main Street, Rochester, NY 14614, and they will supply either the colloidal silver detectors or perform the PAT on a contract basis.

If a manufacturer specifies that a product passes the PAT, you know that the materials used do not contain chemicals that could hurt your photographs. But you do *not* know that the design of the enclosure is safe. An example: a negative envelope made of paper and adhesive, both of which passed the PAT, might still have a center seam that could leave a pressure mark on a piece of film.

What does the PAT not tell us?

- ☐ The effect of the enclosure on color photographs. The temperature and humidity conditions during the test are so extreme that they can produce changes in detectors even when the material being tested is safe.
- ☐ Whether plastics are safe. Because most plastics are insoluble in water, the accelerated aging may not work with them.
- ☐ What extraneous factors (like enclosure design) may affect the stability of the photograph.

It is also useful to list the important criteria for enclosures listed in the ANSI Standard.

For paper cartons, boxes, or containers not in direct contact with photographic material:

- ☐ A pH of 7.2–9.5
- ☐ Alkali reserve equal to at least 2% calcium carbonate

Paper in direct contact with *black and white* photographic material:

- ☐ Made of rag, bleached sulfite, or bleached kraft pulp
- ☐ Alpha cellulose content greater than 87%
- ☐ Free of highly lignified fibers such as groundwood
- ☐ Alkali reserve equal to at least 2% calcium carbonate
- ☐ Minimal sizing shall be used
- ☐ Any sizing chemicals shall be neutral or alkaline
- ☐ Paper shall be essentially free from particles of metal
- ☐ Surface fibers that might offset onto the emulsion shall not be present
- ☐ No waxes, plasticizers, or other ingredients that might transfer to the emulsion; no glassine

This paper also has to pass physical tests for stability, folding endurance, and tear resistance.

Paper in direct contact with processed *color* photographic material:

- ☐ Same as for black and white, except pH 7.0–7.5; 2% alkaline reserve shall not apply

The standard also specifies which plastics are acceptable. They include:

- ☐ Uncoated polyester (polyethylene terephthalate)
- ☐ Uncoated cellulose acetate
- ☐ Polyethylene
- ☐ Polypropylene

Sufficient data do not exist to recommend other plastics. Not to be used are:

- ☐ Chlorinated sheeting (for example, PVC)
- ☐ Nitrated sheeting (especially cellulose nitrate)
- ☐ Highly plasticized sheetings or coatings
- ☐ Plastics of unknown composition
- ☐ Plastics with residual solvents or plasticizers

In general, this standard is a highly useful reference tool, and should be kept handy for constant reference by anyone responsible for storing a large amount of photographic material.

PHOTOGRAPHIC STORAGE PAPER

Only recently has it become possible to recommend with great confidence any single paper for use in close proximity to photographs in storage. Many mills and their agents have advertised medium quality "acid-free" papers that might have a high lignin content or a rough finish, had not passed the photographic activity test, had unacceptably high levels of sulfur present, or were available only with high levels of buffering agents.

That problem has been remedied by new papers from a number of sources made specifically for long-term storage of photographs. These new photographic storage papers come in buffered (to pH 8.0–8.5 with calcium carbonate) and nonbuffered (pH 7.0–7.5) versions. Things to look for in a photo storage paper:

- ☐ Lignin-free pulp
- ☐ High alpha cellulose content (87% or higher)

- ☐ Passes photographic activities test
- ☐ 0.0008% or less of reducible sulfur
- ☐ Nonabrasive surface

Ideally, these papers will be watermarked to indicate presence or absence of buffering.

These papers are opaque white, often cut to stock sizes, and are generally available as an 80 lb text or in envelopes made to standard photographic sizes. You can use them for a variety of purposes in storing photographs:

- ☐ Enclosing single prints (envelopes)
- ☐ Interleaving prints (loose or matted)
- ☐ Wrapping groups of film or paper prints
- ☐ Lining print storage containers
- ☐ Xerography
- ☐ Folders
- ☐ Documentation and labels

Because such papers are relatively new, we expect that any list of suppliers we could give now would shortly be outdated by the addition of new sources. Check on availability with the vendors listed in the back of this book.

Other materials can also be used for interleaving; we discuss that important topic next.

INTERLEAVING MATERIALS

Interleaving sheets protect the surface of *matted* or *mounted* prints from physical abrasion, from contamination by hand-borne dirt and oil, and from chemicals released by neighboring prints that may not be totally cleared of harmful residues. Factors to look for in interleaving materials are first, a chemically neutral composition; then durability; and, for the sake of convenience, as much transparency as possible.

Nonbuffered paper is required for use with albumen, dye transfer, and color prints. The interleaving paper for other photos should be buffered to pH 8.0–8.5 with calcium carbonate.

An interleaving sheet should be put behind the window mat of a hinge-matted print, on top of the print itself. In the case of unmatted or back-mounted prints, interleaving sheets should separate each print from its neighbors in a box or file.

Acid-Free Tissue

This material resembles gift wrapping tissue in appearance and feel, but it has been especially prepared to be acid free. It usually contains a buffer, particularly calcium carbonate. It is very thin, so it does not increase a collection's bulk very much. It is difficult to handle because of its thinness, and it has a high degree of opacity.

Reflex Matte Transparent Paper
This is an acid-free sulfite pulp paper mechanically treated for transparency. The manufacturer states: "The high transparency is achieved without any surface treatment after manufacture and no brightening or clarifying additives are used." This paper was developed specifically for interleaving purposes, and compares well to glassine in handling characteristics. It is available in bulk only from Process Materials Corporation.

Acid-Free Glassine
This is a good interleaving material with a glossy smooth finish and high transparency. Check the manufacturer's specifications and test the glassine itself before using, because most glassine on the market is not acid free. All forms of glassine are, in effect, a kind of calendered tissue paper — that is, tissue that has been passed between steel rollers under high pressure. In addition, it is treated with glycerine to make it transparent.

Acid-Free Bond Paper
Permalife is the most readily available acid-free bond paper. It resembles a conventional sheet of white typing paper, but it is watermarked by the paper mill so that identification is certain. In addition to being acid free, it is buffered and has good handling characteristics such as resistance to crinkling and tearing. Bond is recommended for its low cost, but has the disadvantage of being nearly opaque.

Polyester
The most common form of polyester sheeting is Mylar Types D and S. The recommended type is 0.002 inch (2 mil), and it should be untreated; check your supplier's specifications for these details. Polyester is transparent, ages well, contains no plasticizers, and because of its inherent rigidity has excellent handling characteristics. It provides a barrier to migrating chemicals and moisture. Because it generates static electricity, it will pick up dust and should not be used with prints that have a friable or crumbly surface, like pastel and charcoal media. Polyester is also not recommended where humidity is uncontrolled, because gelatin emulsions can adhere to it and become ferrotyped as they would with glass.

POLYETHYLENE BAGS

When matted prints receive a great deal of handling, they have to be protected. This type of situation will be most often encountered in sales galleries and at art shows, where numerous prints are set out for display and inspection by prospective customers.

The best solution to this problem is the use of open-ended bags made of uncoated virgin polyethylene. These give physical protection by encasing the print in a transparent medium from which it can be readily removed for closer inspection. To keep bulk at a minimum, use a bag that matches the size of the mount.

The disadvantages of polyethylene are minor but significant. This material crinkles with handling, is not completely transparent, has a low melting point, and has electrostatic properties that will attract dust and fine particles. Also, polyethylene becomes brittle with age.

POLYESTER FOLDERS

Unmatted prints need the same kind of protection against abrasion and chemicals that interleaving gives to matted prints. A clear polyester folder that has a self-closing flap along one side and that is open at both ends provides one excellent method of protecting the single print. Polyester has excellent dimensional stability, which is why it is often used for film base; it contains no plasticizers, is chemically neutral, and because of its stability it ages well.

A transparent polyester folder has an advantage over a paper envelope in that the print is readily visible. This reduces the amount of handling it gets during examination. The design of the polyester folder chosen should, however, allow for easy removal, which is why the self-closing flap construction is preferred.

Another similar material currently used for print folders is triacetate, which has the same advantages as polyester except that it becomes brittle with time and can be torn easily.

Nonporous folders should not be used in situations where they might trap moisture.

PRINT HANDLING FOLDERS

You can make or purchase a simple folder for handling and temporary storage of prints. A print handling folder gives you clear visibility for examining the print closely without breathing on it, it provides structural support to prevent folding or creasing, and it gives you immediate access.

The folder consists of a 2- or 3-mil sheet of Mylar (polyester) adhered along two sides to the front of a piece of 4-ply, conservation-grade mount board. Trim the board to match the size of the plastic. The plastic is stuck to it with 3M's Double-Sided Film Tape #415 along the length and the width. This leaves one long side and one short side open so that the print can be slid in place. Round the corners to keep the plastic sheet from snagging and pulling away from the mount board.

Handling folders are simple to make in any size you need, or you can order them by mail. Make or buy a size that is several inches bigger on each side than the prints you expect to put into it. If it is made in stock sizes, it can be used in storage boxes to even out the pressure on the prints.

ENCAPSULATION

A technique developed at the Library of Congress uses polyester sheeting to provide greater protection for fragile or brittle prints and documents.

Encapsulation between two sheets of polyester provides a barrier against moisture, against acidic elements, and — by limiting access to the print surface — against the mechanical damages of handling. Encapsulation seals the print in a safe, sterile environment, with many of the benefits of lamination; unlike lamination, encapsulation can be reversed.

Because each encapsulation is performed individually, this technique provides a suitable way to protect prints of odd sizes that will not fit stock polyester folders.

Materials needed for encapsulation are: polyester film, 0.003 to 0.005 inch in thickness, in sheet or roll form; 3M's Double-Sided Film Tape No. 415, which will be virtually invisible along the edges when burnished; a grid sheet of the kind sold in drafting supply stores; a sharp knife and scissors; a straightedge; an antistatic plastic cleaner like Brillianize; and a brayer or some other kind of burnishing tool.

On the grid lay a sheet of polyester cut 1 inch larger than the print in both dimensions. Put down tape, using the grid as reference, along the outside measurement of the print, taking care that its adhesive will not come directly into contact with the print. The ends of the tape should not touch or overlap, because a small gap will be necessary to allow trapped air to escape. Next, place the print on the area enclosed by the tape, remove the paper liner from the double-sided tape, and overlay a larger sheet of polyester. Take care that fragile emulsions do not come into contact with the tape. Burnish the two sheets together where the tape meets them, and trim all four sides to within 1/16 inch of the outside edge of the tape. The corners of the encapsulation should be rounded, by scissors, nail clippers, or a graphic arts corner-rounder punch. This rounding eliminates corner snagging that can separate the sheets of polyester.

The encapsulated piece can now be treated as a single print to the point of even being matted in this condition. It can also be inspected from both sides.

PAPER ENVELOPES

Most paper envelopes made by conventional means do not meet the requirements for long-term storage of prints; use only envelopes specifically manufactured for this purpose.

Conventional envelopes incorporate a number of deficiencies, in addition to being usually made out of acidic paper. A seam in the center of the envelope body, for example, adds an extra thickness that can leave a pressure imprint across the image area of a print stored inside. The adhesive used to bind the seam, besides probably being acidic, can also intrude onto the print surface if moisture becomes a factor.

Requirements for print storage dictate that a paper envelope should be made of acid-free (buffered or nonbuffered as appropriate) high-alpha-cellulose content paper, contain nonacidic adhesives, and have the seams along its sides. Two designs on the market have, respectively, an ungummed side flap and a thumbcut on one side. Avoid those envelopes on which the thumbcut goes through both front and back of the envelope, because this exposes more of the print to the outside than necessary. The seamless envelope described later for film storage can also be made for holding prints.

A folder of interleaving material can help prevent fingerprints from being transferred to the print during removal; put it around the print before it is placed in the envelope. Paper envelopes protect single prints stored either in boxes or in vertical files.

VERTICAL FILING

Metal file drawers in which prints can be stored vertically provide rapid access to a large volume of prints. They are especially good for storing many prints of the same size, particularly when you have a collection to which frequent reference is made.

To meet the requirements for long-term storage of valuable pieces in a vertical filing system, each print must have its own acid-free enclosure. Paper envelopes of the type already described, or acid-free file folders, serve equally well as enclosures. Divider tabs should also be made of material that meets archival standards. There are no hanging-type file folders currently available made of acid-free paper. Hanging-type files, therefore, should be used for storage only of nonarchival materials like contact sheets and copy prints.

The proper kind of file cabinet for vertical storage has a movable pressure plate at the back of the drawer that can be adjusted to exert an even, light pressure on the back of the file so that the prints remain upright. Otherwise they are liable to slip down in the drawer. Spectacular curling sometimes takes place as a result. Wood file cabinets are generally not suitable unless they have been effectively sealed.

For small quantities of prints, or for collections where frequent access is not important, vertical filing may not be efficient. Another disadvantage is that, because most file cabinets are made for office use, most readily available sizes accommodate prints up to only 10 × 14 inches. Box storage for prints may turn out to be preferable when quantities of prints must be moved periodically.

PRINT STORAGE BOXES

The archival protection of a print does not end with matting or placing a print in a safe envelope. A final housing for a group of prints must be provided, and for this purpose containers of suitable construction and materials should be chosen. Storage boxes designed specifically for prints benefit the collection in a number of ways (Figure 15.1). Besides keeping light-susceptible materials in darkness and safe from many airborne pollutants like dust and

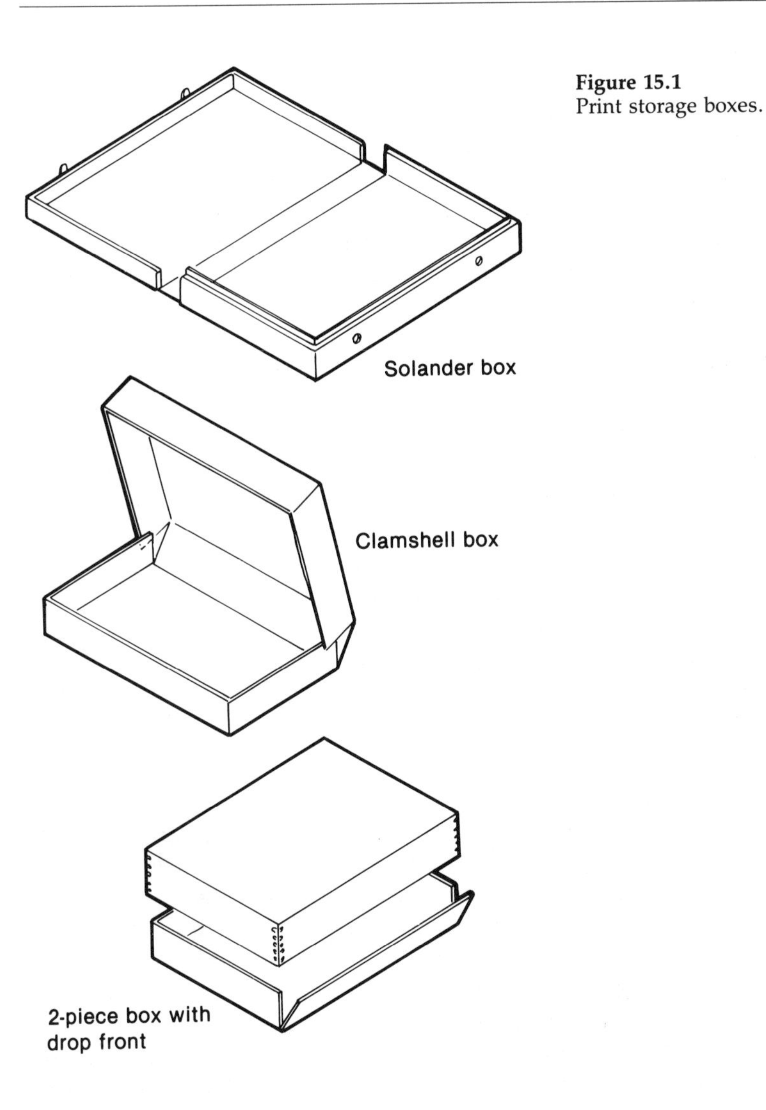

Figure 15.1
Print storage boxes.

sulfur dioxide, boxes provide physical support against flexing and warping, ward off accidental dents and abrasions, keep out many insects, and, with the choice of proper materials, prevent chemical contamination by the immediate environment.

These are rather negative advantages, but boxes have more positive qualities as well. A collection of prints stored in organized form inside clearly labeled boxes of convenient size will be easy to handle and will give the user rapid access to desired pieces. Ease of retrieval in itself greatly increases the value of a collection. Attractively designed boxes enhance the aura of value and beauty that should surround works of art.

Prints stored in boxes, whether matted or loose, should first be interleaved. Each piece should go into a box of the smallest size that allows easy removal, to keep the collection sorted as nearly as possible by size. This prevents sliding and movement inside the box, and equalizes the downward pressure exerted by mats on top. Folders slightly smaller than the interior of the box can provide a suitable housing for loose prints of odd sizes, and will protect against damage from crease marks or warping. Keep in mind that too much movement of mats inside the box will cause corners to snag and possibly tear or scrape the prints.

Do not overfill a container. This causes the box to wear and eventually come apart; it can also create pressure marks on the prints, and makes rearrangement access difficult in handling your collection.

When budget considerations permit, use the shallowest available boxes. Many shallow boxes, compared to a smaller number of deep ones, permit retrieval with less handling per print, and make movement of the lighter boxes easier for people working with them. This lessens the possibility of accidental dropping.

During the past decade an increased number of print storage box designs have definitely advanced the state of the art; so far as conservation is concerned, it has even come to the point where some definite choices about application must be made by the purchaser. It obviously helps maintain the uniformity of a collection if most or all of the prints are stored in the same style of box, so do some careful thinking before initiating purchase of the first storage boxes for a collection.

Solander Boxes

Over the years, solander boxes have become the venerable standard for print storage boxes, and until recently they were favored by museums and galleries with large print collections. The solander is a basswood-sided box with a hinged top and a recessed lip on all four sides to prevent dust entry. Basswood has a low resin content and good structural characteristics. The box is covered with book cloth on the exterior, and should be lined with acid-free paper on the inside (some of the older ones were not so lined). The rigid wood sides permit a solid-working latch to be used, which helps while moving the box, and for prints larger than 16 × 20 inches the box has excellent resistance to twisting and flexing during carrying.

Solanders are easy to access, stack well, and make handsome presentation cases. Their only debits are the use of wood, which some experts question; and their price, the highest of any box.

Clamshell Boxes

These boxes have one-piece construction, with the top and bottom hinged by a flap at the back. The top is made larger all around than the bottom so that it fits down over the sides, giving a double thickness on three sides for extra rigidity and strength. Sides, top, and bottom are usually constructed of binder's board lined with acid-free paper, and adhesives should meet archival standards. Most are covered with book cloth or textured wrapping fabric, and the high-durability Tyvek fabric used for the hinge allows for almost unlimited opening and closing. Where book cloth is used, the material should be coated with acrylic only.

Excellent for large collections where frequent access is a must, the clamshell box provides a reasonable compromise between budget requirements and the need for an attractive, safe housing. Its only drawback is that it is not really rigid enough to house many prints larger than 20 × 24 inches.

Two-Piece Boxes

Drop-front pressboard boxes (commonly called *Hollinger boxes* after their originator) are made of one layer of acid-free board joined at the corners with metal clamps. A separate top fits down over the lower half, holding the drop front closed. The drop-front feature enables the user to extract prints without folding or bending the corners. Usually the exterior finish is a utility gray. These are fine, very low-cost boxes for long-term storage of prints that do not get viewed often; they should not be used for presentation purposes, however, because they are unattractive. The shallowest 1½-inch-deep model should be chosen especially for large sizes, because these boxes do not have much rigidity.

There are also better-quality two-piece drop-front boxes using materials and construction techniques almost identical to the better type of clamshell box. However, these have somewhat limited usefulness when compared to the clamshell design.

Knocked-Down Corrugated Boxes

Boxes have recently appeared made of acid-free board and with a configuration that allows flat print storage. These two-piece boxes are shipped flat, and are assembled with self-locking folds by the user. No adhesives are used to hold the box together. Similar boxes of acidic materials are widely sold, so examine carefully the manufacturer's specifications before purchase. These are strictly for utility storage. Although they will show wear on the finish, they have excellent structural durability and rigidity. Further, because they are also sold in sizes as large as 32 × 40 inches, they provide an excellent method of storing very large prints.

CABINETS AND SHELVES

When a collection of prints has been housed in the right boxes, the next problem is access. If the entire set of storage boxes is simply piled up one on top of the other, access to the lower boxes will be severely limited and there is the danger of actually crushing them if the pile gets too high.

Shelving or cabinet units should be constructed of metal with a baked-enamel finish. Avoid wood shelves, painted or unpainted. Shelves should be easily adjustable, and boxes should be stacked no higher than two high to facilitate removal and replacement at their proper sites. Shelves must be at least as wide as the shorter dimension of the boxes placed on them to prevent overhang and consequent dislodging. The total weight load on the shelves must be taken into consideration when purchasing units, because many of the least expensive units will not hold up under the weight of a large number of print boxes.

Open shelving is most appropriate for collectors and institutions with a separate storage facility. When a relatively small print collection must be stored in a room used for other purposes, some kind of cabinet enclosure is recommended. Closed cabinets provide a number of advantages over open shelving: dust and humidity protection, greater security when fitted with locking handles, and, overall, a more attractive appearance. Cabinets sealed with airtight door gaskets can be fitted with silica-gel desiccant cartridges to absorb humidity during long-term closure. (Silica gel should be in sealed canisters to prevent any leakage of the desiccant into the print area).

Several companies specialize in the production of cabinets specifically for prints and valuable works of art, and these designs go from the quite elaborate to small units designed to hold several portfolio boxes. These, however, are not the only units to consider. Quite a large number of cabinets made by office furniture manufacturers can easily be adapted for print storage.

DRAWER STORAGE

Very large prints can be stored conveniently in the flat file drawers made for keeping maps and architectural plans. This style of storage furniture usually comes in units of four or five shallow drawers that can be stacked with matching units up to four high. These flat files come in two materials: wood, or metal with a baked enamel finish. The wood is not recommended, but if you must use it, line each drawer with buffered, acid-free paper like Permalife. As an alternative, line the drawer with vaporseal paper, a material made of layers of polyethylene, foil, and paper. Most styles can be purchased with either a compressor plate or a drawer-wide dust cover sheet; these accessories keep the prints flat when the drawer is opened so that the leading edge does not catch on the front of the file. When a compressor plate is present, place a sheet of neutral pH board between it and the prints to prevent crease marks.

When using this kind of file, each print should have its own folder of acid-free board in order to equalize distribution of pressure from prints above. The most common type of print abuse found in the use of these files is overfilling of the drawer, so the curator or collector must limit the amount of material that goes into any one drawer.

Cabinets and specialty shelving units are not usually sold direct by the manufacturer, so your best bet in seeking a supplier is to look up the largest supplier of office furniture in the area. Unless the item you need is in stock, delivery times are likely to be up to three months.

STORING DAGUERREOTYPES AND OTHER CASED IMAGES

Keeping daguerreotypes, ambrotypes, and other cased images is greatly simplified by the casings. When these items have been inspected, cleaned, and repaired, wrap them without taping in a large sheet of conservation-quality paper and put into box storage wrapped in this manner. You can write identifying data on the paper wrapper with a pencil (before wrapping) to facilitate acquisition without unwrapping. Suitable containers for a quantity of cased images are any that meet the standards for paper prints. The molded detail of the casings, which is often quite delicate and which adds to their value, should be protected from abrasion. If two or more layers of daguerreotypes are going to be put into one box, put a sheet of 4-ply conservation board or — better yet — acid-free corrugated cardboard between the layers. Because of their small size, safe-deposit boxes in bank vaults make a good housing for small quantities of valuable images of this type.

STORING TINTYPES

Tintypes, which today are usually found loose without a housing, must be handled with care in preparing for storage because of their sharp edges. They should not simply be stored together loose, because their metal corners abrade the images on adjoining pieces. We suggest two methods of storing. One is simply to wrap each tintype separately with a piece of Tyvek, with pertinent accessing and identifying notes on the wrapper. Another method that allows uniform filing and quick visual inspection is to put 4 × 5 sheets of 2-ply conservation board into 4 × 5 polypropylene negative envelopes as stiffeners, and then to insert the tintypes. This method allows most common sizes of tintypes to be put into a single file box or drawer of the kind regularly used for 4 × 5 negatives. You can record pertinent information on the backs of the board stiffeners.

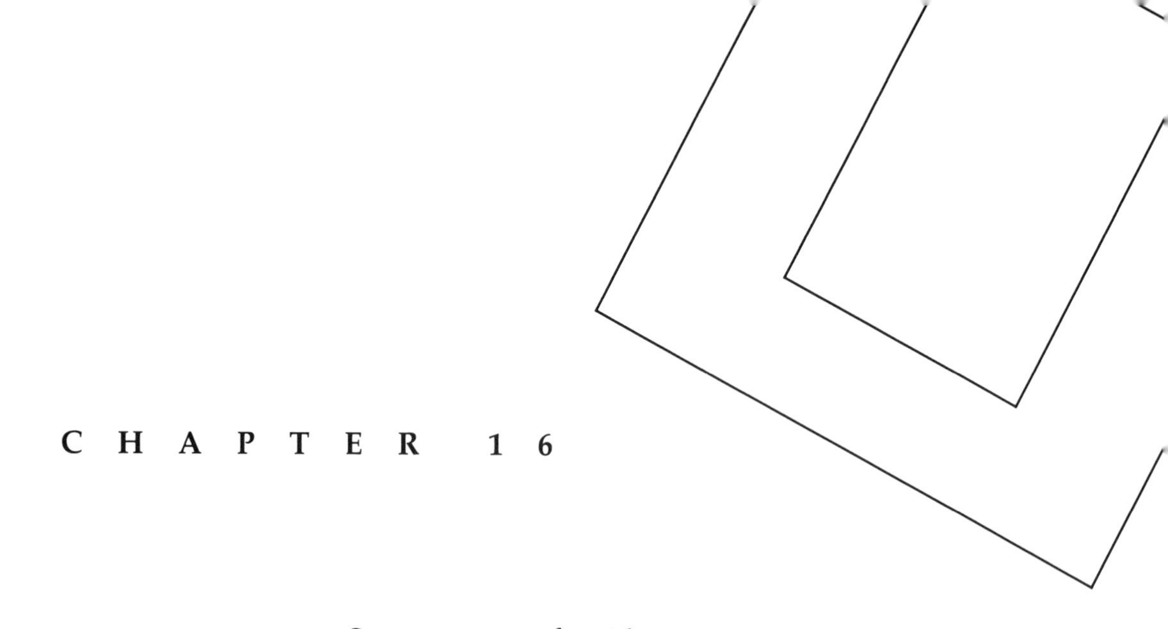

C H A P T E R 16

Storage of Film

Reading the scientific and practical literature on the storage of film, you may be awed by the tremendous number of subtle factors at work on many levels. To look at a simple 4 × 5 negative lying quietly in a paper envelope, for example, it would seem that you are seeing the very essence of passivity, a totally inert object. And yet, in actuality, this same negative and this same envelope, filed away in some dark corner, represent the locus of such a flurry of chemical, mechanical, animal, and vegetable activity that it seems safe to say that investigators have not yet understood (and perhaps never will understand) all these reactions and their interrelationships.

To give an example to demonstrate this point, cellulose acetate negatives (an early safety-base film made between 1934 and 1937) contain plasticizers in the base that volatize slowly. This outgassing of materials causes the film to lose some of its matter and to shrink in size as a consequence. Coupled with this reaction is the tendency of the "subbing" — as the cellulose nitrate adhesive that holds the gelatin to its base is known in the photochemical trade — to give off nitrous oxide gases, which will stain the negative yellow and cause small bubbles to appear under the emulsion.

This is just one example, discovered by photoconservator Jose Orraca [1] of the complex type of interactions that occur spontaneously in photographic materials. What makes it all the more interesting and pertinent is that this activity happens to a material — safety-base film — that previously was thought to be virtually inert when processed to accepted standards.

There are three conclusions to draw from this example. One is that despite the tremendous amount of research already done in the field of film preservation, much still remains to be done. Second, there is probably no single, completely safe method for storing film now known, and there may never be one. Instead, good judgment in light of the known facts and intended uses of the film must determine the storage method. And finally, we can be reassured that there are many general principles already known about film storage that will protect the vast majority of film from these kinds of unexpected damage.

The investigation by Orraca of these cellulose acetate negatives showed that those that had been stored in an air-conditioned room suffered minimal damage, while similar negatives kept in an attic had all suffered serious shrinkage and separation of the emulsion. In short, while we may not have learned everything that can go wrong with a piece of film, we have learned enough to do right by most of them.

MATERIALS FOR FILM ENCLOSURES

Research in the field of film enclosures indicates that the material of which they are made plays a singularly important role in the film's longevity (see [2] for an outline of the standard for these materials). Because of the different physical properties of various materials, some kinds are more suitable for use in specific designs of film enclosures. Irrespective of design, however, different materials have specific advantages and disadvantages.

Glassine

Glassine should *never* be used to house film. This is ironic, because for many years glassine envelopes were favored, and indeed dominated the market for film envelopes in all sizes, in part because they were cheap and in part because they were translucent and enabled the film to be viewed in the envelope. Besides being acidic and containing a volative plasticizer (glycerine), nearly all glassine envelopes have center seams that leave a pressure crease in the center of the image, and they are assembled with hygroscopic adhesives that attract insects and feed fungus growths. Further, glassine's smooth surface in contact with emulsion causes glazing, also called ferrotyping, a condition characterized by irregular splotches with a high sheen. Discard all glassine enclosures on old and newly acquired film, and transfer the film to new enclosures, along with any information inscribed on the old.

Kraft Paper

The earliest commercial negative envelopes were made of yellow-brown kraft paper, and these continue in use to the present day. Because of the residual lignin from the original wood stock, and alum or rosin sizing of a highly acidic nature, these envelopes — and all other paper envelopes of unknown composition — should be discarded. They usually have the undesirable center seam, and in addition they may have been written upon with acidic inks that will eventually mar the film's surface. It will often be found that older envelopes have become stained, or embrittled and crumbled to a stage where they no longer afford even physical protection.

Conservation-Grade Paper

The proper type of paper for negative enclosures meets the same chemical standards as are suitable for photographic prints. It is made from rag, bleached sulfite, or bleached kraft pulp with an alpha cellulose content greater than 87%. It contains no highly lignified fibers (groundwood) or metal particles, and sizing materials are neutral or alkaline. It has no waxes, plasticizers, or other materials that might physically transfer to the emulsion (in other words, it is not glassine).

For storing black and white negatives (especially nitrate-base films) the paper should have a pH between 7.2 and 9.5 with an alkaline reserve of 2% calcium carbonate. For color film, avoid buffered paper; the pH should be as close to a neutral 7.0 as possible.

Paper has several strong advantages as a film enclosure. Its porosity allows the escape of gases produced by decomposition, an especially important factor in keeping nitrate-base films. It accepts writing for identification purposes. Specialized film enclosures can be handmade without expensive equipment, unlike the case with many plastics. Finally, paper is economical.

Cellulose Triacetate

This is an excellent material that is inert around photographic substances. Kodak and others sell triacetate sleeves in standard photo sizes for storing transparencies and negatives. Cellulose triacetate is transparent to allow inspection of film without taking it out of the enclosure. A disadvantage is that this material is electrostatic and attracts dust particles; also, it is hard to write on, tears easily, and gets brittle with age.

Polyester

Virgin, uncoated polyester is sold under several trade names, the most common being Mylar. It is a superior material for negative storage because of its chemical inertness and excellent aging characteristics. In appearance it resembles triacetate, but it costs more. Until recently, polyester sleeves could be made only by special creasing machinery because of its very high melting point.

One type of polyester, Mylar-D, has proved so far to have the most suitable characteristics for a film enclosure. It has been treated (without the addition of a potentially harmful coating) to prevent electrostatic buildup, and it is manufactured in the gauges suitable for working with film. Currently some film enclosures are made with a radiowave welder to create a beaded weld along the closure lines, but the process is slow and prohibitively expensive.

Polyethylene

The most widely available, archivally safe film enclosures are made from virgin polyethylene. These enclosures are not crystal clear, and negatives must be removed from them before making contact prints despite any claims to the contrary. Polyethylene is good material because of its chemical inertness and because of the ease of finding supplies in stock photograph sizes. Untreated polyethylene, however, is electrostatic, and it has another significant drawback. The very low melting point that makes it possible to create heat-sealed envelopes presents a potential hazard in case of fire. Relatively low temperatures that might not even char paper envelopes melt the polyester and fuse it to the film. There is no known remedy.

Polyethylene-Foil-Paper Combination

While this is not a separate material by itself, its introduction by Kodak as a laminate in bags for housing color materials kept in deep freeze warrants a mention. It meets all the current standards for archival storage, with the added benefit that polyethylene stays flexible when cold. This is important for freezer use, because sharp-edged creases at low temperatures could scrape the film's emulsion.

This sandwich material can be bought in envelopes or rolls. The roll material is cut to size, folded over, and sealed on three sides with a hot iron according to instructions supplied with it.

We want to mention here that you should follow proper procedures when freezing and defrosting film in this kind of enclosure. Squeeze out as much air as possible before freezing, so that the amount of moisture trapped inside is minimized. When the film enclosure is taken out of the freezer, allow the contents to come to room temperature before you open it. Otherwise, moisture condenses on the cold film as soon as you take it out, the same way that water forms on a glass containing a cold drink.

Polyvinyl Chloride

We mention polyvinyl chloride (also called PVC) only because some makers of film sleeves have brought out complete lines of film holders in this material. In physical appearance it closely resembles polypropylene, and it is marketed with an express claim that it is safe for long-term storage.

Polyvinyl chloride is not safe for long-term storage. Its volatile plasticizers will make short work of film stored in it, and the effects on color materials will likely be even more drastic. Be extremely wary when you read advertising that fails to specify the type of plastic used to make film enclosures.

Polypropylene

Polypropylene is a plastic that has found increased use among manufacturers of conservation-grade film enclosures. It makes a good substitute for polyvinyl chloride because it offers the advantage of being easy to work with. As a semistiff plastic it has transparency without scratching easily or being brittle, and it is free of harmful plasticizers.

Some types of film enclosures prove more useful in specific applications than others, and so rather than recommend one particular configuration over all others, we offer an overview of the various designs and their strong points to help you make the best choice.

PAPER ENVELOPES

There are three criteria that a paper envelope has to meet if it is to be safe for long-term storage of film. (1) All seams must be on the edges, with the seam flaps folded to the outside of the envelope. (2) An inert, nonhygroscopic adhesive should be used; the only one we know of that meets these standards is polyvinyl acetate, although a dextrin-based acid-free adhesive can be used when keeping negatives in humidity-controlled conditions. (3) Finally, of course, the paper must be acid free and nonbuffered for color materials, buffered for black and white.

Foldover Envelopes

These envelopes without adhesive can be handmade with scissors or razor and straightedge, or they can be purchased (Figure 16.1).

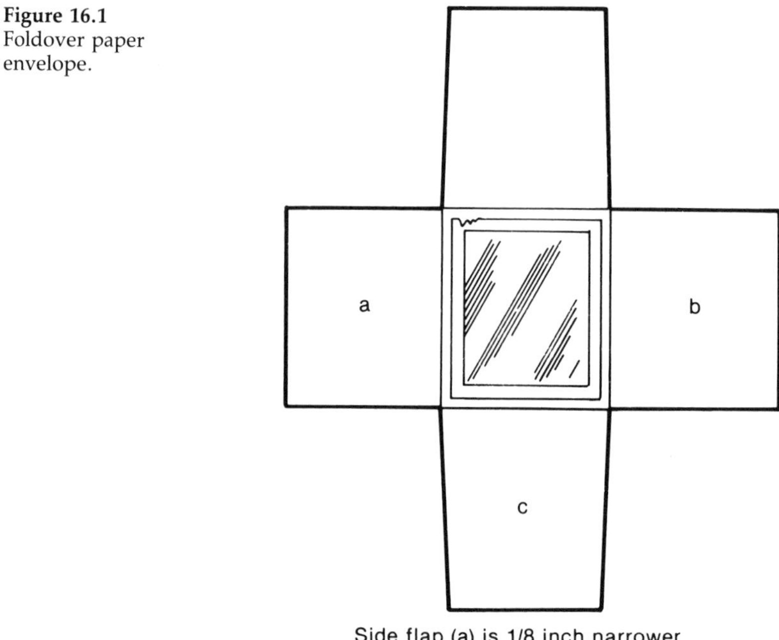
a
b
c
Side flap (a) is 1/8 inch narrower than (b) to allow it to fold under. Bottom flap (c) is tapered and folded in place first. Space is allowed around all four sides of the negative. Top flap is optional, but should be as long as the negative.

Figure 16.1
Foldover paper envelope.

They have the benefit of not incorporating any adhesive, and they do not have seam flaps at all so there is no danger of seams leaving a mark on the film. You can open them fully to lift the film out while holding it along the sides, a method preferred for limiting finger contact with the image area. These envelopes are cut in the shape of a large cross, with the center square ⅛ inch larger than the film along both dimensions to allow room for film to slide in and out without resistance. The inside flap (below the center square) should taper inward toward the bottom to allow for ease of closing, and one of the side flaps should be ⅛ inch shorter than the other, so that it can fit inside the outermost flap. Make creases along the two sides of the center and along the bottom before wrapping the film. A narrow folding flap can be left at the top of the center square for dust protection; it is folded over when the envelope is complete, although it is not always necessary.

This kind of enclosure is most frequently used for storing glass plates. See the section "Glass Negatives" in this chapter for more information on storing them.

Side Seam Envelopes

Side seam envelopes of suitable quality paper can be purchased (Figure 16.2). Both flaps with adhesive should be on the outside. But the most important design factor to look for is that the seams are on the side instead of in the center, where they can leave pressure marks in the center of the image area. The adhesive should be nonhygroscopic if possible, on the order of polyvinyl acetate. Failing this, check with the supplier to make sure that the adhesive is at least acid free and does not contain sulfur-bearing compounds of the kind found in rubber cement. Store film in these envelopes vertically, rather than flat, with the top end open to allow venting from the inside.

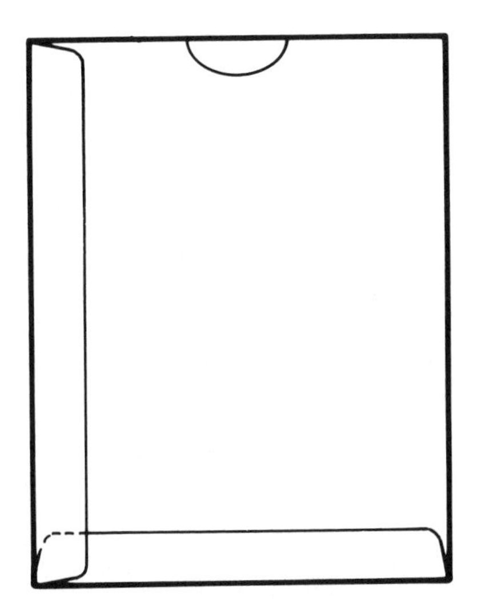

Figure 16.2
Side seam paper envelope. Thumbcut at top is optional.

A suggestion for additional protection: the tendency to reach in and grab the image area of the negative with two fingers when removing it from the envelope can be overcome by making a folder of frosted polyester. Use this folder as an inner housing for the negative.

Paper envelopes in general are the safest type of film enclosure. The paper breathes — that is, it allows moisture to escape — and there is no danger of the emulsion glazing or adhering to the paper. The aging characteristics of paper are well known, and identification on the outside of the envelopes is quite easy. The stiffening provided by the paper serves as an excellent physical support for the film.

There is some resistance on the part of working photographers, however, to using paper sleeves for small negative sizes such as 35mm. This seems to be due to the opacity of the envelopes; in a less than perfectly organized filing system the ability to quickly check the identity of a negative can be an important measure of the enclosure's usefulness. With a huge number of negatives, the problem gets more severe. Another drawback to paper film holders is that small formats cannot be filed readily in binders, a feature that makes some of the commercial systems attractive.

PLASTIC ENCLOSURES

The versatility of plastic fabricating methods means that a richer assortment of designs is available from commercial purveyors, but the expensive machinery involved means also that do-it-yourself designs are almost nonexistent.

Envelopes

Envelopes sealed on three sides and made of uncoated polyethylene are widely sold in camera stores for stock photographic sizes. Some have frosted areas where identifying information can be inscribed. These envelopes are heat-sealed along the sides, which eliminates the need for adhesives. We have seen a suggestion made that sandwich bags can be substituted at low cost, but it would be wise to avoid such false economy. Polyethylene can be coated with various chemicals to improve its handling characteristics, and the manufacturer has no responsibility to inform the public of these changes.

Plastic Sleeves

Heat-sealed on both sides and open at the ends, plastic sleeves are sold in rolls for 35mm and 120/220 film. The film is inserted and then the bulk sleeve is cut to size. To be suitable for long-term storage, this kind of sleeve should be made of polyethylene. Other sleeves in stock photo sizes are sold in triacetate with cement adhesive, and a polyester sleeve has been marketed with the edges sealed by an ultrasound technique and radio waves.

The sleeve design is quite useful for quick removal of the film with a minimum of handling. A negative can be pushed from one open end with a fingertip until enough extends from the opposite side to allow the user to grasp it by the edge. Sleeves also provide for some escape of decomposition gases from nitrate negatives.

Fold-Lock Sleeves
Clear fold-lock sleeves are now being made from polyester and polypropylene. These ingenious designs from Light Impressions employ two crease-folds and the natural rigidity of polyester to create a locking flap along one edge. The other edge is a fold, and both ends are open. There are no adhesives, decomposition gases can escape, and the sleeves are visually attractive.

Binder Page Enclosures
Polyethylene binder page enclosures make fine, easy-to-reference storage systems for negatives. These come in a variety of styles, usually heat-sealed, and any of the ones we have seen on the market meets requirements for long-term storage. Note, however, that this applies only to *uncoated polyethylene;* avoid similar designs made of PVC.

GLASS NEGATIVES

The bulkiest and most fragile of all photographic materials are glass negatives. From the heyday of wet-collodion photography between 1855 and 1880 to the present time, when they are used for very exact astronomical measurements, glass negatives have played an important role in photography. The introduction of flexible-base roll film by George Eastman at the end of the nineteenth century brought the snapshot to the masses, but professionals continued to make their exposures on glass well after this date.

Storage was always a problem with glass negatives, even after the introduction of the dry-collodion process made glass more convenient to use. Their thickness and the large formats usually employed (4 × 5, 5 × 7, and bigger) meant that they were very bulky and required extensive storage space. When stacked loose on top of one another, as often happened when files were no longer being used, their accumulated weight tended to fracture many plates at the bottom of the pile. They were also prey to the agents of deterioration that affect any photographic material, like heat and water damage, separation of the emulsion from the base due to improper subbing, atmospheric pollutants, and mold attacks. Finally, there were the ever active proponents of primitive recycling in the age before ecology became a catchword: the junkmen who scraped the emulsions off and sold the glass to greenhouses and to dry-plate manufacturers who reused the glass already cut to standard sizes. It seems a wonder that any have survived.*

For the collector and curator, glass plates present both a problem

* See Lesey [3] for an account of the intelligent, careful editing and preservation of a large collection of glass negatives.

and an opportunity. Despite the frailties mentioned above, glass has better keeping properties than some later materials like nitrate-base film. It is inert and dimensionally stable, and it does not present a fire hazard. Because the collodion (nitrocellulose) emulsion is very thin, gases due to decomposition can escape with no harm to the image. The rigidity of the glass means that the negatives are unable to start curling as they age.

The ideal way to store glass negatives is to place them in four-flap folders, and to stack them on edge in a baked-enamel metal file cabinet. Drawers in the cabinet should have dividers rigidly attached to their bottoms in order to space the negatives into small groups so that their accumulated weight leaning against one another will not cause cracking. The number per section will vary with size, with a maximum of 10 for 4 × 5 plates, and no more than 1 between dividers for sizes larger than 8 × 10.

Often the original photographers would file their negatives in wood cases with specially designed slots to hold the negatives upright and separate. If metal files of the appropriate design cannot be procured, or if the collection includes only a handful of plates, consideration might be given to reusing these wood boxes. Refinish the interior with a polyurethane varnish or other impermeable sealant, and allow to cure for several weeks before reinserting the negatives.

Before filing glass negatives, it will prove useful to make duplicate negatives. The simplest and least expensive method of doing this is to contact-print them onto Kodak SO-015 Direct Duplicating Film, following the processing instructions given with the film. This procedure gives you a same-sized *negative* without special chemicals or the need to make separate interpositives. When glass negatives are found to be broken while still cased in their original envelopes (check for this possibility before removing *any* newly acquired negative), immediately tape the envelope to a piece of flat board for temporary support, and set aside until more permanent conservation measures can be applied.

To effect a permanent repair to a cracked negative, slit the housing envelope on all three sides with a sharp knife and lift off the top piece. Blow off any dust particles. Cover the negative with a sheet of clean glass that is ¼ inch larger on all sides than the plate. Make a sandwich and turn the entire unit over so that the negative rests on the glass. It will be necessary to reverse the operation if the negative does not have the emulsion side up. Fit the two pieces of the negative tightly together on the glass support. Tape around all four edges with Scotch Magic Transparent 810 Tape, so that the negative is secured to the supporting glass sheet. At this stage, a duplicate negative eliminating the crack can be made on Kodak Professional Black and White Film 4168. Place an oversize piece of film, emulsion side up, on some kind of turntable* under an enlarger, put the glass negative emulsion-side

* You could use a "lazy Susan" condiment rack or an outmoded record turntable.

down on top of this film, cover with a sheet of frosted polyester sheeting as a diffuser, and secure with a sheet of heavy plate glass on top. Make the exposure while turning the turntable (the exact rate is not important), and process according to the instructions. The diffusing sheet and the relative change in light direction obscure the diffraction caused by the crack with minimal loss of image sharpness.

Negatives with multiple cracks probably require the attention of a photoconservator, because the processes required for their restoration are more complex. In these cases, an assessment must be made as to whether the negative is worth preserving.

NEGATIVE FILING SYSTEMS

The biggest challenge facing many photographers or photograph departments is not a technical but an organizational one. No matter how much care is taken in housing negatives, unless the files are arranged so that easy retrieval is possible, negatives might as well have been discarded. The importance of orderly filing cannot be overemphasized.

The widespread availability of personal computers makes them an ideal tool with which to index and track the location of negatives, especially for the small institution or individual photographer who already has one on site. We think that film logs should be created on or moved to disk whenever possible, because the greater ease of working with computers results in more orderly record-keeping.

Many filing systems have been suggested by experts, and variations on any of them could probably be made to meet your particular needs. The easiest methods all have in common the assignment of arbitrary code numbers to each lot of negatives so that the lots can be filed in a recognizable sequence. A code number should include the following information as a minimum: film type, format size, year, and lot number.

A typical code number might be *TX35.78139*. This would be deciphered as follows: *TX* refers to Tri-X film. Other alphabet codes might be *EK* for Ektachrome, *PL* for Plus-X, and so forth. The number *35* means that the roll is to be found with other 35mm negatives. Other format sizes can be designated as *45* for 4 × 5 film, *12* for 120 rolls, and *22* for 220 roll film. This tells the user which collection of film sizes has the desired negative. The number *78* indicates the year the negative was made or acquired. With the addition of an extra digit, the month can also be added, but this is necessary only when a large volume of work is being filed. The designation *139* refers to the lot number, and indicates that this was the 139th roll of 35mm film filed in 1978.

Another digit or initial can be added to indicate the individual photographer who made the negatives, although this information is of little direct relevance in a negative retrieval system. When several different collections of negatives are being filed, this extra digit can also be added in order to refer to a specific collection's source.

Assign access codes to negatives as soon as they are processed. In the case of roll films, the code will refer only to the roll; for easy reference, note an individual frame number along with the code on the back of prints made from that frame.

As negatives are accessed, keep a log in numerical sequence that notes the subject matter or client who ordered the particular job. When separate logs are kept for each format size, they closely correspond to the filing order of the negatives. The same code numbers should also appear on and be used as the basis for filing contact sheets from the negatives. The contact sheets can be used for quick reference to make certain that a particular number refers to the desired image. These can be three-hole punched and kept in binders convenient to the negative files.

Use of a log means that, given the approximate date of the original shooting, a quick glance tells you where the appropriate negative is stored. In case more specific information is necessary, you can keep a cross-index log arranged by subject matter or client. In the case of client headings, the cross-index log would work in this fashion. Clients are given separate pages in a 3-ring binder, and are arranged alphabetically by name. Under each name appears a list of code numbers, with the subject or job description noted to the right. Thus, when a particular client calls up and says a reprint is needed of the photo of "Widget X," the darkroom worker simply turns to the cross-index log under the customer name, scans down the list for the notation that reads "Widget X," and pulls the negative with the appropriate file number.

Negatives and contact sheets should never be discarded by a working photographer. Modern film materials do not require large amounts of space, and their long-term value may go unrecognized for many years. It may be necessary and prudent, however, to cull the negative file periodically for dead material, such as work for customers who have gone out of business, and to place this material in "dead storage" — that is, a separate filing area where little accessing is done. Further disposition of old negatives can often be arranged with local libraries, museums, or historical societies, particularly if the material includes such worthwhile material as record shots of area landscapes or portraits of important local figures.

STORAGE OF NITRATE-BASE FILMS

The flexible roll film responsible for George Eastman's success improved upon the glass negatives that preceded it in that it was light, sturdy, compact, and could be manufactured in large sheets and then rolled to make possible multiple exposures without reloading the camera. Unfortunately, it had a major defect. It was flammable — and not only was it highly incendiary when exposed to heat, but it can, without provocation, burst into flame all by itself. Material like this does not belong in the same location with valuable prints and negatives.

With this in mind, we must acknowledge that there do occur

situations making it highly desirable to keep nitrate-base negatives for periods up to several years. Unique images of historic or aesthetic worth often turn up on nitrate-base film, and immediate copying and disposal may not be practical for financial or other reasons. In these cases, negatives *in good condition* can be kept under proper storage conditions with appropriate safeguards. It should be considered an important objective to copy nitrate negatives while they are still in good condition.

All crinkled, buckled, or sticky nitrate-base negatives should be stored underwater in metal containers and disposed of immediately by burning in an open area.

The problem of handling nitrate film can be conveniently broken down into four areas: identification, assessment of condition, storage, and replacement.

All film negatives made prior to 1950 should be considered suspect unless they bear a manufacturer's label like Kodak's "Safety Film" designation.

Deteriorating nitrate film often has a distinctive acidic odor that should alert you immediately. A simple test to identify nitrate film can be performed with a ¼-inch-square piece clipped from a non-image area of the film. Place the dry film in a test tube filled with trichloroethylene, available from most chemical supply houses. Stopper the test tube and shake thoroughly. Nitrate-base film sinks to the bottom, while other kinds will float.

To determine the condition of identified nitrate-base film, you must be alert to the first signs of deterioration:

- ☐ Brittleness
- ☐ Yellowing of the base
- ☐ Buckling
- ☐ Stickiness

If the film is brittle enough to crack when folded, it has started to decompose.

Yellowing of the *base* (not the emulsion) is a further sign of developing problems. To check for this, cut a small strip from the film edge, moisten with water, and scrape off the emulsion. Holding the base against a piece of white paper gives an accurate idea of the degree of yellowing that has occurred.

As the film decomposes further, it will start to buckle, and the emulsion will become sticky, first in the presence of moisture from the breath and later from absorbed atmospheric moisture.

If the negatives are still in good condition, you can take steps to arrest deterioration. Good ventilation, cool temperatures, and low relative humidity are the most important considerations. Nitrate film in sealed enclosures at temperatures around 100°F can ignite spontaneously; keep in mind that such adverse conditions often occur in an uninsulated attic.

Store nitrate negatives in buffered paper envelopes to allow the gases of decomposition to escape. Plastic envelopes can allow a dangerous buildup of gases and accelerate the aging process. Place the negatives in fresh envelopes upon acquisition, and do not store this type of film in contact with other film. Small quantities can be placed in ventilated metal boxes or cabinets, in a room separate from other prints and negatives. Keep large quantities in a fireproof vault with air conditioning. In either case, the temperature should not exceed 70°F and the relative humidity should be no more than 45%. Lower temperatures are desirable. Smoking and other open flames like pilot lights should not be permitted in the vicinity of nitrate film.

Freezing will greatly retard the deterioration of the material, but several precautions need to be taken. The film must be stored in moisture-proof containers sealed shut in a dry atmosphere because of the sensitivity of the emulsion to moisture. Label prominently all sealed packages as to the nature of the contents, and this precaution should also be extended to the freezer itself. Place a prominent sign on its outside, warning that the refrigeration should not be turned off or otherwise allowed to fail, because a sealed and insulated freezer with a large cargo of nitrate film could easily become a veritable bomb.

Dispose of all nitrate-base film by open burning rather than by discarding with other waste. Follow safety precautions in accordance with local fire regulations; burn only small quantities at a time and do not use enclosed furnaces where a buildup of gases might occur.

The negatives can be copied by conventional methods of printing onto Kodak 4168 Film, either by enlargement or by contact printing. If there is dirt on the film, remove it with film cleaner rather than by water washing. Use a firesafe darkroom with good ventilation. Equipment should be grounded to avoid the chance of sparks or arcing, and — naturally — no smoking should be allowed during the film handling.

An alternative method of rescuing nitrate negatives is described by Eugene Ostroff of the Smithsonian Institution [4]. This procedure, in brief, consists of dissolving the nitrate backing away from the emulsion, which has been taped to a piece of glass. The emulsion is then cut away from the glass and transferred to a subbed sheet of Kodalith film, and coated with Acryloid b72 sealant. The method requires great skill, a fairly large investment in materials, and an extremely well-equipped laboratory. There seems, reading the article, to be a high probability of destroying the negative. Furthermore, it is difficult to tell whether Ostroff feels the method is widely applicable. At one point he states: "The method is relatively slow and the cost of labor — a specially qualified technician is needed — and materials is high, consequently the approach is not cost competitive with photocopying." Later he concludes: "The method can be applied on a mass production basis, and makes it practical to handle large quantities of material and thereby achieve high quality at low cost." We suggest that those who feel that this might be a potentially useful technique consult the article itself, where complete, illustrated instructions are given in a lucid fashion.

REFERENCES

1. Robert A. Weinstein and Larry Booth. *Collection, Use, and Care of Historical Photographs*. Nashville: American Association for State and Local History, 1977, pp. 193–194.
2. American National Standards Institute, IT9.2–1988. "Requirements for Photographic Filing Enclosures for Storing Processed Photographic Films, Plates and Papers." New York: American National Standards Institute, 1988.
3. Michael Lesey. *Wisconsin Death Trip*. New York: Pantheon, 1972.
4. Eugene Ostroff. "Rescuing Nitrate Negatives." *Museum News*. September–October 1978: 34–42.

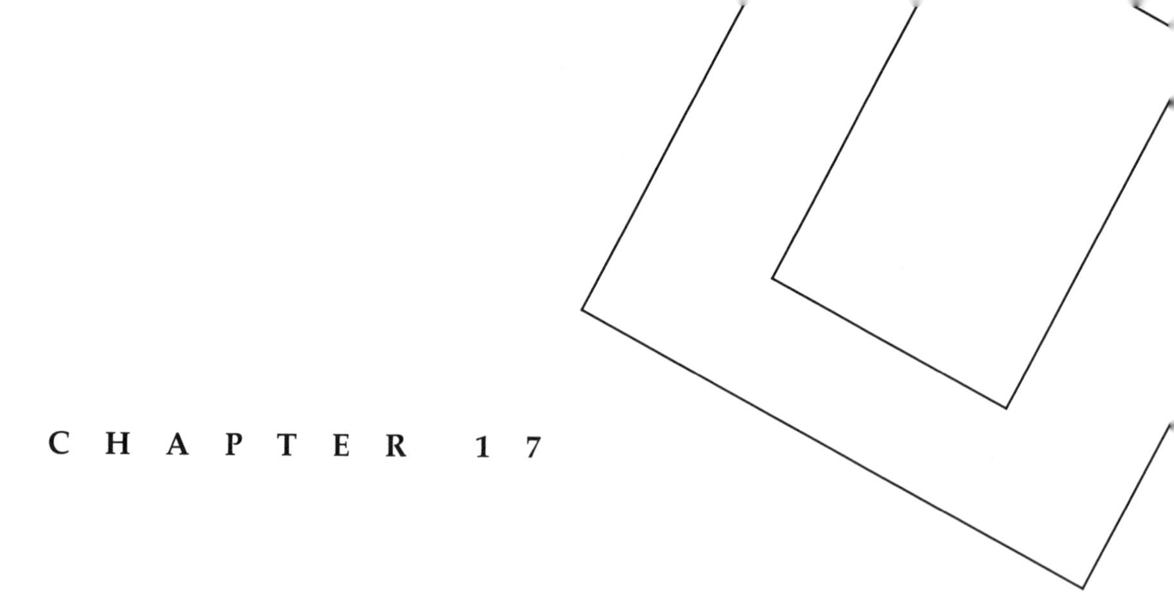

CHAPTER 17

Caring for Color

Color photographs present a number of problems for the curator and collector, not the least of which is the inherent instability of the medium. Given the current status of the technology, few if any color photographs can be called permanent. This is particularly true of modern media, most of which create images by the formation of dyes during development. Even if an economical method of making completely stable color photographs were to come onto the market tomorrow (which will not happen), our problems would not be solved, because we already have half a century of commercially produced color photographs. Many already reside in collections and archives.

Color print stability is not accomplished by modification of the processing, although variations in processing may affect stability. Producing a color image by modern photographic techniques demands strict adherence to the manufacturer's instructions. Even minor variations in the developing chemicals can cause major shifts in color. This means that there are no "archival processing" techniques for color prints in the same way that you can use toning, for example, to protect black and white prints. Thus, responsibility for preserving color imagery falls on the owner of the print. We cannot look to the darkroom for answers to preserving color photographs.

The difficulty of preserving color photographs does not mean that the cause is hopeless. We already have a number of techniques (discussed on the following pages) to conserve the imagery, and more are on the horizon. It may very well be that some type of computerized digitization — much like digital audio tape — will in the long run prove to be the best way to preserve a color image. Then, of course, we will probably run into some problem with *that* storage medium, like the redox pimples that currently afflict some microfilm copies of newspapers; however, let us leave that worry for another time, and talk about what we can do now.

IDENTIFYING COLOR PHOTOGRAPHS

To preserve color photographs we need to understand what they are and how they evolved. From the very first, photographers wanted to make pictures that recreated the beautiful hues and shades that came through the lens. This led to some fantastical efforts — like sensitizing paper and toning prints with juice squeezed from flower petals — methods that smacked more of alchemy than photoscience.

Various methods were developed to mimic real color. Most involved applying some kind of paint or gilt to an underlying monocromatic image. It was quite common, for example, for photographers to apply red paint to the cheeks of an ambrotype portrait and to daub a little gilt on a necklace to suggest gold. During the forties and fifties of this century, it became common to sepia-tone black and white studio portraits made on silver gelatin paper and then to add color to eyes, skin, hair, and clothing with light washes of transparent oil paints. The effect created this way, while sometimes garish to the knowledgeable eye, convinced many people that they had bought a genuine color photograph. These methods are not what we mean when we talk about color photography.

THE HISTORY OF COLOR PHOTO MATERIALS

Additive screen photos were made according to a theory first proposed in 1869 by the French scientist Louis Ducos du Hauron in his book *Photography in Color*. He suggested that it shoud be possible to make a color photograph with a single exposure (instead of through separations) if a single screen ruled with alternating lines in the primary colors (red, green, and blue) were used as a filter.

Joly Plate

The first application of Ducos du Hauron's idea was attempted by a Dublin physicist named Charles Jasper Joly. A process he commercially introduced in 1896 required binding a black and white film positive into perfect register with the ruled glass screen that had been used to make the original negative. The resulting transparency was backlit for viewing. Because the emulsions of the time were not yet fully panchromatic, this method produced results that were less than realistic.

Autochrome

The first widely available color medium was the Autochrome technique, introduced to the market in 1907 by the Lumiere brothers. It did not use ruled lines. Their process created a positive transparency on a glass base with an image composed of small points. These points were actually extremely fine grains of potato starch that had been dyed orange, green, and violet. The effect is quite lovely, looking rather like a pointillist painting. Autochromes in all standard glass plate sizes were made until 1932. Many are found in holders that provide for backlighting so the image can be viewed.

Similar Processes

A number of imitative processes for making color transparencies enjoyed modest success. Dufaycolor, first appearing in 1908, eventually became more popular than Autochrome in the 1920s because of greater speed and more accurate color. It was used until the 1940s. Finlay Colour, a technique from a British firm, was marketed intermittently from 1908 through the mid-1930s. It used an incorporated rectangular screen with a precise grid rather than the random pattern of Autochrome.

Tri-Color Carbro

A commercial process invented in 1905 by Thomas Manley and introduced to commercial use in 1919, tri-color carbro was used through World War II (and in some cases even later) for magazine and other forms of display advertising. Because they are made with pigments, carbro prints exhibit excellent color stability. They were never used to make mass market snapshots because all variations of the process are complex, difficult, and expensive. In tri-color carbro, a special camera made three separate black and white negatives of the same scene, each exposed through a different color filter. Each negative was then printed onto a specially prepared gelatin bromide paper, which was dyed the complementary color of the filter using a pigmented tissue. The print maker then used transfer paper to assemble the three color images onto a single sheet. There are several keys to identifying tri-color carbro prints: they are usually studio photographs with an evident commercial orientation; the colors are vivid, bright, and strong, and exhibit little sign of fading; the subjects are usually scenes like a touring car of a vintage not often seen in color pictures; the images are frequently larger than snapshot size, because the expense of printing color on a press could only be justified for a large advertisement.

Dye Transfer

The dye transfer technique introduced by Kodak in 1946 largely supplanted tri-color carbro. It also creates prints that are known to be relatively stable. The complexity of making prints by dye transfer means that it is used primarily by studio photographers working for publication or by serious artists. The technique involves making three gelatin matrices — one each for the cyan, magenta, and yellow dyes — usually from separation negatives, and frequently with film masks to control contrast. The gelatin matrices are then soaked in their respective dyes, rinsed, and squeegeed one at a time onto a sheet of gelatin-coated paper. The dyes transfer to this base from the gelatin matrix to create the print. Registration is critical. Such prints can frequently be recognized because of the ragged overlapping of the dyes around the edges of the print.

Chromogenic Materials

Processes that create colors directly by development are interchangeably called tripack color or chromogenic color. No matter what you call them, all consist of three layers of color-sensitive emulsions laid on top of one another; all three are exposed and developed simultaneously, and the development process creates or adds dyes to the emulsion. Chromogenic materials can be divided into three categories: the reversal process, the negative-positive process, and the dye diffusion transfer (or instant) process.

Reversal Processes

Reversal processes are so called because the materials actually go through two stages: first, development of a black and white negative image, and the second, its conversion ("reversal") into a color positive.

Kodak's Kodachrome, the first chromogenic reversal process, was brought on the market in 1935 as a movie film and the next year as a 35mm slide film. The film had to be developed by Kodak in three separate color developers. (Kodachrome prints, introduced in 1941, were made on direct positive paper, and are often brand and date stamped.)

Agfacolor slide film, released in Germany the same year as Kodachrome film, had its color couplers in the film and could be processed by the user. Ektachrome film, introduced in 1940, used the same technology as Agfacolor.

Negative-Positive Processes

Negative-positive processes yield color negatives that are then used to expose film or (more usually) paper that also has a negative-acting emulsion. The two negatives reverse one another to produce a positive image. The chemistry is more complex than reversal processing, but color negative-positive processing resembles the use of camera negatives to make prints in black and white photography. Of course, in addition to reversing the densities, colors are also reversed during printing to make their complementary colors.

Kodak introduced the first color negative-positive process in 1942 with Kodacolor. Because the color couplers were added during processing, these materials had to be developed by professional labs, and many early Kodacolor prints are labeled with the name of the process.

Since that time, many companies have produced negative-positive processes, including Agfacolor and Fujicolor. In the absence of documentation or identifying stamps, it is frequently difficult, if not impossible, to determine what kind of process was used in making a particular print, although some media are known to be less stable than others. When you acquire color photographs, find out from the source what materials were used to make them. Keep a record of it; this material may prove invaluable to a conservator.

A couple of words here about terminology in the photography market. The suffix *-chrome* on a film name indicates a film for making positive transparencies (e.g., Kodachrome). The suffix *-color* refers to a negative film for making prints (e.g., Fujicolor). The one exception (from the 1930s) is Agfacolor. *Type C* prints are those made from a negative; *Type R* are made from transparencies. These terms have passed into generic use, though they originally referred to a specific type of Kodak papers and chemicals manufactured between 1955 and 1959.

Dye Transfer Diffusion Processes

This rather elaborate terminology refers to "instant" photographs. The most common amateur versions are the Polaroid SX-70 prints and (until recently) Kodak's Instant Print Film PR 10. The SX-70 prints, introduced in 1972 in a 3⅛ × 3⅛ format, now come in a variety of sizes. This and the Kodak film, taken off the market because of a patent infringement ruling in the mid-1980s, both contain all the light-sensitive materials and processing chemicals in a plastic packet, one side of which shows the image.

The original instant color process introduced by Polaroid in 1965 is called Polacolor. After it is exposed in the camera and passes through a set of rollers in a film pack, the Polacolor paper negative is peeled away from the print and discarded. Polacolor prints, which have superior color stability compared to other materials, are now available primarily in a 4 × 5 format for use as a proofing material by professional photographers, and the vast majority are discarded after a quick viewing. Other sizes for amateur use have been largely supplanted by the SX-70 film.

The rise of instant photography marks the return of the singular image, the one-of-a-kind print like the daguerreotype. With these prints, there is only one original, a factor that the collector and curator must keep in mind when cataloguing and exhibiting.

THE HISTORY OF KODAK COLOR PRINT PAPERS [1]

Throughout the history of color photography, product changes by Eastman Kodak Company have paced the photofinishing industry that served the amateur market and much of the professional market. Changes in terminology and trademarks often reflected changes underlying the structure of the papers and chemistry being used. This historical survey of color papers from Kodak demonstrates many of the changes that have taken place in the broad field of popular photography.

1942–1954

During this period, Kodak photofinishing papers had to be processed at Kodak facilities in Rochester, New York, or Palo Alto, California.

Type I Kodacolor Paper (1942) This six-layer paper had a conventional-order yellow/magenta/cyan coating on a fiber base. Type I used high amounts of gelatin and silver compared to later papers; processing required seven solutions and four washes. Borders (and some of the image) tended to yellow because of a reaction caused by the magenta coupler. The magenta dye in particular faded quite rapidly in the light, and fading overall was markedly faster than in modern color materials. All of these early processes were quite unstable because the processing used the very fugitive ascorbic acid.

Type II Kodacolor Paper Type II resembled Type I but had an ultraviolet-absorbing layer.

Type III Kodacolor Paper This replaced the previous version, had improved magenta dye stability, and another ultraviolet-absorbing layer. The yellow layer was moved adjacent to the base to improve a problem with mottling.

1955–1967

During this period, independent labs could perform photofinishing for the first time using Kodak materials.

Kodak Color Print Material, Type C (1955) Type C resembled Kodacolor. Differences included a different yellow coupler, increased ultraviolet absorber, and a new color developing agent (CD-3) that gave improved dye stability.

Kodak Ektacolor Paper, Type 1384 (May 1958) Type 1384 replaced Type C for photofinishing ("snapshot") use, although Type C continued in production for professionals. Initially it required the seven-step P-122 process, which was gradually shortened and simplified over time. Type 1384 exhibited lower staining, improved cyan dark-keeping qualities, and absorber dyes for increased sharpness.

Kodak Ektacolor Paper, Type 1583 (1962) This was an improved version of Type 1384, with the difference that it had a new ultraviolet absorber to give lower staining and improved dye stability.

Kodak Ektacolor 20 Paper (1965) Ektacolor 20, introduced in conjunction with a new five-solution EKTAPRINT C process, could be processed more simply, quickly, and inexpensively. More important, the pre-exposure keeping qualities of the magenta layer had been improved to allow labs to give more accurate color, and changes in the shapes of the color curves gave a more achromatic gray scale (read: more lifelike color). New couplers gave better yellows, greens, and reds, and better stability for yellow and magenta dyes.

Kodak Ektacolor 20 Paper, Type 1870 (1967) Type 1870 incorporated another new stabilizing added to the magenta layer.

1968–1984

The major change during this period was the introduction of a resin-coated (RC) base to replace the fiber-base papers.

Kodak Ektacolor 20 RC Paper, Type 1822 (1968) This had the Type 1870 emulsion on a glossy RC base. Kodak felt that since face-side polyethylene absorbed ultraviolet radiation, it increased magenta stability.

Kodak Ektacolor 30 and 37 RC Papers (1971) These papers featured new emulsions, a new hardener, and a new magenta coupler; changes in sensitometry gave better neutral and color reproduction. Ektaprint 3 Chemicals, introduced at the same time, provided significant ecological benefits. This process cut wet time to 8 minutes and 3 solutions. Ektacolor 37, originally introduced for professional use, replaced Ektacolor 30 as demand for its textured finish increased.

Kodak continued to change the chemistry for these papers during the 1970s. Ektaprint 3HC (1975) improved efficiency, while the Ektaprint 2 Chemicals of 1976 reduced the processing stages to two, cut the amount of chemistry required, and offered various other efficiencies.

Kodak Ektacolor 74 RC Paper This had a new textured surface (E) that replaced the predominant photofinishing "silk" surface. Changes in the emulsion reduced staining, increased speed, and gave better tone reproduction.

Kodak Ektacolor 78 RC Paper (1979) Ektacolor 78 gave greater color saturation and improved contrast, in response to demand generated by competition in the amateur market.

Kodak Ektacolor 78 Paper, Type 2492 Type 2492 featured longer emulsion runs for the convenience of photofinishers, and offered greater resistance to fingerprints.

Kodak Ektacolor 78 Paper, Type 2524 (1982) Type 2524 had increased speed for higher productivity and optimum sharpness.

Kodak Ektacolor PLUS Paper (1984) A new cyan coupler improved dark-keeping quality; magenta light stability was increased by the use of a new stabilizer in processing. It also featured certain technical improvements such as an improved D-min and reciprocity characteristics. This paper can be processed in either Ektaprint 2 Chemicals or using Ektaprint 2 Developer Replenisher to minimize use of chemicals.

FACTORS THAT AFFECT THE STABILITY OF COLOR MATERIALS

Whether we are talking about negatives, transparencies, or prints, most color changes happen along three paths.

The first process — *dark fading* — occurs around the clock and proceeds inexorably 24 hours a day. Dark fading does not require any sort of radiant energy to drive it, and is a function entirely of ambient temperature and relative humidity. In other words, even if you put your photographs into a sealed, lightproof container they will undergo some change.

The second change that occurs — *light fading* — is the result of photochemical events in the emulsion that are induced by ambient light. In addition to the duration of the exposure, the intensity and the spectral quality of the light that strikes the emulsion determine the rate of change.

Both light fading and dark fading affect the magenta, cyan, and yellow dye layers of the film. What makes the changes they produce especially disturbing to the viewer is that the dyes fade at different rates, so that the photograph changes color.

The third process that alters a color photo — *staining* — is less of a problem than it used to be. On older color papers such as Kodacolor Type I, by-products of the magenta coupler left in the emulsion caused yellow staining in the border and highlight areas of the print. Contemporary emulsions typically have less problem with this. Little can be done to remedy this problem except copying or cold storage for prints that show susceptibility to it.

Dr. Donald Koop of Kodak has reported some interesting tests on the perception of fading.* He took a series of Ektacolor prints that had been subjected to prolonged irradiation to produce various degrees of fading. Koop showed them to a large population of nonprofessional viewers, who were asked to rate them for quality. He found that a surprisingly large number of people did not judge a print to be of unacceptable quality until it had undergone — as measured by objective criteria — significant fading. Dr. Koop suggests that subjective factors influence perception and are important criteria for evaluating the importance of fading. The average viewer, he said, might not notice so much that a print had faded if it showed the face of a loved one. We are not certain what to make of these findings, because it is difficult to factor them into the same equation with footcandles and dye densities. For Kodak to argue that the masses don't care about fading may be a case of taking lemons (dye fading) and making lemonade (public indifference). This may be an overly harsh assessment, however, and we should keep in mind Dr. Koop's central point: that as a practical matter fading may not be so severe a problem as testing methods suggest.

KEEPING QUALITIES OF CONTEMPORARY MATERIALS

In the first edition of this book, we presented a list of projected life spans for color materials based on the published research of Henry Wilhelm, an independent researcher and expert in the field of photographic conservation.

We did this with some misgivings. First, this research had not been widely duplicated. Second, we had no way of telling when

* D.A. Koop, "A Relationship Between Fading and Perceived Quality of Color Prints." *Second International Symposium: The Stability and Preservation of Photographic Images* (Society of Photographic Scientists and Engineers); pp. 335–349.

RESISTANCE TO DARK FADING

Most Stable
Cibachrome
Kodak Dye Transfer
Fuji Dyecolor
Polaroid Polacolor
Kodak Kodachrome

Excellent Stability
Kodak Ektachrome (E-6)

Good Stability
Kodak Ektachrome (E-4)

Moderate/Poor Stability
Kodak Ektacolor
Kodak Vericolor
Kodak Ektachrome (E-3)
Eastman Color Negative

and how ingredients in the materials listed might be altered by the manufacturers; in this case we might unfairly stigmatize a worthy product or mislead you, our reader. Third, we cannot say with confidence that a particular photograph will last more than fifty years in dark storage. After all, color photography has barely been around for fifty years.

The situation today has not much improved, and there remains a paucity of reliable data. Some comparative testing goes on, but for various reasons little of it is published. Manufacturers test the competition's film for proprietary purposes; other testing, by ANSI, for example, is done for reasons that have little or nothing to do directly with the individual user. (The ANSI committee writing a color stability standard needs to determine how the available color films actually keep in order to write the standard, but they are not a consumer testing organization.)

On the other hand, it seems unfair of us to proclaim that we know of extreme differences in the stability of color materials but then to coyly shy away from saying what they are. We do so, however, with reluctance, subject to the same reservations expressed above, and we would ask you to note that this comparison relates primarily to *dark fading*, and tells you only about the emulsion; other factors — such as the RC base — may have equally important effects on the real-life keeping qualities.

Again, these comparisons are compiled from research presented by Henry Wilhelm in various forums (although of course we take full responsibility for its format and presentation here). Comparative data are not yet available about Kodak's Ektar 25, Ektar 1000, or Fuji Photo Film Company's Reala.

CONSERVATION TECHNIQUES FOR COLOR MATERIALS

There are quite a number of things that can be done to extend the life of color photographs. Many of the procedures we recommend here are the same ones that apply to most other photographs, but there are some differences when it comes to color.

Light Management

Irradiation by both visible and ultraviolet light is the most important cause of light fading of color photographs.

☐ *Ultraviolet Filtering*
Light sources such as windows and fluorescent lights can be filtered with materials available from suppliers of archival materials. Framed prints can be protected with ultraviolet-filtering glazing (i.e., UF-3, OP-1).

☐ *Print Rotation*
Limiting the display time of a print by rotating it with other photographs in the collection not only reduces its exposure to light but adds to the variety of the display.

☐ *Dark Storage*
Prints not on display are best kept in light-tight enclosures or storage boxes.

Temperature and Humidity Management

Humidity makes possible the chemical and photochemical changes in fading; high temperatures drive them.

☐ *Air Conditioning*
Air conditioning not only controls humidity and temperature, but with suitable filtration equipment it can remove damaging industrial pollutants.

☐ *Cold Storage*
Freezing will slow or stop most of the chemical changes involved in fading. Materials to be frozen can be put in a pouch made of foil/polyethylene/paper layers, which is then heat-sealed according to the manufacturer's instructions. All materials in the pouch have to be clearly indexed; the pouch must have no pinholes that could let in moisture; the contents must be allowed to come to room temperature before it is opened.

This technique is expensive, time-consuming, and labor intensive. Consequently, many institutions are using freezing only for materials like negatives that require limited access, and for other uses are moving toward cool storage.

☐ *Cool Storage*
In this technique, a storage area adjacent to an air-conditioned work area is kept quite cool (50°F–60°F), and the relative humidity in both rooms is monitored constantly to make sure that prints being taken from storage do not cross a dew point that would cause condensation.

Copying

Copying, already covered in another chapter, is an excellent way of preserving the information in an image, and adequate as a way of recording the appearance of a photograph. Of course, it does nothing to protect the original.

☐ *Separations*
Because they are made on silver emulsion black and white film, continuous tone separations — the kind used for making dye transfer prints — are a precise and archivally acceptable way of making copies of a photo. These separations are not the same as the ones made for printing on an offset press.

☐ *Black and White*
Straight black and white copies of color photos will preserve the image, and are often quite suitable in situations where there are no aesthetic considerations.

☐ *Transparencies*
Copying onto Kodachrome or Ektachrome (E-6) transparency film (preferably large format) has a number of advantages. The material is relatively stable, simple to use, and easy to view for retrieval.

HOW TO MAKE LONG-LASTING COLOR PHOTOGRAPHS

The right choice of materials at the start can go a long way toward making color photographs more durable. The photographer who intends his or her work to last should make every effort to pick the most durable media possible. Of course, when you are acquiring photographs for a collection, that choice is not yours to make, but dupes and copies can still be made on durable materials. For an integrated approach, we suggest the following:

Slides and Transparencies

☐ Kodak Kodachrome 25 (for dark storage)

☐ Kodak Ektachrome with E-6 processing (for projection and large-format work)

Kodachrome, while slow, makes an excellent general-purpose film because of its vivid but realistic colors. It has good dark-keeping qualities. When your slides will be projected, Ektachrome resists the fading created by the bright light of the projector.

Prints

☐ Ilford Cibachrome Glossy Print Materials (CPS)

☐ Dye transfer prints

The glossy Cibachrome, with its polyester base, has much better keeping qualities than the RC-based matte finish Cibachrome. Prints can be made directly from slides with this material. For more information, see "The Stability of Cibachrome Prints."

Kodak's standard recommendation for stable color — the dye transfer print — gives excellent color saturation without being garish or too contrasty.

RECOMMENDED STORAGE CONDITIONS FOR COLOR PHOTOGRAPHS

Cool	50°F to 60°F with no crossing of a dew point during handling; minimal temperature cycling
Dry	30% to 40% RH
Dark	Light-tight enclosure. Limited display time. All light sources UV filtered, and light levels kept to recommended minimums.

Instant Prints

☐ Polaroid Polacolor-2

This material resists both light and dark fading. For some people, the fact that Polacolor requires a large format camera is a limitation.

Color Negatives

☐ Duplicates on Kodak Ektachrome with E-6 processing and appropriate filtration

Making a dupe allows the original film to be frozen without having to be thawed repeatedly for printing. An Ektachrome "negative" will keep for many years in conditions of controlled temperature and humidity.

THE STABILITY OF CIBACHROME PRINTS

Here we look at the use of Cibachrome Glossy Print Material (CPS) for making long-lasting color prints. (See the chapter on mounting for the special mounting requirements of this material.) Cibachrome is unique, because — unlike the chromogenic method of forming dyes from color couplers during processing — it has stable azo dyes incorporated into the emulsion when it is manufactured.

Cibachrome is what is known technically as a silver dye bleach material. This is a direct positive process for making prints from slides and transparencies onto a material that incorporate azo dyes. The dyes are bleached in proportion to the amount of light that reaches them during the printing exposure. Silver compounds and the other chemical by-products are taken out by a final washing and fixing step.

Ilford makes an extensive line of Cibachrome materials, including translucent and transparent display films. These are used primarily for trade show and advertising displays; print materials such as we are interested in are the Cibachrome II and Cibachrome-A II. These have a high gloss surface with the emulsion coated onto an opaque white polyester base 7 mils (.007 inch) thick. To prevent curling, the backs of these materials are also coated with a layer of gelatin that reacts with humidity at the same rate as the emulsion on the front. A low contrast version is available.

The high gloss Cibachrome materials are the most dimensionally stable color print materials currently on the market. The same emulsion with a (semimatte) "Pearl Surface" is also available on a resin-coated (RC) fiber base, but this material does not exhibit the same keeping qualities and dimensional stability as the polyester-based Cibachrome.

In an excellent publication on mounting, displaying, and projected life expectancy [2], the manufacturer states that "one of the most important factors in maintaining Cibachrome stability is to offer protection from high humidity." Conservation of Cibachrome prints should be planned with this consideration foremost.

Storage and Handling Recommendations

Ilford specifies that for archival storage, Cibachrome prints should be kept at 30%–50% RH, with conditions never exceeding 60% RH and not cycling daily between extremes. Storage temperatures should be 50°F–70°F not cycling more than 7°F daily and not to exceed 86°F.

To protect against fingerprints, you must always handle Cibachrome prints with cotton gloves. The surface shows the oil of fingerprints in a particularly unsightly fashion, and long-term damage will result. Remove fingerprints with film cleaner and a soft cotton glove. Never use paper towels, tissues, or antistatic cloths. In severe cases, you can rewash the print, but because the emulsion is very sensitive to abrasion when wet, this creates the risk of removing part of the image.

LAMINATING CIBACHROME PRINTS

Ilford indicates that, unlike chromogenic color materials, Cibachrome shows increased (3 to 10 time more) resistance to fading when laminated with polyester sheets. Lamination protects it against ultraviolet radiation and humidity. This suggests that anyone making a Cibachrome print would be well advised to follow Ilford's directions for laminating as a standard procedure when creating the final print.

Laminating a photograph is a radical step, and we confess that we felt some misgivings when we learned about this advice. It would preclude any future chemical restoration of the photograph. Two factors, however, seem relevant. One is that no restoration techniques — aside from copying — are currently available for color photographs, nor do we know of any that are likely to be developed. Color images do not respond to traditional techniques of paper conservation. The other consideration is that Cibachrome prints do not react well to wetting after processing. Ilford states that attempts to rewash or refix only a month after the print is made are likely to lead to the emulsion peeling. This alone would make any restoration difficult or impossible.

What are the potential dangers of lamination that we can identify at present? The important ones seem to be yellowing and discoloration, brittleness, and separation of print and laminate.

Yellowing Definitive research has not been done on this subject. Prints and laminating material sometimes yellow when exposed to high levels of illumination. This condition may be aggravated by high humidity and temperatures.

Brittleness Polyester and polypropylene are known to have excellent keeping properties when not exposed to stressful conditions. Irradiation may cause brittleness, and the adhesive could be suspect in this case. Again, relatively little is known.

Separation (delamination) The major precaution to be taken here is proper mounting procedures, especially including predrying the print and using adhesives that have been stored according to the manufacturer's direction and applied before their expiration date.

Cold Laminating Films

Laminating films, which are clear plastic supplied in rolls with an adhesive on one side, must be applied using professional cold laminating machines. Ilford states that: "Polyester and polypropylene films provide the best overall moisture and light fading protection and the best aging characteristics of available laminates." Polyvinyl chloride and acrylic sheeting are not appropriate when you want the print to last a long time.

When using laminating films, the edges must be wrapped with the film to seal against moisture. The film does not protect completely against humidity, and the finished print should be protected against conditions that might cause condensation.

Recommended cold laminating polyester films with UV protection include:

S-75 UV HI-GLOSS	FILMOLUX
PermaGard ILP-7000 (Gloss)	MACtac Permacolor
Printshield UV Gloss	Seal
Glossy Overlam	CODA

To prevent separation, prints should be predried by being held for several days in a room at 35%–45% RH, if possible, and never mounted at levels above 55% RH. Humidity levels can be tested with a device such as a sling psychrometer.

In conjunction with lamination, Ilford suggests that the print should be cold mounted with a film adhesive onto a suitable nontextured substrate such as aluminum sheeting.

Heat Laminating

With this technique the print is contained between two layers of polyester film, and a ⅛-inch border is left on all sides to keep out humidity. Again, the process requires professional equipment with rollers or heated shoes designed specifically for this use.

Films for heat laminating are:

Durafilm	Graphic Lamination, Inc.
Duraseal	Meteor Photo Company
Thermashield	Seal

The same considerations about humidity apply, with the added problem that an overly moist print may exhibit red spotting when heated for lamination.

ALTERNATIVE TECHNIQUES FOR PROTECTING CIBACHROME

There are two alternative techniques also recommended by the manufacturer for protecting Cibachrome print materials. These are intended primarily for outdoor display, require professional expertise, and will be of minimal interest for the individual photographer or collector.

Embedment

Several companies can embed Cibachrome materials in liquid plastic reinforced with fiberglass. This gives exceptional protection from light fading, ultraviolet damage, moisture, and pollutants. On the down side, it mutes the colors and creates a textured surface. Suppliers of this service are Key Color Lab (Canada), Pannier Graphics, and Watco ID Systems. See Appendix A for addresses.

Face Mounting

Two companies offer kits with a special clear adhesive for mounting Cibachromes to polycarbonates (Lexan) and glass. These require a laminating machine with a roller system, but they do not diminish the colors or alter the surface appearance. Because the adhesive contains an ultraviolet-filtering agent, Ilford claims this technique offers excellent protection. For more information, contact Fotographika, Inc., Clear Mount Industries.

PROJECTED FADING RATES FOR CIBACHROME

Ilford is the only manufacturer of color materials to publish projected fading rates and life expectancies. We admit that, as the manufacturer of relatively stable materials, this is to Ilford's advantage, but other companies would do well to follow suit.

In trials that gave the results in which we are most interested, Ilford tested gray patches with a density of 1.0 to determine the time that they would take to show a 10% loss of density. (In most prints this degree of change can only be detected with a reference print for comparison.) Conditions were moderate: 45%–55% RH, and maximum temperatures of 86°F. Light levels corresponded to those in an office environment; prints were protected from direct sunlight by hanging at least 7 feet from a window, with a higher level in the same environment caused by spotlights on the ceiling above. For the test, it was assumed that lights are on for 12 hours a day. (Light levels could be kept lower in museum conditions, and prints are not necessarily exposed to light for 12 hours every day.)

		Theoretical Value in Years	
Cibachrome Material	*Protection*	*Normal (500 lux)*	*Medium (1000 lux)*
Glossy Print	Unprotected	3>	2
Glossy Print	Glass	5	3
Glossy Print	Polyester Laminating	14	7
Glossy Print	Embedment	27	14
Glossy Print	Face Mounting	27	14
RC Pearl Paper	Unprotected	2>	1>
RC Pearl Paper	Glass	4	2>

These figures are not really happy news. When you read them carefully, they indicate a relatively short life expectancy for color photographs on display.

However, more optimistically, 500 lux is about two times the optimal light level for display in a museum. So, if you laminated a print, kept illumination levels down, displayed it for only a third of the time, and kept it in dark storage the rest of the time, you might project a period of 105 years before major changes took place. That's not exactly as good as an oil painting, but it's not writing on sand, either. (These are theoretical values, primarily relevant to fading; other factors do and will influence actual print life.)

STORAGE OF COLOR SLIDES

Color slides are a wonderful way to take pictures. Because they use transmitted light for viewing, slides have a longer contrast range than do prints. That means that they are more brilliant. When they go up on the screen, they are so much larger than prints that the effect can be overpowering. Grain is nearly eliminated because slide dyes do not clump like silver when forming the image. And the color of Kodachrome has become synonymous with "vivid."

By now most of our photographic record probably consists of slides, so the preservation of slides has become an important part of photographic conservation. Back in the 1930s the combination of Kodachrome film and cheap Argus 35mm cameras swept the world of amateur photography like a storm. Probably nothing like it had been seen since George Eastman promised members of the new American middle class that they could become photographers with the pressure of a single digit.

The slide revolution spread, and we now have an entire audiovisual industry that daily cranks out slides by the millions — literally. Maintaining slide libraries has become big business. We would like to think that this will mean increased demand for more durable slide-making processes, and corresponding benefits to the noncommercial user.

STORAGE METHODS FOR SLIDES

The most convenient, least expensive method to store small collections of slides is to put them into binder pages free of polyvinyl chloride (PVC). These, and more elaborate professional systems working on the same principles, give adequate protection while allowing rapid access. The ultimate in "dead storage" is to freeze the photographs in sealed bags.

Slides can be left in the boxes as they come back from the processor and tossed into a drawer. When the boxes are polystyrene, this is a viable option, but storage becomes cumbersome. Storage in Kodak-type Carousel trays will enhance the probability of survival, especially when an organized slide show is kept this way. These circular trays keep the slides separated so air circulates freely, give physical protection, and provide a crude method of filing. But one major benefit of slides is that they are compact, and 80 slides in a carousel take up the space of hundreds stored by other methods. Retrieving individual slides will be difficult, in that they cannot be seen without removal one by one from the tray.

Binder pages offer an inexpensive step forward from the tray in both quality and convenience. But watch what you buy! Some binder pages on the market are made of polyvinyl chloride (PVC), a thick, limp plastic containing a heavy dose of plasticizer that breaks down to form hydrochloric acid. Even if you are not an experienced chemist, you know that you do not want hydrochloric acid on your slides. Further, its flexibility lets the PVC slump against the surface of the slides, where it sticks. The result is local glazing and ferrotyping, splotchy patches similar to those caused on prints contacting glass in the presence of moisture. (To remove these splotches, take the slide out of its mount, wash in distilled water, dip in a dilute solution of Photo-Flo, and hang to dry in a dust-free environment. When dry, remount.)

Binder pages that are not made from this chlorinated PVC come from Franklin Distributors. Franklin's Saf-T-Stor pages are made out of *semirigid polypropylene.* These molded pages measure

9½ × 11¼ inches and will hold twenty 2 × 2 cardboard, metal, or glass mounted slides. The protective backing allows air to circulate, and it diffuses the light for easy viewing even without a special illuminator. The slides are open in front for unobstructed viewing. The molded construction allows the pages to nest one on top of the other to save space and to lock out dust and insect intruders.

Saf-T-Stor pages have multi-ring holes punched along one edge so that they can be put into nearly any kind of binder. If you decide to use binders, again, do not get the chintzy kind covered with plastic: most contain PVC. Instead, choose a clothbound one, preferably lined with Permalife or with Tyvek. Buy one with D-rings to prevent distorting the pages because of pressure. Make sure that you never overload the binder, and that you keep it in a slip case to control dust.

Saf-T-Store pages are good for long-term storage because they allow air circulation around the slides to prevent fungus growth. For slides that get more frequent handling, Light Impressions makes binder pages of *flexible polypropylene* under the trademark Slide Guard. These cover both sides of the slide and are not rigid.

Safe polyethylene binder pages are also available. Print-File and Vue All, for example, make *polyethylene* pages with twenty 2 × 2 pockets on each. These pockets protect the front of the slide from finger contact better than Saf-T-Stor pages. There is a tradeoff: the flexible polyethylene can slump against the slide surface and create glazed splotches.

Binder pages can be converted to hanging files with attachments that clip along the long side or that can be inserted into pockets on the long side (Figure 17.1). These allow the pages to be hung in Pendaflex-style drawer frames available at stationery stores. All you need to make a complete storage system is the addition of a metal file cabinet with baked enamel finish.

Bulk Storage

It is often necessary to store large quantities of slides. There are a number of options for doing this that range from very economical to extremely high-quality storage enclosures. Choose an enclosure

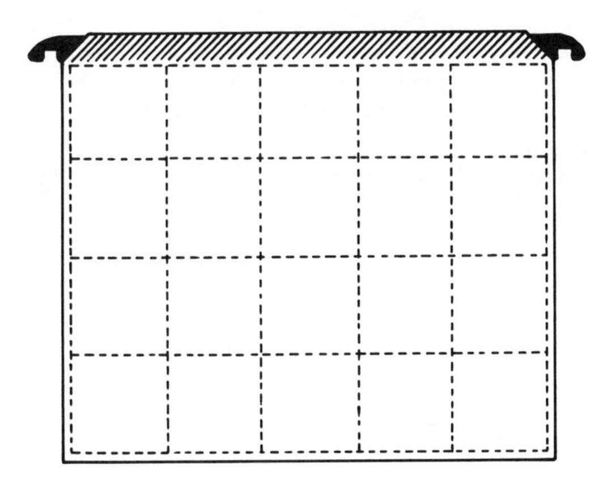

Figure 17.1
A binder page converted for a hanging file.

that meets all the standard criteria for archival housing of color materials.

- ☐ No wood, varnish, or shellac
- ☐ Baked enamel finish on steel preferred
- ☐ Paper materials of conservation grade (neutral pH, nonbuffered, lignin-free)
- ☐ No PVC; polystyrene, polyethylene, polypropylene acceptable

Paper boxes: the simplest, least expensive enclosures for bulk storage of slides are paperboard boxes available from archival supply houses. These hold slides stacked on end in rows, and each file can be inserted into a drop-front box specifically designed to accommodate a series of such files (Figure 17.2).

Metal and plastic boxes: these can be purchased with built-in trays for filing slides in groups, and hold between 600 and 1200 cardboard-mounted slides. Some are moisture resistant (Figure 17.3).

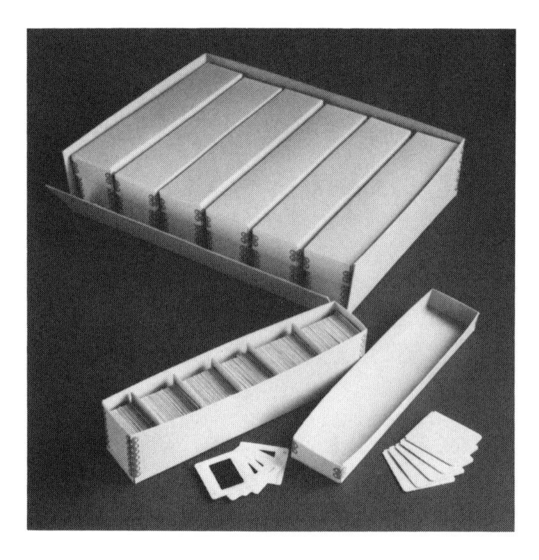

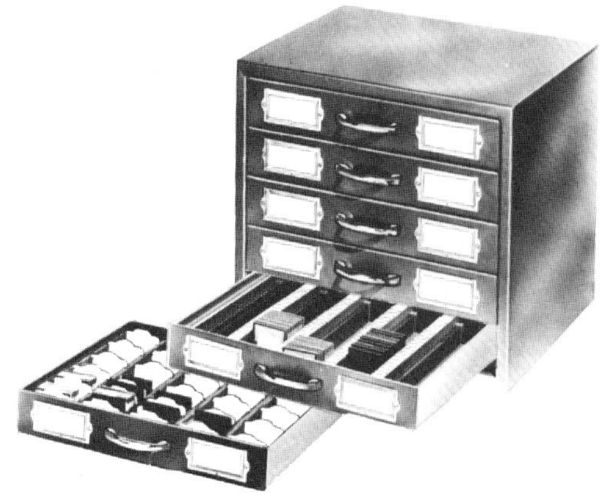

Figure 17.2 (upper left)
High-density slide storage in archival boxes.

Figure 17.3 (upper right)
Plastic and metal boxes for group storage.

Figure 17.4 (left)
Metal cabinets with drawers for individual or group storage.

Metal files and cabinets: these come in quite a variety of configurations for bulk storage. A very simple one of steel has removable plastic drawers that will hold 200 cardboard mounts. The ultimate in plain vanilla bulk storage is the Neumade metal cabinets, which are stackable two-, three-, or five-drawer units. They are sturdy, rigid, and have numbered position slots for indexing (Figure 17.4). The Luxor Slide Storage Units are much more luxurious — stackable drawers on rollers with black polystyrene trays for holding individual slides upright. Luxor also makes a slide bank filing and viewing center that allows you to pull out large pages of slides in front of a vertical light box. All these units are made of steel with a baked enamel finish.

Binder pages and baked enamel drawer files together offer a satisfactory solution to most slide storage problems. The inherent tendency of color slides to fade during long-term storage, however, cannot be attested simply by good physical protection with inert materials. For this, the slides must be kept cold and dry.

For a long time, Kodak literature suggested that the only possible way to overcome the long-term instability of color materials was either to make color separations or to freeze the originals.* Because separations are made on black and white film that does not fade when correctly processed, Kodak maintained that this would protect the image. We know of no organization or individual seriously contemplating this as a method of storing any large quantities of still color photographs.

Freezing holds out a somewhat greater chance of success. The recommended method uses heat-sealed bags with three laminated layers of paper, foil, and polyethylene. The paper serves as a structural support, the metal foil provides a moisture barrier, and the polyethylene interior, which stays flexible when cold, protects the photographs inside from physical damage by the foil. The larger bags each take several slide pages.

Place material to be frozen in the bags in a dry environment, with around 25% to 30% relative humidity. Seal the bags with a household iron, at the "cotton" setting, run along the edge. When removing from storage, a necessary precaution is that the entire contents of the bag should be allowed to reach room temperature before opening. Along with identification of the contents on the outside of the package, add a warning notice to this effect. You will probably find that the waiting time to reach room temperature will vary from 2 to 4 hours. (When color prints or large transparencies are housed in these bags, they need interleaving to prevent them from sticking together.)

Freezing may not be the ideal solution to the problem of keeping your color materials. Immediate access is impossible, and the bags,

---

* Much corporate literature on color stability reads as if it were written by lawyers concerned about product liability, and then put into lay language by the public relations department. This curious hybrid style helps neither the consumer nor the conservator.

once opened, cannot be reused. Any defects or pinholes in the moisture barrier can let moisture inside. In freezers with uncontrolled humidity levels, ice and even slime on the outside of the bags can make identification difficult and can cause the paper housing to deteriorate. These are all factors to consider when planning long-term storage.

The best way to handle slides would be the creation of a humidity-controlled system of freezing. Though some facilities like this have been built by federal agencies, the capital investment required is outside the realm of most individuals or small institutions.

PROJECTION AND HANDLING OF SLIDES

Projection significantly reduces a slide's probable life span. The heat and intense light emitted by a projector lamp give the slide a double whammy that accelerates fading. The problem can be worse for glass-mounted slides. How long does a slide last when it is projected? Kodak states that

> For most viewing purposes, pictorial slides made on properly processed Kodak Color Films will be acceptable through 3 to 4 hours of total projection time. This is true when the slides are used in an Ektagraphic or Carousel Slide Projector that is equipped with a tungsten-filament lamp and has an unrestricted air circulation, even if the projector is operating with the selector switch set at HIGH. With slides containing noncritical pictorial content, such as line drawings and charts, the viewing life of the slide may be considerably longer than the time stated here [3].

This assumes optimum storage between projections, short projection intervals, and careful processing. Kodak does not state the criteria it used to decide what is acceptable "for most viewing purposes," and to us its estimate seems a bit optimistic.

There are a number of steps that you can take to extend the viewing life of a slide.

☐ *First, do not project the original.*
Instead, for projection use duplicate slides made from a master, and keep the original in permanent storage under optimum conditions. This practice is commonly followed by large institutions that maintain "active" and "master" files, but there is no reason it cannot be adopted by smaller organizations and individuals. Duplicates can be made with an inexpensive bellows unit for any of the common "systems" 35mm cameras, or duplicates can be made by photoprocessors.

Kodachrome slides last longer in storage, but Ektachrome slides hold up better under projection, so it is a good policy to shoot masters on Kodachrome and have the dupes made on Ektachrome. When the duplicate starts to fade, replace it with another copy made from the master slide on file. As an alternative to duplicating, consider shooting at least two copies of each slide when possible. Use one for a file copy and the other for projection.

- ☐ *Limit projection time to short intervals.*
  Projection for 5 minutes straight causes more rapid fading than projection for 10 30-second intervals. Twelve straight minutes of projection with a 250-watt bulb will seriously fade a slide. Published studies disagree on the best maximum time, but recommendations vary from 1 full minute to 15 seconds; 15 to 30 seconds allows plenty of time for viewing. If you are just showing shots of the kids in front of the Winnebago, audience reaction might be a good reason to make it even shorter.
- ☐ *Use the low-intensity setting for the projector bulb.*
  This reduces the amount of light and heat that hits the slide.
- ☐ *Use the heat-absorbing glass.*
  Most modern projectors have a heat-absorbing glass shield between the lamp and the slide gate. Check that your projector has one, and never remove it. Make sure that the air circulation intake for the projector fan is not obstructed, because blockage leads to rapid buildup of heat inside the projector.

Safe handling procedures go along with good projection practices. The most obvious one, of course, is to handle slides only by the edges of the mounts. A fingerprint in the center of the film leaves acid oils that stain the picture and that can provide a feast for fungi. Avoid getting dust and grit on the film. Hypo dust from darkrooms is particularly bad, and causes the same damage on color slides as it does on black and white film. It is a good idea to use the thin cotton gloves sold for just this purpose in camera stores when sorting and mounting slides, but do not be guilty of false economy here: when the gloves start to fray, replace them, or else the lint and dust particles they generate can make the problem worse. You can use one of the specially constructed light tables or viewing systems while sorting. Whatever surface the slide sits on, make sure that it is clean and free of sharp objects that might puncture the film. When someone who seems unfamiliar with slide handling participates in editing, take the time to instruct that person in the proper way to hold slides.

Because some older color processes were even more unstable than are modern materials, slides on these stocks may have faded despite the best storage and handling techniques. Kodak has two films to correct the particular kind of color fading known to occur in dark storage. They are Ektachrome Duplicating Film 6121 and Ektachrome Slide Duplicating Film 5071. Instructions may be obtained by writing Eastman Kodak, Customer Technical Services, Rochester, NY 14650. These films have a color balance designed to eliminate the need for elaborate filtration to bring back the original hues.

GLASS MOUNTS

While glass mounts do not have a proper place in long-term slide conservation, they can be useful when strict control cannot be maintained over slide handling. Obvious examples would be when a library or audiovisual department keeps a set of slide shows for lending to groups and classes, or when commercial presentations must be screened by inexperienced projectionists.

Glass slide mounts are made by Gepe, Quikpoint, Emde, Lindia, Weiss, and Agfa. Quite a number of collections already contain slides mounted between pieces of glass. Unlike the open paper mounts that seal with adhesives, most glass mounts snap together and hold the film in place by little closures around the edges of their masks. A very thin sheet of glass comes already joined to front and back parts of the mount. The material is usually plastic or aluminum, with plastic more common.

Possible benefits claimed for glass mounting are increased film flatness during projection, protection against handling dangers like fingerprints, and buffering from changes in environmental humidity. Mounting in glass represents an additional expense in preparation of slides, and it seems worthwhile to ask whether it usually justifies the cost. We think not.

The heat to which an open-faced slide is suddenly exposed during projection causes it to "pop" from a flat to a curved shape. This means that the center or the edges of the image suddenly go soft after a few seconds of projection. There are a number of ways to deal with this problem other than glass mounts. Kodak Ektanar "C" series projection lenses for Ektagraphic slide projectors compensate for the film plane curvature as part of their optics. Preheating slots in projectors and automatic self-focusing devices further help eliminate focus problems caused by slide popping.

Scratches and fingerprints can be prevented by handling with lintless cotton gloves and by restricting access to only those people who show their willingness to handle the slides by the edges. Additional protection against fingerprints can be provided by Kodak Film Lacquer (this has to be done before mounting) or by acetate slide sleeves.

Glass mounts sealed around the edges with Mylar silver tape do buffer against fluctuations in humidity, but equal protection can be provided by air conditioning or storage in metal cabinets with desiccants.

Recent research indicates significant reasons to avoid glass mounts. If slide film is mounted or stored in a moderately humid environment, heat from the projector lamp actually causes steam to form from the trapped moisture and leads to fogging of the screened picture. In conditions of extreme heat, the film can actually melt to the glass. Permanent buckling and warping of the slide inside the glass is not uncommon, and negates the supposed film-flattening benefit. Greasy precipitates on the glass, whether from film plasticizers or silicon-based wetting agents, have been reported in conjunction with warping inside glass mounts.* And, finally, Newton's rings compound the problem. These amoeboid-shaped splotches with rainbow hues appear somewhere in the middle of the picture; even "anti-Newton's ring" glass mounts are not always successful in combating their appearance.

An alternative to glass mounting is to put each slide inside an individual sleeve. Kimac Co., Saf-T-Stor, and Light Impressions all sell these inexpensive items through audiovisual suppliers. Slides in these sleeves get protected against some handling abuse, and they can still be inserted into the binder pages.

FUNGUS, LACQUER, AND SLIDE CLEANING

Dust and fingerprints can be removed from slide surfaces by conservative treatment methods. Either cans of compressed gas like Dust-Off or clean, dry sable brushes will take off dust. Fingerprints and oily smudges come off when the slide surface is lightly swabbed with a Q-tip dipped in Kodak Film Cleaner.

Fungus growth on slides, both color and black and white, constitutes a serious problem in humid climates where average relative humidity equals or exceeds 60%. The reason that the problem is so severe is that in addition to making opaque splotches on the slides, fungus makes the film emulsion water soluble. Misguided efforts to wash off the fungus can actually remove parts of the image and leave bare, clear patches on the film. Never use any type of cleaning solution that contains water when fungus growth is suspected. If at some time you have successfully taken fungus off a slide — say, with film cleaner — pencil that fact on the slide mount, because the emulsion stays water soluble after treatment even though the fungus is gone.

The best antifungal treatment is prevention. For small collections in humid climes, one answer is to store slides in a sealed metal box containing a desiccant like silica gel. Use binder pages to allow air circulation around the slides. Seal the box corners with solder, and install a plastic gasket. Replace or bake the silica gel in an oven when it gets moisture saturated. From time to time open the box in a dry, air-conditioned room to allow an exchange of air.

Lacquer is an additional way to inhibit damage caused by fungus. Because the fungus has difficulty penetrating the lacquer film, it cannot convert the emulsion to water solubility. New slides can be coated with Kodak Film Lacquer as part of the mounting process when you mount your own. Just follow the label directions, which are quite simple.

Prior to 1970, when Kodak discontinued the practice, all Kodachrome slides were treated with lacquer. For thorough cleaning

* Many of these problems were covered by Christine L. Sundt [4], who feels that these are reasons to limit the use of glass mounting to high-use collections.

of old slides, this lacquer can be taken off. (This method is safe for slides that have been attacked by fungus.) Scratches that did not get through the lacquer to the emulsion will come off with the lacquer. Check the processing date stamped on the mount of any Kodachrome slides you want to clean to determine whether lacquer was applied during processing.

Remove the slides from the cardboard mount. To do this, peel the mount apart from the edge until the slide film is exposed. Use cotton gloves or tweezers to pick up the film and transfer it to a clean sheet of paper. Mix 15 ml nondetergent household ammonia with 240 ml shellac-thinning alcohol (available at hardware stores). Agitate the film in this solution at room temperature for a maximum of 2 minutes, and then hang to dry in a dust-free environment. You can recoat the film with film lacquer before remounting. This solution also cleans dirty old slides quite effectively. To remount the slides, you can buy snap-together plastic mounts like the ones made by Gepe, or use cardboard Kodak mounts that can be heat-sealed with a household iron. Camera stores or mail order houses carry both types. Remount the slides so that the emulsion faces the right way — that is, toward the screen during projection.

This information applies equally to black and white slides. Though they resist fading better than color, monochrome slides need the same protection as do color slides and other film. We might add that the ease with which they can be processed and mounted, and the widespread availability of low-cost 35mm copying stands, suggests that a large reference collection of black and white slides would be an excellent solution to the problem of user access for large collections that have to restrict direct handling of their materials by patrons.

REFERENCES

1. Ronald L. Heidke, Larry H. Feldman, and Charleton C. Bard. "Evolution of Kodak Photographic Color Negative Print Papers." *Journal of Imaging Technology* 11: 93–97, 1985.
2. Ilford Photo Corporation. *Mounting and Laminating: Cibachrome Display Print Materials and Films*. Catalog No. 7929. Ilford: Paramus, NJ, 1988.
3. *Kodak Ektagraphic Slide Projectors*. Kodak Publication No. S-74. Rochester, NY: Eastman Kodak Company, 1977, p. 159.
4. Christine L. Sundt, "The Glass Mounted Slide: Causes and Effects of Heat Damage." Unpublished paper delivered at the Mid-America College Art Association, October 1980.

CHAPTER 18

Family Photographs

A collection of family photographs can give you a great deal of pleasure. You can also use it to explore the events and patterns of your personal life. Gathering together family pictures sparks insights and creates connections in ways that no single image can. You might see, for example, in two different photos taken decades apart how strikingly a daughter and her mother resembled each other when they both turned thirty, or — and this might be even more interesting — how similar their husbands looked. This is the strength and attraction of a well-managed family photo archive.

You probably know in your heart of hearts that your photographs should be carefully mounted in neat and uniform rows on the pages of your family album. But right now you have them jumbled in shoe boxes on a closet shelf, stuffed in a drawer someplace still in the original processing envelopes, and thumbtacked to a bulletin board. Just the idea of having to do something about them makes you nervous.

Relax. You probably haven't done as badly as you think. In our experience, more damage is done to photographs by overly aggressive attempts to take care of them than by neglect. And if you approach the care of your photographs as a project for pleasure rather than as an onerous, compulsive task, you will probably do a better job of it anyway.

Most people automatically associate family pictures with a photo album. An album does give you an excellent way to arrange, store, and display your pictures, all at the same time. We suggest, however, that you think of it as one tool among many for creating a family photo archive. Envelopes, boxes, and file cabinets can also be useful for your toolkit.

We should mention in passing that while we use the term "family" throughout this chapter, the same principles and techniques apply equally to any small collection of photographs.

ORGANIZING YOUR PHOTOGRAPHS

This part is a lot of fun. Arranging your collection gives you a chance to see what you have and to start making the connections that tell you more about yourself and your family.

How do you start? Cases differ, of course, but most of the same methods work equally well for organizing your photographs, whether you have recently found your grandfather's collection in the attic or you want to gather together all the snapshots you have taken since 1970.

Try this working approach. Start by getting everything in one place (Figure 18.1). Dig out all your photos from their hiding places and put them together. Ask other people in the family if they would like to give you any old pictures they don't need anymore. Explain what you intend to do and promise to take care of their photographs, and you will often find that your relatives part with them with surprising alacrity. After all, they probably feel just as guilty as you do about not taking adequate care of them.

You need to get some kind of workspace. A large table that you can leave undisturbed for periods of time is best. You might set up a trestle table in some out-of-the-way corner with good lighting. If your collection is housed in crates and boxes that have accumulated a lot of dust and dirt, you may need another, separate place for your initial examination and unwrapping.

Allocate yourself a reasonable amount of time for the work. Avoid rushing; it spoils your pleasure in the job. It also tempts you to handle things carelessly, to mount prints with the most convenient method, and to dump photographs higgledy-piggledy into boxes.

Start by evaluating all material. Examine the condition and note the format, medium, and state of the enclosures. This first sorting tells what kind of material you have to work with and indicates

Figure 18.1
Often, family photographs are found in a variety of places and enclosures. Start by getting everything together.

the scope of your project. It enables you to identify potential problems and to see what media are in your collection. You will find this information useful, because different media will have different storage and display requirements. Most collections (see Figure 18.2) of family photographs can usually be divided into these media:

☐ Contemporary prints

☐ 35mm or other small-format (126) slides

☐ Instant (Polaroid or Kodak) prints

☐ Black and white (silver gelatin) prints

☐ Studio portraits

☐ Snapshots

☐ Film negatives

If your heritage goes back a ways, you may also find some pictures in older media:

☐ Cased photographs

☐ Daguerreotypes

☐ Ambrotypes

☐ Tintypes

☐ Albumen cabinet prints

☐ Cartes de visite

☐ Glass plate negatives

A few words of caution:

☐ If you do have some cased images, do *not* take them out of their case. Turn to the section on nineteenth-century media to learn how to care for them. Removing them from their case to put them into an album will damage them and sharply reduce their historical value.

☐ If you find glass plate negatives still in their paper envelopes, do not pull them out. The glass support may be broken even though the picture-bearing emulsion is still intact, and pulling them out may tear an emulsion that could still be printed. Instead, support the envelope on a piece of glass and cut around the edge of the envelope. Lift off the paper to check the condition of the contents. If the negative is broken, tape another sheet of glass on top to make a sandwich and set it aside for expert treatment.

Even though these are your own photos — or *because* these are your own photos — we suggest that you wear light cotton gloves

a

b

c

d

e

f

Figure 18.2
These are the different media you are likely to encounter in a collection of family photographs:
a. contemporary prints
b. 35mm slides
c. black and white (silver gelatin) prints
d. studio portraits
e. snapshots
f. film negatives

g

h

i

j

Figure 18.2 (continued)
g. cased photographs
h. tintypes
i. albumen cabinet prints
j. cartes de visite

when handling prints. It protects them from finger oil, and you will find it also promotes an attitude of respect (Figure 18.3).

Prints that are brittle or that are on deteriorating mounts should always be supported with a piece of board underneath them or carried in a rigid folder. Prints that are 11 × 4 inches and larger also require some kind of support.

Now you need to decide what to save and what to discard. This question seems to come up automatically. Our rule of thumb: discard very little. One thing that you must watch for is deteriorated nitrate film (see the chapter on film storage). Copy it if possible before discarding, but get rid of it as soon as feasible.

Make note of any identifying or historical data on enclosures before throwing them out. For example, a print might come in an envelope that you decide to get rid of. Check to see whether it has dates or the names of people, places, or studios on it, and make a record of these to keep with the print. The preferred approach — especially if we are talking about an institutional collection — is to keep the envelope in a separate container with a cross-reference to the photographic artifact it held. You have to decide whether time and money can be allocated to this, or

Figure 18.3
Photographs in deteriorating mounts can be supported in an acid-free rigid folder with a clear polyester cover while being handled. See "Print Handling Folders" in Chapter 15.

whether you should allocate your resources to more critical tasks in managing your collection. But keep in mind that if you get rid of a picture of Aunt Mylene that she hated because it didn't flatter her, it may turn out to be the only picture anybody in the family had of her. Pictures that now don't seem especially interesting often acquire increased meaning with the passage of time.

EVALUATING THE WORK OF YOUR PREDECESSORS

You may very well inherit photographs from somebody in your family who has already created a family collection. This means that you have to evaluate the work that has already been done. If everything is in good condition, you are that much further ahead. But how can you be sure? Look for the following problem areas (Figure 18.4):

☐ Albums with "magnetic" pages

☐ Vinyl (PVC) pocket pages

☐ Prints mounted with Scotch tape, glue, or rubber cement

☐ Album pages with acidic paper

☐ Poor album design
  Broken or creased prints
  Mechanical stress from pressure

☐ Staples, paper clips

☐ Loose photos

☐ Housekeeping problems: dirt, dust, grease, bugs

☐ Moisture damage
  Waterstains
  Mold and fungus

☐ Writing on back (especially with ballpoint pen)

Any one of these merits correction. It is usually sufficient to transfer the photographs to a housing of acceptable design and materials.

a b

c d

Figure 18.4
Typical kinds of damage found because of improper handling:
a. Albums with "magnetic" pages use a medium-tack adhesive that can harden to the point of causing damage.
b. Prints mounted with Scotch tape or other adhesive tapes.
c. Rubber cement and other glues.
d. Information written on the back of the image area with ballpoint pen.

The most common and most intractable problems come from the adhesives people use to mount photographs. Albums with "magnetic" pages are probably the worst offenders. These have strips of slow-drying, medium-tack adhesive running diagonally across the page. People use them because they are convenient. They can simply put the print on the adhesive and then drop a plastic cover sheet over the whole page; the albums accept a mix of any size prints, and never let them fall out. But over time the adhesive turns color, hardens so that the print can't be taken out of the album, and then starts to hydrolize the paper and bleach the photographic emulsion. If this is caught in time, you can take these prints out of the album by gently unpeeling them from the page. It may be possible to take residual adhesive off the back by swabbing lightly with a Q-tip or cotton swab dipped in a solvent; a trial on your least valued prints can tell you what will work. Apply small amounts, constantly changing the applicator, and do not soak the print.

Not all prints that are badly mounted can be remounted. If the paper of the page tests acidic, however, and the print shows signs of sulfiding or fading, you will want to take action. Sometimes you find photographs stuck down with water-soluble adhesives. This is good, because you can simply immerse the page in a tray of lukewarm water until the glue releases, and then dry the print in the correct way (see Chapter 1 for information on processing prints). Always check, before immersion, to make sure the photographic emulsion does not stain or dissolve. Test for this by putting a drop of water on a nonimage area of the photograph, blot it lightly, and inspect for any changes.

When an adhesive proves intractable, it may be necessary to use a solvent to release the print. This kind of sophisticated technique should be performed by a professional photo conservator, and so is not covered in this book. Because of the many variables and dangers of working with solvents, we suggest that you do not attempt it without thoroughly researching the health hazards and potential danger to the photograph.

Paper and RC prints can be cleaned by brushing lightly with a clean camelhair brush. Remove dirt or grime from the back by applying eraser powder and moving it around with a gentle circular motion. Work from the center out, and reapply the powder until it no longer discolors. Never apply it to the front of the print. Clean the face of the photograph by dipping a Q-tip or cotton ball in Kodak Film Cleaner and swabbing the emulsion lightly. Move the cotton in a circle until it no longer picks up dirt and grease. (Do not use this technique on prints that have been colored with oil paints.)

THE FIRST SORT

After your initial evaluation, your next step is to arrange everything. When you look at the mishmash in front of you, the prospect of doing this probably seems daunting. Press on. The fun is about to start. And there is a simple and logical way to bring order out of chaos.

Sort by size Every print that's the same size goes together into a stack. Black and white prints are separate, of course, from color. Common sense tips: don't separate prints from negatives if they are still together. (See Figure 18.5.)

Sort prints that came from the same roll of film This is easier than it sounds, and it is a useful identification method. Look at the width of the margins, the style of scalloping on the edges, the dates applied by the photofinisher, paper marks on the back of the print. Any variation indicates that the prints were made at different times and probably did not get made from the same roll of film. There are more minor, telltale differences between prints than you might think possible. (See Figure 18.6.)

Once you get your prints grouped this way, you can often use

Figure 18.5
Start organizing the collection of loose pictures by arranging them according to size.

Figure 18.6
It is often possible to find prints that came from the same roll of film by looking at the margin width, scalloping style on the edge, and the date applied by the photofinisher. Look closely.

Figure 18.7
Prints from the same roll can be *temporarily* sorted into clear envelopes.

Figure 18.8
Label and sequence the prints into larger groups by chronology.

Figure 18.9
While getting ready to make the album, you can store the prints in a box, with dividers to indicate groups.

a face or place that you recognize in one print to infer the identity of other subjects.

Put prints from each roll into a container You can best use a uniform-sized bag of clear plastic to make the sorting easier. The top-sealing type made for refrigerator storage is excellent (for sorting — not for long-term storage). (See Figure 18.7.)

Sort the photographs into larger groups by chronology e.g., 1950 to 1960. Store in large clear envelopes and place in boxes. Place divider tabs between the groups to equalize the pressure on the prints and to mark the chronology. (See Figures 18.8 and 18.9.)

By now, you have already learned a great deal about your collection of photographs. This procedure forces you to examine each image, and it also gives you a good idea of how much detective work you have to do to identify the subjects.

Look at this first sort as merely a working arrangement. It doesn't have to determine the sequence in which you finally house and display your photographs. You might start by trying this two-step approach: (1) Arrange photographs by the media (start with daguerreotypes and work through contemporary color). (2) Sequence the individual images chronologically.

Maybe you need a more precise scheme of organization — dividing the images thematically by subject matter. You could start with these categories:

☐ Studio portraits

☐ Formal group photos

☐ Records of important occasions: weddings, graduations, etc.

☐ Holidays

☐ Family members and relatives

☐ Friends

☐ Recreation and vacation

☐ Places

☐ Pets

You might do best to separate these into blocks of time, perhaps by decade. Get together all your 1950s studio portraits, formal group photos, and on through the pet pictures, then the same for pictures from the 1960s and 1970s on down to the present day. Alternatively, you might pick a more personal chronology: Period 1 might be the time before your grandparents came to this country, Period 2 your parents' childhood, Period 3 their adult lives before they met, Period 4 their wedding, and Period 5 might be your own childhood. The divisions will be obvious to you.

Your next objective is to positively identify and label all subjects.

Documenting the people, places, and dates of a photograph adds greatly to the pleasure that other people will experience when they look at your collection. However, this will be an ongoing process, one that you are unlikely ever to complete fully. New information may come in nearly every time you show the collection to somebody.

The final (and most time-consuming) stage is to put everything into its final housing. This may mean mounting into a photo album, inserting into slide storage pages, storing in sleeves and boxes, or filing in a folder. The question is, what do you choose?

CRITERIA FOR SELECTING HOUSINGS

It helps to start by looking at specific needs and their possible solutions.

Type of use Will you use your pictures for browsing, casual shared looking, or reference, or do they merely need to go into dead storage? An album is best for browsing, a file cabinet for reference, and boxes for seldom-accessed material.

Future expansion In terms of future additions, is the collection closed or open? If you choose to put your pictures into a photo album, but expect to acquire more pieces in the future, the obvious answer is the type of binder that allows you to insert additional pages.

Physical size of photographs 35mm slides can be stored in boxes or binder pages. Glass plates go into file cabinets or the wooden boxes in which they may have originally been stored (after the wood is sealed to protect the emulsion). They should be wrapped in special enclosures. Large paper prints need the protection of portfolio-style boxes or drop-front boxes.

Ease of retrieval A file cabinet affords the fastest and most precise retrieval of prints and negatives. An album is not as quickly accessible, and boxes (unless the contents are thoroughly indexed) tend to be even slower. If you have a small collection through which many people will look at specific pictures, albums are not the best way to house it.

Budget and size of collection Hand mounting individual prints in an album is feasible only for small collections. Once a collection includes several thousand pictures, it may become necessary to find another way to house prints so that you can insert images into the collection in a standardized way. This usually means boxes or file folders.

When you have more photographs than money, but you still want to be able to look at some of them frequently, try this approach. Select the pictures you most want to see, buy one good-quality binder album with removable pages, and store the rest of your pictures in envelopes in acid-free boxes, As time passes, plan to acquire more albums and expand the number of photo pages.

Film storage Negatives can either be stored adjacent to the prints or filed separately with a cross-referencing system. If you want to store negatives adjacent to the photographs, one easy way is to put the negatives in page sleeves (these are available for all negative formats from 35mm through 8 × 10) and interleave them in a binder with prints in plastic pocket pages.

The materials of all housings should be of archival quality. We've covered this subject in detail already, but to recap briefly: material should be nonacidic and lignin-free, the housings should be built to protect photographs against mechanical damage and moisture, and their design should enable the photographs to be removed for examination without damage.

MAKING A PHOTO ALBUM

The first step in making a photo album is to determine your objectives. You may wish simply to have quick and efficient storage, in which case pocketed pages would be a good approach. A more studied and thematic album can be created by using large pages with many photographs attached with mounting corners. You may wish to display images and documents at the same time, which can be done with pocketed pages or on blank pages using corner adhesives. Another reason for making an album is to create a master layout for a video tape; then you would want to allow plenty of space between the images.

Certain images definitely do not belong in an album. Among the ones we think of first:

☐ Stereo cards made on curved boards

☐ Cabinet cards with curved backs

☐ Cased images (daguerreotypes, ambrotypes, etc.)

☐ Glass plate negatives

All of these, in addition to the images you choose not to show frequently, can be put into storage boxes, envelopes, or file folders.

You can pick from a variety of photo albums. Some of them offer certain benefits that might not seem immediately obvious.

Bound Album with Paper Pages

These are usually manufactured like a high-quality hardbound book. They sometimes come with leatherbound covers that are highly decorated and very impressive in appearance. There are a number of things to look for in this format.

☐ Quality of the paper.

☐ Interleaving sheets. There should be a sheet between all the pages, to prevent photographs coming into face-to-face contact.

☐ Spacers built into the binding. These hold the pages far enough apart to compensate for the additional thickness added to the book by the insertion of the prints.

- ☐ Slipcase, box, or latch to hold the book closed.
- ☐ Weight and thickness of the paper pages. These should be thick enough to bear the weight of the prints without tearing.
- ☐ Adhesive that may come into contact with the print. There can be adhesive pages in many types of albums, and they should always be avoided, no matter what the quality of the album is otherwise.

Bound Albums with Plastic Pages (Pockets and Sleeves)

These are mass produced, often inexpensive, and usually easy to use. They vary in both the quality of the plastic used and the design. Many times they are limited to one format of print per book.

Factors to consider when purchasing:

- ☐ Type of plastic. You need to determine what plastic is used. Mylar and polyester are a good choice.
- ☐ Pocket size. Pockets that are too small cause the photo to bend and make removal difficult, possibly leading to further damage. Pockets that are too large let the contents slide around and make it possible for them to fall out.
- ☐ Mounting style. Can the page accept prints mounted front and back?
- ☐ Quality of binding and heat-sealing on page. These need to be durable enough to stand up to use over time, or else the album should be used only temporarily for sorting.

Spiral Binding (Plastic and Wire)

This kind of album is at the low end. You will want to verify the material used for the pages in the same way that you would for other albums with paper or plastic pages. Many of these albums have adhesive ("magnetic") pages and are totally unacceptable.

The overall design does not repay the effort of organizing and putting a large number of prints in the book, because the pages can and will wear out where they join the binding. They cannot be replaced.

Wire Binding with Straps

A wire staple is attached to a paper card stock, with a space to allow a plastic strap to thread through and hold the pages together. This design has significant advantages:

- ☐ It opens flat and stays flat while the pages are turned.
- ☐ It is expandable.
- ☐ Both pocketed pages and blank pages are available.

Make sure that the plastic and/or paper meet conservation standards.

Hinged Albums

These albums have a thin hinge designed into each page and into the front and back covers. The hinge is attached to the page by taping.

This type of album design is popular among wedding photographers, and is available in many sizes with many types of preformatted pages.

The hinged design can be expanded, but this usually requires that you have the manufacturer's instructions, because it is not obvious how it works. This type of album is not very adaptable to multiple formats.

The major advantage to a hinged album is that it lies flat open. This puts very little stress on the binding and on the photos mounted in the book. As with other albums, you need to check the quality of the materials used.

Post Binder Albums

This kind of binder is commonly used to make scrapbooks. Each page is drilled with a predetermined pattern of holes that the posts fit through. The posts hold both the pages and the covers together. These binders provide a number of important benefits:

☐ They can be expanded by changing the post size and adding pages

☐ Spacers can be added between pages to accommodate the thickness of the photographs

☐ Pages can be worked on outside of the book

☐ Quality of the page paper can be customized easily

There are also some drawbacks:

☐ Pages do not open flat

☐ Pages have stress points both at the post holes and where they fold when the book is opened

Ring Binders (Three or More Rings)

There are a variety of ring binders to choose from. They can be troublesome from the point of view of the materials used.

☐ Vinyl covers are common

☐ Pages are also frequently a problem, and finding out the manufacturer's specification for materials is often difficult

Some types of rings are better than others for use in an album. The three major types of rings are the standard, circular ring mounted on the spine; the D ring, mounted on the back cover; and the oval ring, usually on the spine. The standard and oval rings allow the album to get overfilled. When this happens, the covers expand past the parallel. This puts pressure on the pages and on the prints. The D ring is the preferred design for album use, when mounted to the back cover. The rings keep pressure evenly applied to all pages.

When turning pages in a binder, take care to turn them slowly, one at a time. When finished, turn all pages back to the original position to prevent the rings creating pressure marks.

Page closers — rounded, ridged plastic inserts — can be put in the front and back of a 3-ring binder. These may cause sloppy handling and a false sense of ease when moving the pages back into place after viewing, and for that reason are not recommended.

We can think of quite a number of reasons to use binder albums. They are adaptable and easily expanded; they make an excellent tool for organization; they can be changed easily; and pages for them come in a variety of formats.

- ☐ Blank pages with cover sheets
- ☐ Pocket pages for film, with your choice of pocket sizes to fit everything from mounted 35mm slides through negatives — 35mm, 2¼ formats, 70mm, 4 × 5, 5 × 7, 8 × 10
- ☐ Pocket pages for prints of any size
- ☐ Special designs for storing documents

Binders come in multiple sizes. The most common has 8½ × 11 pages in a 9 × 12 binder. Another common size has 10 × 11 pages in an 11 × 12 binder, and of course many larger sizes are available on a specialty basis.

Multi-Ring Binders

These are popular with professional photographers and models as a presentation system. They are commonly available in large formats and can be effectively used as a scrapbook, where the large size is advantageous.

You will not find a great choice of specialty pages for such binders.

MOUNTING PHOTOGRAPHS ON BLANK PAGES

An important consideration when mounting your pictures on blank pages is the weight of the paper. In order to support the prints securely, this needs to be either a stiff card stock or a lightweight board. Bond or text-weight papers do not provide sufficient support.

No single method of attaching the photos to the page works equally well for all applications. Corners, for example, work well with small photos, but they can cause problems when used with larger photos or prints on thin and fragile paper. The techniques used for mounting pictures in an album should meet the same criteria for other types of archival mounting: it must be possible to reverse the process and remove the object from the mount without damage, preferably without the use of chemical treatment (i.e., solvents).

Avoid attaching your photos with the following methods:

☐ "Magnetic" album pages

☐ Cellophane (Scotch) tape

☐ White glue (Elmer's, etc.)

☐ Rubber cement

☐ Paper clips or staples

Accepted techniques for mounting include:

☐ Paper hinges with starch paste

☐ Encapsulation

☐ Tapes

☐ Mounting corners

☐ Dry mounting

☐ Mounting strips

☐ Certain specific adhesives

☐ Die-cut album pages

We described paper hinges and dry mounting previously. Apply the same consideration when using these in albums as when mounting prints in mats. Paper hinges are best used for light, small, and valuable pieces. When dry mounting in an album, we prefer heat-reversible tissue.

Encapsulation

We discussed encapsulation in the chapter on matting, but there are some differences when it is done for an album. Encapsulation for albums can be done in two ways.

Double-Sided Encapsulation: The print is enclosed between two sheets of Mylar and sealed on all four sides with 3M's Double-Coated Film Tape No. 415. No taped border should be closer to the print than ⅛ inch. This is an excellent technique for valuable pieces, because

☐ The print can be viewed from both sides

☐ It is completely protected from the environment

☐ Encapsulation makes the print rigid enough for mounting with photo corners

Single-Sided Encapsulation: Place the print on the page. Apply tape around the edges (no closer than ⅛ inch) and cover with a trimmed piece of Mylar. With this method, it is critical that the backing paper be of conservation grade.

Tapes

Acceptable tapes include a high-quality, acid-free paper tape and Filmoplast tape. These have an adhesive that can be reversed in water. Use these tapes to make hinges or to hold paper corners to the page.

Several cautions apply when using tape. First, cellophane tape is not acceptable, because it is corrosive, cannot be easily reversed, stains the print, and quickly yellows. Second, do not tape directly over the front edges or corners of the print.

Mounting Corners

Mounting corners meet many important criteria for use in albums:

- ☐ Relatively easy to use
- ☐ Completely reversible
- ☐ No adhesive contacts the print
- ☐ Clear corners allow the entire object to be seen

Figure 18.10
Mounting the photographs on album pages using photo corners.

Paper corners made of acid-free paper are available from several suppliers, or you can make them with a pair of scissors. These have to be attached with paper tape. Another type of corner made of clear polyester is available, backed with No. 415 tape, and this is also an excellent choice. (See Figure 18.10.)

The only drawback to corners is that, because they hold the print by a relatively small area, they might overstress fragile paper and cause part of it to break off. Physically weak prints can be encapsulated before mounting with corners.

Mounting Strips

These plastic (Eastman Kodacel) channels have a nonacidic adhesive on the back edge. They can support thick pieces like albumen prints that are already adhered to mount board or heavy card stock. The mounting strips can be cut to size with scissors. The print can be easily taken out of the channel.

Adhesives

There are two brand-name adhesives that you could use to hold snapshots or other items of somewhat limited value.

Gudy-O-Stickers: These transfer tabs of a neutral pH adhesive to the back of the print. They are a convenient way to stick snapshots onto the page.

Glue Sticks: Certain glue sticks contain a fairly neutral adhesive that is akin to wheat paste. Make sure that you buy the kind with a white paste, rather than one with a clear glue. The clear glue is acidic and attacks the photos.

Die-Cut Pages

Some album pages come with angled slots already cut so that you can tuck the corners of the prints into them. This is a suitable way to mount prints, provided that you take care not to bend or crease the paper while inserting the print. They have the following drawbacks:

- ☐ Pages accept only one format
- ☐ Prints can be mounted on only one side of page
- ☐ Print corners are not visible

LABELS AND IDENTIFICATION

Knowing about the subjects of your photographs adds greatly to the value and interest of an album, and you should identify them with whatever information is available to you. You don't need to provide reams of data, but pertinent material — dates, place names, names of the subjects, occasion pictured, name of photographer — are all useful. In planning and labeling your album, it helps to look at it through the eyes of somebody a hundred years from now.

Use a hard pencil rather than ink to make your notes. Graphite does not interact with photographic emulsions, does not smear or run, and will not fade over time.

Some situations that call for labeling are best met by putting information on an enclosure like an envelope. Make such information descriptive enough to put the photograph back with the housing if they get briefly separated.

SCRAPBOOKS

For housing nonphotographic materials, a scrapbook is an excellent adjunct to a photographic album. Generally, the larger-size binders make your work easier, because many of the materials you work with — like diplomas — are relatively large, especially when compared to snapshots. You will find that encapsulation is an excellent tool for handling the many papers that may be in fragile condition. Single-sided encapsulation is an excellent way to lay down a wide range of materials on a single page.

Experience shows that family papers can be broken down into these categories:

☐ Letters

☐ Diaries and journals

☐ Greeting cards

☐ Financial and legal documents

☐ Printed memorabilia (e.g., news articles)

You can use the same techniques to organize these papers as were used for sorting photographs: separate the material by type, or arrange chronologically within type.

Some documents are better housed in boxes, especially when there is little reason to look at them frequently. Such enclosures should generally be made of buffered board. It helps to do an inventory or index to keep track of the documents that you have in your possession.

MANAGING THE FAMILY PHOTO ARCHIVE

The first principle of managing your family photo archive should be the Black Hole Rule: what goes in never comes out.

To put the same idea even more bluntly: always a borrower and never a lender be. It seems that the more pictures you have in your collection, the greater the number of friends and family who want to "borrow" one. You need a fair amount of tact to avoid ending up with lots of pictures of people you no longer speak to.

The best solution to this problem is to offer to make copies at cost for anybody who wants them. This eliminates idle requests from people who say they want to make a copy but never get around to it, while it enables you to take care of the people who are sincere. Of course, it means you have to keep your negatives organized and accessible. One fortunate thing about copy requests:

most of them will be for recent photographs, which means that there is a better chance of the negatives still being available. In the case of older photographs, you will have to go to a photo studio or a processor to have a copy negative and a print made.

You might make it a project to draw on the memories of older family members to identify and document your photographs. A relatively painless way to do this is to sit with them in front of a tape recorder and to go through the pages of the album, asking them about the people in the pictures. Assign an identification number and title to each album, write page numbers on each page, and you have an easy way in the course of your conversation to index the stories to the pictures. You might phrase your inquiry like this: "We're here in Album Three, the Fifties, and on page sixteen we're looking at the image in the upper right. . . . Can you tell me about the man behind the wheel of the car?" That's the general idea.

HOW DO YOU HANDLE THE PHOTOGRAPHS ALREADY IN AN ALBUM?

You have acquired a photographic album that seems to be in reasonably good condition. What do you do with it?

You have a number of options. For one, you can simply shelve it. To protect it, we suggest that you wrap it in conservation-grade wrapping paper and tie it with string (Figure 18.11). Albums that come without binder sleeves also benefit from being wrapped with tissue and placed in a storage box to stabilize their environment (Figure 18.12). This kind of conservative treatment is suitable in a number of situations: for example, all the prints are permanently mounted and cannot be removed; the photographs do not need immediate intervention to save them from disintegration caused by the album itself; or the album as an artifact has sentimental, historical, or aesthetic value.

On the other hand, the old album may be damaging to photographs. It might have acidic paper, which you can be sure of if it's made of the old style black construction paper. Mounting corners may be falling off, or the album may be so overstuffed with material that pressure is creasing the prints.

In this case you also have a number of choices. Again, you can simply put the album on the shelf or wrap it and store it in a suitably designed box made of buffered board. This is often a good interim choice, if nothing else can be done immediately.

Assuming the original album has no inherent aesthetic or sentimental value, you can remove all the photographs and remount them in a new and safer style of album. Here you might wish to photograph each page, if this is feasible, to document the original sequence of the photographs. In any case, copy and transfer any identification found in an original enclosure.

Perhaps you find that the pictures are permanently mounted onto black paper bound into the album, and they are already fading and sulfiding. If this is the case, interleave the pages with 2 mil Mylar sheets (Figure 18.13). The 2 mil sheets are thin enough

Figure 18.11 (left)
An older photo album can be wrapped in paper for better protection during long-term storage.

Figure 18.12 (below left)
Place the album in an archival box to stabilize its environment.

Figure 18.13 (below right)
The photographs in this album are permanently mounted onto poor-quality paper that will cause fading or sulfiding. Interleaving with thin, 2 mil Mylar sheets trimmed to fit the book can slow the deterioration, because the emulsion will no longer be in direct contact with the paper.

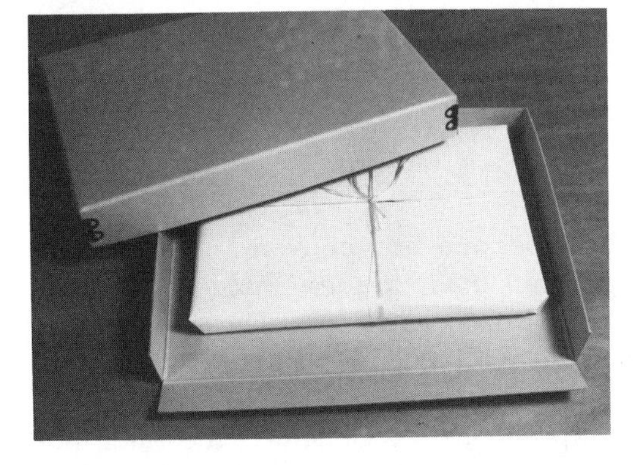

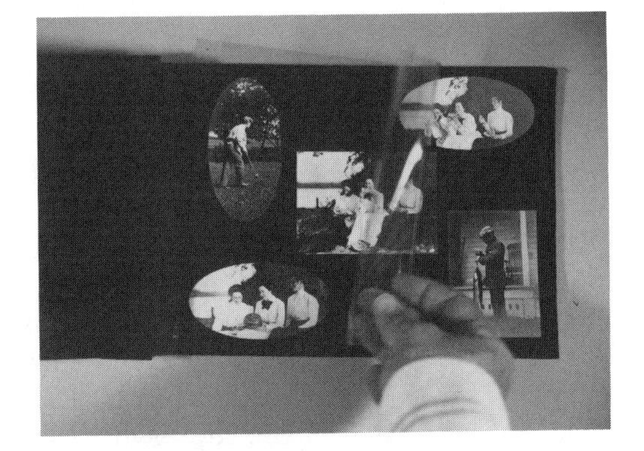

to fit into the album without forcing the pages apart. They protect the emulsion of the photo from touching the back of the facing page, while the dry mount tissue will provide a barrier against the migration of acids from the back. Copy photos made with a polarizing filter are an effective backup step in this case.

A photographic album may come into your possession in such circumstances that you know little or nothing about the people and places shown in the pictures. It is still a good idea to make a note of the date and circumstances under which you acquired it (to establish a provenance) and where you think it might have come from. Write this down and file it with the album. Others may eventually be able to research it more completely, and the information you supply may give them the clue they need to make their identification positive.

CHAPTER 19

Nineteenth-Century Photographs

DIRECT POSITIVE PROCESSES

The daguerreotype, ambrotype, and tintype were the most common early images, and portraits made in these three media introduced photography to its first mass audiences. These techniques share a central characteristic with Polaroid and other kinds of instant prints. Each is a direct positive process. In other words, the light-sensitive plate exposed in the camera is the same one that is the final picture. No negative, no enlargement, and no duplicate prints resulted; each photograph is unique. Duplicates can be made only by copying.

The value and historic importance of direct positive photographs range from priceless to nearly worthless. It is important to distinguish among different media, in order both to care for them properly and to determine their probable value. In general, daguerreotypes are the earliest and most valuable, while tintypes fall on the other end of the scale, many having value only as curios or historic artifacts.* Here we will describe how to tell the three kinds of direct positive images apart and what features make some more important than others. We will also tell how to inspect and clean them.

Direct positive photographs have proved to be remarkably durable. Given the meager knowledge of chemistry possessed by their makers and the long, often arduous provenance of these articles, this is amazing. Several factors, we think, contribute to their survival. For one thing, the metal and glass substrates on which the photographs were made did not trap processing chemicals, and a quick wash was usually sufficient to remove any residues. The hardness of the photo plates themselves also helped. You cannot just crumple up a sheet of copper and toss it into a wastebasket

---

* This may change. Not too long ago some people kept daguerreotypes around the living room solely because they were quaint.

like a scrap of paper. Finally, the photographers of the time usually put the finished image into a case before giving it to the customer; the housing helped protect it from damage.

DAGUERREOTYPE IMAGES*

The daguerreotype was introduced to the world in 1839 by one of it inventors, Louis Jacques Mandé Daguerre. The French government had bought his rights on the condition that he give the process to the public. "Daguerreomania" swept the world as thousands of scientists, amateurs, and businessmen took up the process. From 1840 to 1860 this difficult and exacting process was the dominant medium of photography used in the United States. Despite widespread dissemination of a "fast" portrait lens after 1844, most daguerreotype images were rather stiff studio portraits; some exterior and candid "instantaneous" views remain, though they are relatively rare.

Numerous inventors sought with varying degrees of success to improve upon Daguerre's process, most by methods meant to increase the sensitivity of the plate; but after 1840 the basic outlines of the process stayed substantially the same. A thin sheet of copper was plated with silver and then polished to a mirror finish with increasingly finer grades of jeweler's abrasives. Often these plates were purchased commercially rather than finished by the photographer. The plate was then sensitized in dim light by inserting it into a vapor chamber with a dish of heated iodine or bromine. Next it was put in a camera and an exposure made to produce the latent image.

Development took place by holding the plate over a heated dish of mercury. Mercury vapor amalgamated with the iodine-sensitized silver more densely in the highlights than in the shadows. This created a visible image. The plate was fixed with a solution of sodium thiosulfate that changed unexposed silver halides into soluble silver thiosulfate.† This thiosulfate was removed by a water wash. After 1840, gold toning to stabilize the image and to create warm colors was usually practiced by daguerreotypists as a final stage.

For presentation to the customer, the daguerreotypist put a decorative brass mat on top of the plate, covered both with a sheet of glass the same size as the plate, and sandwiched all three into a metal border known as a preserver. This assemblage was then inserted into a case.

Addition of color was frequently offered as an option. Sometimes various metals were galvanically plated onto the image to create beautiful delicate hues, but more commonly daguerreotypes

* The two most authoritative books on the history of the daguerreotype remain those by H. and A. Gernsheim [1] and Beaumont Newhall [2].

† At first Daguerre had used a sodium chloride (table salt) solution to stabilize the image, but later he took up Herschel's suggestion to use sodium thiosulfate as a fixer.

were hand-colored. Stencil or brush application of dry pigments onto a tacky gum arabic finish were used to give a more lifelike appearance to a finished portrait.

The surface of a finished daguerreotype is physically quite fragile. Any abrasion removes the image bearing silver and mercury amalgam. For this reason, the surface of a daguerreotype should *never* be cleaned, especially by any method that involves wiping or the use of polishes. It is common to see otherwise valuable daguerreotypes that were damaged or ruined by attempts to rub off dirt, sometimes with silver polish.

The cases that protect daguerreotypes are often remarkable pieces of photographica in themselves. Early ones were made of wood covered by embossed paper, morocco leather, or papier-màché. Later manufacturers used molds to impress decorative motifs into a heated mixture of wood fibers and adhesives — the first commercial application of thermoplastic molding.

A typical case has a front and a back of equal size, hinged to open like a book. Cases sometimes hold two images, but it is far more common for the front to be covered on the inside with a dark, embossed velvet. The case was usually held closed by a brass hook. Some rare variants encompass a magnifying lens of low power held in a card in front of the image, and stereographic daguerreotypes had their own special viewing cases, with two lenses and a dividing card.

A daguerreotype should never be separated from its case. Loss of the case is often a cause of destruction of the image itself and will hinder efforts to date and identify the image. During the early decades of this century it was common for antique dealers to discard the photograph and sell the case as a container for cigarettes. Thrifty photographers also liked to discard old portraits and insert new ones.

Identification

It is quite common to hear all cased nineteenth-century images referred to as daguerreotypes. A little experience in comparing them to examples of other media like tintypes makes it easy to distinguish the true daguerreotype. Its chief characteristic is a silvery, highly reflective finish that makes it difficult in certain lights to see the subject. This reflective quality led early observers to aptly dub the daguerreotype the "mirror of nature." A generally conclusive test of its nature is to hold a sheet of white paper with writing on it at right angles to the front of the picture. The writing appears reflected on the face of the image, reversed, and the image, instead of appearing positive, will be a ghostly negative. This easy test makes it possible to determine the medium without taking the picture from its case. If the image is already outside its case, you can identify it by the copper backing. A magnet held against the back of a case also helps to differentiate between an ambrotype and a daguerreotype; but be sure to apply the magnet only to the back, or to the glass, never to the image itself.

Few processes are capable of rendering detail more exactly than the daguerreotype. Under a magnifying glass, such minute details as the leaves on a tree appear distinctly and with a clarity that has to be seen to be believed. With the greater magnification of a microscope, actual beads of the silver-mercury amalgam can be seen rising from the polished silver surface.

Although millions of these photographic jewels were produced for an admiring public each year during their heyday, the process passed quickly out of vogue once cheap and easy processes appeared that gave similar results for a lower price.

Cleaning and Care of Daguerreotypes

Few of the dangers that beset photographs on paper are likely to faze a daguerreotype. It is not easy to casually dispose of a sheet of copper, particularly when it is snugly ensconced in a tough little case. Moisture does not rot it, nor do insects (at least any we know of) graze on mercury deposits.

The underlying silver-plated copper sheet does not absorb and retain chemicals, so even a quick rinse removed most thiosulfate residues that might fade the image. Long and undisturbed storage even in relatively hostile conditions seems to do minimal damage to the sturdy daguerreotype; you might even say that a quiet neglect seems its best friend.

Indeed, aside from being discarded, the most dangerous thing that can happen to a daguerreotype may be "help" from friendly hands. Most of the worst damaged images we have seen were cleaned at some time, usually by mechanical means like rubbing or wiping. The fine dispersal of mercury droplets that creates the image will not stand such direct, rough treatment. There is good reason to question the wisdom of any attempts at restoration, even in the most extreme cases of deterioration.

The most common cause of daguerreotype deterioration is grime. Even if the image itself has survived in immaculate condition, it appears dull, lusterless, and lacking in detail when obscured by dirt on the cover glass. Dirt and tarnish on the brass mat will worsen its appearance and exaggerate the effect of any problems on its surface. Often, cleaning up the case makes an old daguerreotype look almost new. Taking the daguerreotype out of the case is an extreme measure, so this should be done only when absolutely necessary.

To clean the case and housing:

- ☐ Remove the daguerreotype by gently inserting a dull, unpointed blade (a kitchen knife works quite well) between the edge of the preserver and the cloth or leather trim around the outside. This causes the assembly of preserver, glass, mat, and copper plate to pop out as a unit.
- ☐ Open the preserver frame by gently bending back the flaps of soft metal that wrap around the back of the copper plate. Set the preserver aside and slit open the tape that holds the glass to the copper. Often, this tape will have disintegrated already.

Always take care to avoid sliding the glass or mat across the daguerreotype surface. Either one can abrade the image.

☐ Remove old tape from the glass and the back of the copper plate with a sharp razor. Always handle the copper plate by the edges to avoid marring the surface.

☐ Set the daguerreotype to one side.

☐ Wash the cover glass in a solution of Ivory Flakes or similar mild soap. Avoid detergents, which can leave a film. Rinse well and dry. For extra cleanliness, you can also use washing soda after the soap. Conditions of high or wildly fluctuating humidity can cause a condition known as "weeping glass," in which chemicals from the glass itself will form small crystalline deposits on the surface. These usually come off during washing, and a recurrence can be prevented by careful sealing and humidity control. Often, the inside glass surface has acquired such a coating of grime as to largely obscure the image, and the effects of cleaning will be quite remarkable.

☐ Clean the brass mat by wiping it with a clean, lightly dampened cloth, and then brush the high-relief features with either a jeweler's brush or an ink eraser. Wipe thoroughly again to remove all grit.

☐ Reassemble the daguerreotype, mat, and cover glass. Clamp together with clothespins on the two short sides, prior to retaping. Taping prevents airborne pollutants from penetrating the case. Guard against scraping the surface of the daguerreotype during this step.

☐ Retape the edges of the pinned assembly. Either Scotch Magic Transparent Tape or a paper tape can be used. The paper should be either Japanese tissue or Permalife bond, joined with a polyvinyl acetate adhesive. For ease of taping, use two sections and overlap them at the corners where they meet to provide a completely airtight seal.

For replacement of cracked or broken glass, use the lightest glass available, usually picture weight.

While the daguerreotype is out of the case, this is an ideal time to make copy photographs. Document the image both ways, with and without the frame. At the same time, write down all data you find inside the case for a permanent record, so that later workers do not have to open the case again. The metal flaps on the preserver do not tolerate much bending before they break. If you find no data, record this fact also.

Two serious causes of daguerreotype deterioration on the image proper are silver sulfide tarnishing and mold growths.* Other

* Swan [3] offers an excellent analysis of the causes of daguerreotype deterioration, based on electron microscopy and gas chromatography.

causes may be responsible for blemishes, but in addition to these two, the major problems are likely to be evenly distributed black spots caused by fixer crystals from unfiltered solutions, and green crystals of copper salts caused by puncturing the silver plating.

Mold growths and the black measles formed by fixer crystals cannot be removed; their only treatment is good storage in a cool, dry environment to prevent further aggravation of the condition. Mold growth is often very disfiguring, and its appearance can be aggravated by removing the tarnish so that the growth contrasts more strongly with the cleaned areas.

Tarnish and copper crystals can be removed by a thiourea bath, though the copper will usually reappear in short order. This kind of cleaning falls into the realm of restoration rather than of preservation, and should be undertaken only by those with developed laboratory skills. We do not recommend it. There is always some risk of image loss in any chemical treatment; in addition, a cleaned daguerreotype tarnishes more rapidly than one with the patina of age still on it. Chemical treatment may remove any coloring agents on the daguerreotype, and can impair the historic value of the image. This should be taken into account before you decide to proceed with cleaning or restoration.

Daguerreotype tarnish is usually caused by airborne sulfur resulting from removal of the cover glass, failure of the tape seal around the edges, or cracking of the cover glass.

AMBROTYPE IMAGES

Ambrotypes were popular in the United States from 1854 to circa 1870. They were invented in 1852 by the Englishman Frederick Scott Archer, also the inventor of the glass negative. The ambrotype achieved its peak of popularity in the United States just before the Civil War. The hazards of combat, the mails, and the economic vicissitudes of the conflict made the cheaper, unbreakable tintype (introduced in 1860) more popular with separated families and soldiers' sweethearts than the fragile glass ambrotype.

In appearance, ambrotypes superficially resemble daguerreotypes, though they lack the fine detail of the earlier technique. Archer took advantage of the fact that a very thin underexposed wet-collodion negative on glass backed with a sheet of black paper or varnish appears as a positive. At first a sheet of black paper or a piece of black velvet was used behind the image, but later other methods were adopted. Sometimes the back of the ambrotype was painted with paint or varnish; at other times, the ambrotypist used a deep red glass (called *coral glass*) for a substrate.

A patented method used after 1854 (but fortunately not very common) called for pasting a cover sheet over the front of the ambrotype with balsam resin. Usually, when these are encountered it will not be possible to remove the cover sheet.

Identification

Superficially, the ambrotype resembles a daguerreotype, because it is usually found in a fancy case with a brass mat, glass cover sheet, and preserver. Another point of similarity is that most ambrotypes were studio portraits.

The most apparent difference is that the ambrotype fails to reflect light in the same fashion that a daguerreotype does; for example, when a sheet of white paper with markings is placed against the image, the marks are reflected only by a daguerreotype. The tonal scale of the ambrotype is shorter, flatter, and much coarser than that of the daguerreotype, a difference that readily appears to the experienced eye. Again, using a magnet often helps; it is useful to carry one of these to antique stores and junk shops where you are not allowed to take apart the case to make a positive identification.

Cleaning and Care of Ambrotypes

An ambrotype can be disassembled for cleaning in the same fashion as a daguerreotype. Often, cleaning and rebacking will prove even more spectacularly successful. The same precautions have to be taken when cleaning an ambrotype as were outlined for the daguerreotype. For example, movement of the mat across the surface of the collodion emulsion has the same distressing effect of removing part of the picture. Make copies both before and after cleaning, and keep the ambrotype image itself separate from other elements in the picture package during their cleaning.

Begin by removing the preserver frame from the case and removing any disintegrated particles of tape. The ambrotype should have been mounted with the emulsion on the side of its glass backing that faces the viewer, so that the image is reversed from left to right. This made for a sharper and more contrasty image, and allowed application of colors to the emulsion. However, check to make sure that the emulsion is actually on the forward side; it may later have been incorrectly mounted during a cleaning.

After setting the ambrotype aside with the emulsion side facing up, clean the cover glass and mat. Avoid chemicals other than soap and water, because the collodion emulsion may interact with residues you trap inside the case. A jeweler's soft brass brush or an ink eraser, as before, will clean the mat; again, of course, all pieces of grit should be blown free.

Proceed to clean the nonemulsion side of the ambrotype with a piece of clean dry cloth, cotton batting of the type sold for first aid purposes, or a piece of cheesecloth. If serious deterioration has actually set in, only a trained technician should attempt restoration of the emulsion. Little can be done for a flaking emulsion.

And now for the step that will seem like magic the first time you see it: replace the backing. When the old backing of an ambrotype has faded or started to disintegrate, the image's visual integrity suffers to the point where it may seem hopeless. Many times, simply rebacking makes an entirely new, fresh picture appear. (This only works for ambrotypes with separate backings; those on coral glass will not be affected.)

Like any silver halide emulsion, a collodion negative can be adversely affected by a wide range of chemicals in small amounts.

It is foolish to introduce potentially harmful agents from paint, black paper, or cloth into proximity with the delicate picture. Instead, we suggest that you use a piece of polyester-based sheet film, fully developed after total exposure to light and then fixed and washed to archival standards, should be used for the new backing material.

When the film has been cut to fit and the cleaning is completed, put the picture package back together and tape it in the same fashion as a daguerreotype. The film, instead of being taped to the back of the glass, can simply be inserted in the case before the picture package goes in. This makes for a more securely fastened package.

As previously mentioned, make a permanent file record of all information found on the inside of the case.

TINTYPE IMAGES

The tintype (also properly called a ferrotype, from the Latin word for iron) gained widespread popularity after 1860 in the United States as a cheap, quick way to send portraits home. The tintype process combined features of the daguerreotype and ambrotype to make a rough-and-ready kind of photograph at a price that allowed people of modest means to possess portraits of families and friends for the first time.

The technical progression from ambrotype to tintype bears remarking, for in many ways the tintype was merely a tougher and cheaper version of its precursor. A tintype was made by coating a sheet of iron instead of glass with the collodion emulsion. To provide the black background needed to reverse the collodion negative image, the iron sheet was "japanned" with a coat of black varnish. The tintypist bought a quantity of these prepared plates, which he coated with the wet collodion emulsion just prior to exposure, inserted in the camera to make the picture, and then developed, fixed, and washed.

When multilens cameras came into widespread use around 1860, it became common practice for multiple images to be made on a single ferrotype plate. After processing, they were cut apart with tin shears into discrete pictures. The finished pictures were usually about 2½ × 3½ inches in size, although many tintypists also specialized in "gems," exceedingly small pictures sometimes numbering as many as 36 on a 5 × 7 plate.

In general, tintypes are not considered to have great value, with the obvious exception of unusual views and portraits of famous people. Contemporary photographers considered the tintype something of an aesthetic abomination because of the haste and general lack of care that characterized its production. Many regular photographic galleries of the time did not even offer tintypes, and separate establishments catering to the lower classes specialized in their manufacture.

Early tintypists sometimes mounted their works in cases similar to those used for daguerreotypes and ambrotypes, but in keeping with the cost-cutting attitude that dominated their production, later tintypes were mounted in die-embossed paper sleeves, usually with an oval opening cut in the face. Many of these paper sleeves have disappeared.

Identification

Because most tintypes will be found loose, their identification generally poses little problem. The metal on the back will be iron instead of the copper found on daguerreotypes. Often, the corners of the plate were clipped to facilitate insertion into the paper sleeve.

Their flat, muddy whites, similar to those of the ambrotype, their low contrast, poor definition, and generally dog-eared appearance make tintypes a rather depressing family of images. If mounted under glass their general resemblance to ambrotypes can mislead, and you may have to unmount them to make a positive identification.

Unlike ambrotype and daguerreotype photographs, the tintype rarely exhibits positive-negative reversal unless the backing is very dull. After 1870, a chocolate-brown varnish producing more life-like coloration came into vogue. In addition to its muddy, flat appearance, this brown color readily betrays the image's medium, and can help date it.

Care and Cleaning of Tintypes

Tintypes most often turn up loose, without any kind of mount. These images, despite their age, generally have little value save as historic curios. The only exception would be some of the very rare outdoor views. Given their current minimal worth, an economical method of storage should be adopted. The best method of storing loose tintypes is to insert them individually in polyethylene or polyester sleeves of the kind widely used for negative storage. These sleeves allow visual inspection, protect against abrasion, and keep out moisture without releasing potentially dangerous chemicals.

The same method can also serve for storing tintypes found in their original sleeves. To keep the edges of the sleeve from being broken off in the plastic enclosure, it is an excellent idea to sandwich the sleeve and the tintype between two pieces of acid-free conservation board before putting them into the polyester enclosure.

Elaborate restoration methods have not been developed for tintypes, due to the fact that the image can be suitably recorded by proper copying techniques, and often improved upon by increasing the contrast. You can clean a tintype by immersion in a mild soap solution like Ivory Flakes, completed with a rinse in distilled water. Since rust can be a problem, dry thoroughly with a hair dryer after rinsing.

A crease caused by bending a tintype can never be completely removed. Weinstein and Booth [4] suggest bonding a plate to good-quality board of the same size with polyvinyl acetate adhesive, and then using a mounting press to flatten the tintype plate and to join it to the backing board.

THE ALBUMEN PRINT

By far the most popular printing medium from 1850 to the 1890s was the albumen paper print. Pure rag paper was used as the raw stock, and it was coated with the albumen (protein) obtained from egg whites. Before 1872 this coated stock was sold to the photographer unsensitized. It was made light-sensitive by treatment with a silver nitrate bath, and then printed out from a glass negative by exposure to sunlight. After 1872, sensitized paper displaced the coated stock, although the basic printing procedure remained unaffected. The demise of albumen printing came about because of the introduction of gelatin emulsion and carbon and platinum printing papers. Both a change in aesthetic tastes and a concern for stability were factors in promoting the change.

The future survival of existing albumen prints seems only a dim possibility. James Reilly, the leading expert on albumen printing techniques of the current generation, estimates that 85% of these prints already exhibit serious discoloration, and that "possibly in another 75 years, not a single albumen print will at all resemble its original appearance"* [5, p. 109].

After 1855 many photographers were extremely conscious of the need to thoroughly wash their prints to remove thiosulfate residues. However, the deterioration endemic to albumen photographs arises from the interaction of egg proteins with silver during sensitization. A certain amount of the silver seems to lock onto sulfur-bearing side groups of the protein molecules. This silver cannot be removed by any technique except treatment with potassium cyanide, which is highly poisonous and which also bleaches out the less stable colloidal silver of the photographic image. This unexposed silver remains behind after fixing and washing to become silver sulfide, creating the highlight yellowing characteristic of most albumen prints.

Deterioration caused by this process cannot be slowed or reversed by means now known. The only feasible conservation measure is further research. One phenomenon that has already materialized is that albumen prints have a particular sensitivity to the buffering agents used in some board to make it acid free. The yellowing actually speeds up in situations where albumen prints have been mounted with buffered board, so it is a matter of some importance to use only acid-free, nonbuffered boards and papers in their presence. Increasing supplies of these kinds of materials should be available in the next few years.† In the meantime, if a

* Reilly [5] has published the most authoritative treatment on the history, chemistry, and technique of the albumen process, and is an indispensible reference work for anyone handling nineteenth-century photographs.

† Light Impressions is one of the few sources for this kind of nonbuffered, acid-free board and paper that we know of, though other suppliers are in the process of entering the market.

source cannot be readily found, some kind of suitable plastic enclosure such as Mylar would be a good interim measure.

Albumen prints can also suffer from the same kinds of damage that affect other silver prints. Residual fixer complexes will cause generalized image fading, as will sulfiding induced by atmospheric agents like sulfur dioxide. (Albumen prints cannot be treated for sulfiding of the image by the "bleach-and-redevelop" method used on conventional silver reprints, because the silver locked in place by the albumen proteins also redevelops as a fog.)

The mount boards and glues used in the nineteenth century by the original photographers to mount albumen prints are a major cause of their deterioration. It was common practice during the heyday of albumen prints to mount all prints on hard board made of fancy lining papers over an acidic core of wood pulp containing a high percentage of lignin. By now the breakdown products of lignin have had the chance to work through the back of the print to attach themselves to the silver in many photographs. Short of attempting to dismount the print and risking its destruction, the best conservation measure is to store the prints in an environment that retards chemical activity: cool, dry, and dark.

OTHER HISTORIC PRINT-MAKING TECHNIQUES

Many methods of making photographic prints have been invented during the last century and a half, and it would require several large volumes to give the details of all the processes when they are known. The brief descriptions that follow will not substitute for a complete study of the subject, but they give enough information to aid in identification.*

In the case of a print produced by any of the photomechanical processes such as gravure, there is no reason that the print should not be treated exactly as would an etching or any other kind of ink print. In the case of a print produced as a silver image such as the kallitype, the same care should be taken as with a silver gelatin print; for example, efforts should be made to isolate these types of prints from sources of sulfur contamination. And some methods of printing — carbon, platinum, and gum bichromate — while photographic in nature, have proven to be as stable as ink-based prints.

Salted paper prints (1840 to mid-1860s) were the original paper prints. Upon close examination the image looks to be embedded in paper fibers, because there was no baryta coating to fill in the spaces between the fibers and provide a smooth surface. Prints are typically red-brown if they have not been toned, or purple if they are gold-toned. Subject matter was typically a landscape, and the in-camera original was usually a paper negative. Antique salted

* The definitive book on historic print-making techniques is Riley's *Care and Identification of 19th Century Photographic Prints* [6]. In addition to a wealth of technical data, it contains a practical flowchart for identification (also available separately), with color reproductions of all the major processes.

paper prints were not mounted on standardized mounts, and typically show some deterioration: yellowing, fading around the edges, and bleached-out highlights. The process enjoyed a revival by pictorialists around the turn of this century, and these later images may be in better shape.

Platinum print or platinotype, from around 1880 to the 1930s, has a very subtle tonal range and was positive printed on uncoated papers with matte finish. Its characteristic color is silver-gray, although it can also be found in warmer tones up to a fully sepia color. One of the "permanent" photographic processes, platinum printing was forced out of existence by the rising cost of its active metal, although some contemporary photographers occasionally use the process.

Woodburytype (in French, *photoglyptie*), from 1865 to the 1890s, was the most exquisite form of photomechanical reproduction. It used a gelatin relief matrix in a hydraulic press and was a proprietary process, so most examples are easily identified. The color is often chocolate brown, though violet-brown is also common. A distinct, if subtle, relief image can be seen on the print surface, and examples are always mounted or tipped onto a book page or card. Because of the nature of the process, it is always a paper positive.

Gum or gum bichromate, from 1858 through the 1920s, was primarily employed by the Pictorialists for multiple-exposure prints utilizing different colored pigments. Most often, though not invariably, it is found on matte finish paper; the process produces a positive. Colors can range from a flat black, due to the use of carbon pigment, through entire rainbows of color. "Realistic" color like that produced by modern color processes was not often attempted, however.

Kallitype, vandyke, or *brownprint,* announced in 1889, is based on the work of W.W.J. Nichol. Variations on the process were numerous, although it never achieved widespread popularity because of questions about its permanence. Kallitypes resemble platinum prints, both in their use of the light sensitivity of ferrous salts to create the image and in their long tonal range and excellent shadow detail, but the image is composed of metallic silver. Originally a print-making medium, it is now used for proofing printer's litho negatives. An untoned print will be a rich brown, although gold toning was used to get a purple color. Kallitypes are often found on matte finish paper.

Photogravure, 1879 to the present, is a photochemical process used both for original prints by photographers and as a commercial reproduction technique. Close examination of the image shows either a halftone screen composed of square dots or a very fine grain. Single prints, as opposed to those found in publications printed on large rotogravure presses, often show plate marks around the edges. Color depends on the ink chosen by the printer and the quality of the paper substrate, which varies widely. A compressed tonal range was popular with Pictorialist photographers who printed in gravure, though later workers such as Paul Strand have also favored a style with rich black shadows.

Cyanotypes or the *ferroprussiate process*, invented by Sir John Herschel in 1842, was revived by a French concern, Marion & Cie, in the 1870s as the *blueprint process*. The chief characteristic of prints made by this process is their bright blue color; the ferric salts used for the process can be coated onto any substrate, especially cloth, so it is not uncommon to find many unusual examples of this process. It is usually quite stable as long as the substrate endures, but may fade because of exposure to light or alkalis. A common use is to produce architect's plans, hence the generic term *blueprint*.

Carbon prints, patented by Alphonse Louis Poitevin in 1855, did not become fully practical until Joseph Swan announced the transfer process using his carbon tissue in 1864. The stability of the pigment makes these prints another of the permanent photographic processes, but it can be hard to spot them (aside from the lack of deterioration) because of their resemblance to conventional silver-gelatin prints. It may be possible to spot grains of pigment beneath the gelatin transfer tissue and the paper substrate with the aid of a microscope; and if a different color pigment than carbon has been used, this will be a strong indication.

Carbro or *ozotype*, developed in 1899 by Thomas Manly and marketed after 1919 by the Autotype Company, was also a carbon-type process. A conventional bromide print was used to sensitize the carbon transfer tissue.

Oil, bromoil, and *transfer prints,* 1904 through the end of the Pictorialist movement, were favored by Robert Demachy and his followers because the process allowed the creation of a painterly look through the use of hand manipulation. The process depended upon the ability of a gelatin matrix that had been differentially hardened by exposure to hold varying amounts of oil-based ink. A major identifying characteristic is the print's manipulated look and the appearance of ink on its surface. Highlights often lack much detail, and the brushwork gives the print an irregular, grainy appearance.

Printing Out Paper (POP) processes could have either a collodion or a gelatin emulsion. Printing out papers were developed by contact printing with exposure to sunlight, and the action of the sun caused the actual formation of the image without a chemical developer being necessary. The remaining silver was removed by fixing. All of them were widely used for commercial portraiture. *Gelatin POPs* (late 1880s through 1920s) were usually gold-toned to reddish- or purplish-brown hues. They can be glossy or matte finish, have a thick baryta layer that obscures any paper fibers, and are subject to the same types of deterioration characteristic of other silver images. *Collodion POPs* came in both glossy (late 1880s to 1920s) and matte (1894 to 1920s) versions. The glossy collodion prints are nearly indistinguishable from gelatin papers and have the same type of coloration. Matte collodion prints have a cooler, more neutral hue because of the combined gold and platinum toning they usually received. Because the baryta coating is thinner, the paper fibers can be seen more clearly with a microscope. These prints have excellent image stability. After the turn of the century, they were frequently mounted on dark, square boards instead of cabinet cards.

Mixed media prints have been produced almost since the inception of photography. They present special problems in identification, problems that are often simply insoluble without information from the photographer. For example, it was a common practice during the Pictorialist years to overprint gum on top of platinum prints to get added depth, detail, and color. Various processes of hand coloring and toning were also used and were often as individual as the photographers themselves. In these cases, little can be done without extensive historical research.

Anyone handling a variety of very old photographs should also be aware that many experimental processes — that may even have appeared briefly as a commercial venture — have since disappeared almost without a trace. Examples include anthotype (made with flower juices), energiatype, chrysotype, and chromatype.

Crayon prints are a variation on mixed media photographs. This is the technical name for a type of large print — usually portraits — produced from the time of the Civil War through the beginning of this century. These are handworked images created on top of weakly developed prints made on matte surface paper. This practice compensated for photographers' inability to control the tonality and densities of developing out paper, and because of the labor involved was supplanted as soon as enlargement became a practicable darkroom technique. Though the medium may vary from charcoal to wax crayon to pencil, you can recognize a crayon print because of the handwork evident throughout the surface. In addition, these prints can have a rather startling appearance created by the discrepancy between the lifelike pattern of the photograph and the somewhat wooden abilities of the photographer-artisan. They were usually made on highly acidic paper and often show a great deal of damage.

Collotype (less frequently called *albertype, phototint, heliotype,* and *photogelatin*), invented in 1868, was a printing method that used the differential absorption of ink by a gelatin colloid to create the image. It gives a very close approximation of the qualities of an albumen print, particularly when coated with an overlay of sizing. It was very widely used during the nineteenth century as a method of reproduction, and until recently most movie posters were printed by the collotype method. The image has an appearance of exaggerated sharpness, one that is quite distinctive, and certain identification can be made by close inspection that reveals a worm-like pattern of ink caused by the reticulation of the gelatin. Be careful not to confuse this with Fox Talbot's *calotype* process for making his paper negatives.

REFERENCES

1. H. and A. Gernsheim. *L.J.M. Daguerre, the History of the Diorama and the Daguerreotype*. 2nd ed. New York: Dover, 1968.
2. Beaumont Newhall. *The Daguerreotype in America*. 3rd ed. New York: Dover, 1975.
3. Alice Swan. "The Preservation of Daguerreotypes." *AIC Preprints, 9th Annual Meeting, Philadelphia, PA 27–31 May 1981*. Washington, DC: American Institute for Conservation, 1981, pp. 164 ff.
4. Robert A. Weinstein and Larry Booth. *Collection, Use, and Care of Historical Photographs*. Nashville: American Association for State and Local History, 1977, p. 165.
5. James M. Reilly. *The Albumen and Salted Paper Book: The History and Practice of Photographic Printing, 1840–1895*. Rochester, NY: Light Impressions, 1980.
6. James M. Reilly. *Care and Identification of 19th-Century Photographic Prints*. Rochester, NY: Eastman Kodak Company, 1986.

CHAPTER 20

Inspecting and Reframing Old Prints

THE IMPORTANCE OF PERIODIC EXAMINATION

As it hangs in its frame, any print can undergo drastic changes that will not always be seen by the casual observer. The same holds true of prints stored in cabinets, boxes, and drawers. This is a most important point: serious damage can be done to prints hanging in plain view, and can still go undetected unless an informed, concerted effort is made to determine what changes are occurring.

One reason for this is that the human eye does not possess the ability to make comparisons from memory. When a print begins to fade gradually, we have no way to tell that it is fading until we lift up the window mat to look at the protected areas around the edges of the print, where fading probably will not have occurred. Other deterioration takes place in the same way: yellowing paper, for example, might not be seen until it is quite advanced.

Much print damage does not appear at the front of the print until it is complete and irreversible. In this category can be included the staining that takes place from the back forward in prints improperly backed with corrugated cardboard or untreated wood. In this case, the staining shows up last on the front of the paper, after affecting everything beneath.

Mold and fungus grow quite happily between the interstices of the paper, and remain invisible until their damage has nearly been completed. Until you see the distinctive foxing produced by them, it is impossible to tell that they are present. They can, however, be detected by their distinctive odor when the frame is opened. Some types of paper-eating insect also live undetected in the confined world of a framed print, chewing and burrowing their way through both cellulose and various kinds of mucilage.

Small amounts of moisture that condense within the frame cause the print to adhere to the glass, if it is in direct contact with the surface. This adhesion goes unnoticed because it does not change the image itself.

These are the arguments for periodic inspection, including unframing, of important prints. However, what does "periodic" mean? This remains a matter of discretion and, for public collections, of economics. The ideal system for examining and storing prints would be reusable metal frames of stock sizes, with prints matted in stock sizes. After display, the prints are stored matted in safe containers, where a complete inspection would consist of carefully leafing through each container looking for any problems under each mat.

But let us be realistic. Many prints are put in permanent frames, or come to us in this fashion. They may need to stay in those or similar frames for various reasons. Given these limitations, it is our feeling that most framed prints should be taken out of their frames at least once sometime between every 5 and 15 years. Excessive handling, of course, can do some damage, but not nearly as much as lack of vigilance. Any print that has been newly acquired should be opened as soon as possible, even when newly framed, if only to determine that safe materials were used.

WHAT TO LOOK FOR

The key to successfully opening an already framed print is to stop and look very carefully before taking any other step. At this stage a great many things can be learned that cannot be reconstructed once the entire assemblage has been taken apart. It helps to organize your perceptions if comprehensive notes are made at each stage, and these notes can later be used to provide a permanent record. It also helps to have a form on which you enter data in a systematic manner. This form can be used to authenticate and catalogue pieces, and as the basis for any insurance claims that might later be made.

Initial examination should cover both good and bad points about the condition of the print. In addition to determining what damages the print may have incurred, it will help to determine the nature of the beast, so to speak. You will look for clues to the origin of the print, its probable history, and pertinent facts about how it was originally or later framed. In addition, you will want to determine both the medium of the print and the ways that medium was used.*

On the negative side, you will be looking for three types of

* A systematic procedure for determining the precise photographic medium of a print has been published by the Canadian Conservation Institute in the technical bulletin series [1]. This technique should find widespread use among institutions both in establishing working methods and in training new workers in the conservation field. A flowchart with samples of photographic media that is also very useful in identification is included in Reilly [2].

damage: mechanical, chemical, and organic. These categories apply to the framing assemblage and to the print itself. *Mechanical* damage covers areas like broken glass; a broken, chipped, gouged, or dented frame; punctures, tears, scrapes, or abrasions; folds; and water damage, especially stains and adhesions to the glass. *Chemical* problems include a wider variety of possibilities, ranging from fading or yellowing of the paper base to stains from outside sources such as backing materials. Also in this area fall crumbling or embrittlement due to high acid content, adhesives on mounting materials that have hardened, fading of dyes in the emulsion, and hypo residues. During the course of the entire examination, make an effort to determine exactly where and when these problems arose, so that they can be corrected. *Organic* damage can be equally problematic, but fortunately arises from fewer, and usually simpler, causes. Unless your print has large and ominous teethmarks on it, or some other unlikely damage, we can usually confine these organic agents to mold and to paper- and glue-living insects.

With the print still in the frame, start the examination from the back. There you can find the first signs of problems that have deeper implications, as well as helpful information about the date and provenance. Look for framer's labels, signatures, or identifying legends, and examine the condition of the dust cover. Has the dust cover crumbled and opened the interior to the outside environment? Does it have water marks on it?

Turn the piece over, and examine the frame itself. Is the glass intact? What kind of frame is it — wood, metal, gesso and gilt? Does the frame come from an assignable period? Is there dirt or insect debris visible inside the glass? Can you tell the medium of the print, as well as its date and subject matter, without opening the frame? And, since you are going inside, are there any visible signs that might give you pause about removing the print from its surrounding environment? Look very carefully at points where the print might be in contact with the glass; little differences in coloration or texture might hint that the print has formed an unnatural union with its glazing material. In short, look at the print not as a picture but as a very fragile physical object, and look at it carefully.

WORK AREA FOR OPENING THE FRAME

Cleanliness and precision count for everything. Keep the work area spotless, and always take care not to handle the print or frame roughly. It would be a shame for an image to have survived for many years, only to expire at the hands of one who wished to help it.

A good work surface for the framed print should be slightly resilient to prevent denting or scraping the exterior of the frame. We have used a kind of flexible foam that can be purchased by the square yard, but any flexible rubber mat would work as well. Another suitable work surface is sheets of Styrofoam covered with kraft paper.

The tool kit that needs to be assembled is not long, and consists of readily available materials.

1. Disposable-blade knife	removing paper backing, hinges
2. Pliers	taking off screw eyes, brads
3. Spatula	lifting print, mat, and glass from frame
4. Small hammer	knocking apart frame
5. Masking tape	holding broken glass in place
6. Sponge and acid-free blotter	application of moisture
7. Acid-free folder	holding print when removed from frame
8. Small brush	lifting dirt from print, removing insect remains and mold
9. Sheets of thick glass and acid-free cardboard	working support for print outside of frame

A strong, glare-free source of illumination proves a real boon when you examine the print, both inside and outside of the frame. If possible, the work table should have a shielded, neutral-color fluorescent fixure several feet overhead.

OPEN THE FRAME AND REMOVE THE PRINT

After examining the outside of the frame, rest the piece face up on the work table. Lay masking tape down the entire length of any cracks in the glass facing to secure the glass in one piece. Turn the piece over and remove all hanging attachments on the back, such as the wire and screw eyes; discard these until they are of value.

By hand, tear the dust cover to find out what is inside the back, and then trim carefully around the edges. Do not just stick the knife in and start hacking away, because this can result in serious problems. As a general rule, never insert a knife into any part of the frame where it is not possible to make a visual examination first.

Use the pliers to pull out the retaining brads all the way around the edges of the backing board. Discard the brads.

At this stage it is important to determine whether the glass, mat, and backing can be removed as a unit. Lift up one end of the frame and with the fingertips push gently on the glass to see whether it moves easily in the frame.

If the glass moves easily and is not broken, use it as a support to lift out the entire piece. Set the frame aside. If the glass is broken, take the print and mount out from the back — unless you suspect adhesion to the glass. In the latter case, tape the glass together on the front and rest both ends on the frame face down on some supports (a pile of books will do). Cut a piece of cardboard the same size as the inner face of the molding and use it as a support to lift out glass, print, and mat.

You may encounter instances where the glass has been set into the frame and attached independently of the mat and print. Use a dull-bladed spatula to shovel out the print and mat. Remove them as a unit, without disturbing any possible attachments.

At each step in the process, look carefully for significant evidence about the print's origin, materials like captions or signatures, and signs of deterioration like mold or insect remains.

The glass, print, and mat, once clear of the frame, should be laid face down on a clear area. If there is no indication of adhesion, move the mat and print separately to a clean sheet of paper, where further investigation will proceed.

It is sound working procedure to assign a job number to each frame when a large quantity is being examined. This number is then put on or with each part of the frame when it is laid aside for storage. Save any materials that bear on the history of the piece, like framing labels or notes on the back.

OPENING THE MAT

If the piece taken out of the frame was lucky enough to have a mat, it may be necessary to proceed with great caution in removing the mat. Some antique mats from the nineteenth century are quite ornate and of historic interest in themselves; when possible, save and reuse with a protective inner mat cut to protect the print.

Do not lift any of the sheets of paper in the mat or open the mat wide until the method of adhesion has been determined. Carelessly pulling the print away from the mat can destroy the print. In ideal cases, the print will be hinged into a mat constructed to the highest conservation standards. If this is the situation, you need not even take the print out of the mat when there is no sign of deterioration. Very gently, wearing white lintless gloves, lift apart the sheets of paper in the mat and print assemblage and peer between to determine how the unit was put together. Most likely you will find that the print has been joined in one of these ways.

☐ Stuck directly to the backing with spots of glue

☐ Glued overall to the backing

☐ Joined to the back of the window mat with spots of glue, with a continuous trail of glue around all edges, or with tape

☐ Sandwiched between window mat and backing with an adhesive around all edges of the mat, or

☐ Not directly attached to the mat in any way

It is so difficult to predict all the different ways that prints are found attached that no uniform method of working can be prescribed. Instead, we suggest the following working methods to be adapted as individual circumstances require.

When the print has been joined to the backing with spots of glue, work with the print face down and gently lift the backing board away from it so that a razor-sharp knife can be inserted to cut away the backing directly behind the glue. Essentially, what you are doing is cutting the spots of glue off the backing, along with some attached material. If the surface of the print seems fragile or likely to be disturbed by laying it face down, support the edges with blocks.

Once the backing has been lifted off, remove the remaining material by one of these methods.

- ☐ Scrape the remaining particles off with the flat edge of the blade, and with a nearly dry, clean sponge wipe away the remaining fragments and glue. Small amounts of organic adhesive can be allowed to remain if they do not seem to be affecting the print.
- ☐ Lightly moisten remaining pieces of board with a drop of cold water, and with fingers or tweezers pull parallel to the surface toward the center of the glue spot. Proceed slowly and stop if there seems any danger of tearing the paper.
- ☐ Where neither of these methods work, you can try the following, after testing a corner of the print's paper support for staining with a drop of warm water. Place a piece of acid-free blotter that has been lightly moistened on the stuck part, and touch it with a hot tacking iron to drive steam into the adhesive. Slowly roll the blotter back after moistening again, and lift off. With the knife scrape away softened board and adhesive, and repeat the process with fresh pieces of blotter until the back is clean. These same techniques can also be used to remove paper tape, cloth tape, and other kinds of attachments where a water-soluble adhesive joins the print to the backing.

When the print has been joined over the entire surface of the backing board and there seems to be little or no damage to the print itself, the backing may be allowed to remain. In cases where you can see that deterioration has occurred because of this condition, the only remedy is to entrust the care of the print to a paper conservator, who may or may not be able to restore the print. Be warned that the cost will probably be high.

If the face of the print has been stuck to the back of the window mat with glue, proceed in the same fashion as when removing the spot-joined back.

When tape has been used, cut through the tape away from the edge of the print and lift the print away to rest on a sheet of protective paper, face down. If it is water soluble, moisten the tape with cold water and, holding your hand quite low, pull the tape toward the center of the print.

If the mat has been stuck together in a sandwich, work from the sides with a scalpel to pry apart the two pieces. Do not cut into the mat vertically to the surface. If possible, push the knife gently into the gap between the two pieces and lever them apart, a small section at a time if necessary. Proceed all the way around the edges before attempting to lift them apart.

Pressure-sensitive tapes like masking tape and cellophane tape do not have water-soluble adhesives, tend to stain the print, become brittle with age, and should be removed (except when they have been used for mending tears, which means a visit to the paper conservator). Very brittle tape can be removed by flaking off pieces.

Pressure-sensitive tape and its adhesive can also be taken up by "dry cleaning" with one of four solvents: heptane, petroleum benzine, toluene, and acetone, in order of increasing strength and activity. These solvents are dangerous to touch or breathe, and should always be used in well-ventilated areas. They also represent a fire hazard, so treat them accordingly. Wear impermeable rubber gloves when using them, because some can be absorbed directly through the skin.

Test the solvent you intend to use before its application by putting a small drop on a blotter, which is then touched firmly against an insignificant corner of the print. If any part of the print is affected, or discoloration occurs on the blotter when it is lifted away after one minute, choose another solvent. Always work from the weakest to the strongest solvent. To avoid staining, the smallest possible amount of solvent should be used.

To remove the tape, bleed a small amount of solvent around the edge with a wad of cotton batting and allow it to soak in for about a minute. Lift a corner of the tape with the knife and pull the corner back using a pair of tweezers. Repeat this until all the tape has been pulled off.

MOLD

Sometimes when a frame is opened, the distinctive musty smell of a fungus infestation will greet you. Irregular splotches — mold — may or may not stain the print; in any case, the smell is a certain clue to the existence of this problem.

Fungi are very tough multicellular plants that thrive in dark, moist spots and that require a minimum of nutrients. They can leave permanent, irremovable stains on paper. The traditional practice of museums in treating mold infestations has been to place the affected print in a sealed cabinet with dishes of thymol crystals, which are then heated 3 hours a day for 4 days with a low-wattage light bulb. However, thymol is not safe for fumigating photographs, despite its efficacy as an antimold agent, because it can soften the emulsion of the print.

There are, luckily, several less drastic measures that you can take to combat mold. First, discard all contaminated material in the frame; this means everything but the print. Mold spores are numerous, tough, and invisible. Second, remove the print to a sunny area, preferably outdoors; with tweezers, a small brush, and a can of compressed air like Dust-Off, blow away any visible mold spots. Expose the print on both sides to half an hour of direct sunlight; the ultraviolet radiation will kill most of the remaining mold. Finally and most important, the best conservation measure in fighting a recurrence of the mold is to keep the print in a low-humidity environment. Mold needs moisture; deprived of moisture, the mold will simply go dormant. Mold control is as good an argument as any for storing prints in an air-conditioned environment.

CASE STUDIES

The following three case studies in examining prints show the benefits of periodically taking prints out of their frames. We chose at random three prints in our possession. Although two of them seemed to be in good condition and one had an obvious minor problem at first glance, in all three we found problems that would only have become progressively worse with time. Although we tended to take a pessimistic view to start with, even we were somewhat surprised at the extent of these problems.

In passing, we mention two things that we did not find in these examples, but that will frequently be encountered by anyone examining large numbers of framed pieces. The first is severe edge burn, a kind of browning caused along the lines where an acid-core window mat has been in direct contact with the paper support of the print. Most often this appears as a sharply defined rectangle the same size as the window mat opening. The edges of the print will usually remain pristine. *There is no real cure for this*, except by cutting a somewhat smaller window of good-quality board. Some conservators might wish to undertake a deacidification treatment of the entire sheet to arrest the further action of the deposited acids.

The other common problem we did not come across was staining from corrugated cardboard backing. This happens when a print has lain directly against the cardboard, and it shows up as a series of brown parallel lines approximately ⅛ inch wide. You might consult with a paper conservator about this, but at present there is no easy, effective way of salvaging a piece that has suffered this kind of damage.

WORLD WAR II PORTRAIT

Original Condition

The photograph is a World War II-vintage studio portrait of the father of one of the authors, Laurence Keefe, Sr., found in the original frame in an attic (Figures 20.1–20.7). Initial examination showed that the print seemed in good condition, although the thin paper mat and the backing paper had become discolored and embrittled. There was no evidence of water damage or adhesion to the glass. The fact that the print had been matted, even with an extremely thin die-cut paper mat, probably had helped prevent adhesion.

The mat had two areas of gray tint around the border, which appeared to be an applied watercolor. The proportions of the mat were "off" in that the side borders were very narrow and the bottom excessively wide.

The print was found in an inexpensive wood frame with an underlayment of ochre color gilded with silver paint. Exterior dimensions were 9⅝ × 14⅛ inches. It was decided after examination to discard the frame, which was without much historic value, and to reset in a larger frame to give more pleasing proportions to the mat.

Interior Examination

Interior examination of the photograph revealed a number of potentially harmful conditions, but little deterioration. There were no signs of mold or insect attack. The die-cut mat had been hand-trimmed along both edges and the top, indicating that the pho-

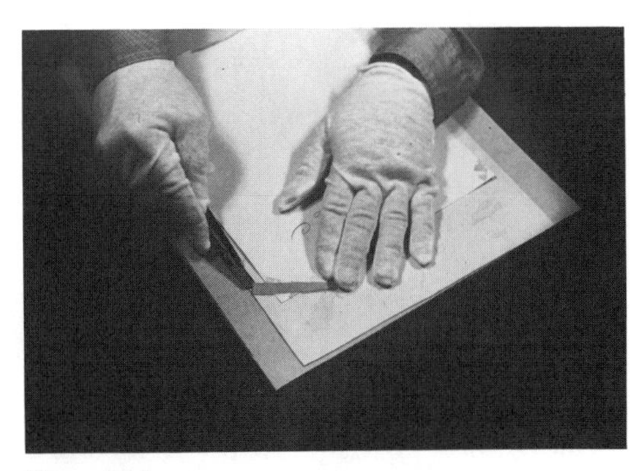

Figure 20.1
World War II portrait. Straps on the back of the print were carefully removed with a sharp knife and, where needed, with moisture.

Figure 20.2
To turn the print and mat over together, a piece of corrugated cardboard was put behind the print, already lying face down on another support.

Figure 20.3
The entire assemblage was turned over as a unit while the sides were held firmly. This technique is especially useful for old and fragile prints and should be used whenever there is doubt about the strength of the paper.

Figure 20.4
The print was uncovered, and . . .

Figure 20.5 (above left)
. . . the mat was removed, leaving the print supported on the cardboard.

Figure 20.6 (above right)
The print is now ready for careful examination, temporary storage, and mounting in preparation for reframing.

Figure 20.7 (left)
The portrait has been matted properly and inserted into a new frame.

tograph had originally come in a folder mat of the type commonly used by commercial portrait studios. It was probably framed at a later date. The print had been joined to the back of the die-cut mat with short strips of kraft paper, stuck diagonally across the corners with animal glue. Various signs of old adhesions were found on the top and the bottom of the mat, but there was no indication of what they had been joined to.

The print and mat had been held in place inside the frame by a piece of chipboard directly in contact with the back of the print. This chipboard was, in turn, nailed into the frame with a number of small brads around the edge, but it was not stuck directly to the print or window mat. The chipboard, upon testing with a pH-indicating pen, proved very highly acidic, and indeed already showed signs of flaking and starting to crumble at the edges. It presented the most serious immediate danger to the print, and it was promptly discarded.

The photograph itself was removed from the mat by cutting through the paper straps at the corner; it was then slid onto a sheet of clean board, where most of the remaining straps were removed with moisture and a sharp knife. The back of the print bears the imprint of a rubber stamp identifying it as the work of Neisner's, a department store studio active in downtown Rochester, New York, during the last five decades but now out of business.

When the mat had been removed, we turned the print over and found that the negative had been printed flush to the borders on a sheet of 8 × 10 paper. Some discoloration and fading around the edges appeared where the mat had overlain the paper, but it was not serious enough to warrant remedial attention.

The print had been made on brown-tone paper with a lightly textured surface, very similar and perhaps identical to Kodak's Portralure with a G surface. The paper had not become particularly brittle and showed no signs of discoloration or fading in the major image areas.

Evaluation

On the whole this portrait survived in very good condition, considering the primitive framing techniques used originally. Positive identification of both subject and photography studio was possible from identifying marks on the photograph and frame unit, although such information could also have come from family members. No particular care had been taken to preserve the framed photograph, other than to leave it undisturbed in an attic.

This photograph does provide an excellent example of the importance of periodically examining old prints. Allowed to remain in its original frame for much longer, there is no question that the fading around the edges and contamination from the acidic chipboard backing would have caused it to deteriorate quite rapidly, perhaps even spectacularly, when moved to a different environment. This process could have accelerated given the presence of large amounts of atmospheric moisture, when the print came down from a relatively dry, dusty attic to a normal living room environment.

Reframing

The portrait was placed in a brown 100% rag acid-free mat hinged onto a backboard of conservation board. The brown of the mat closely matches the khaki color of U.S. Army uniforms, and provides a fitting surround for the portrait. The hinged mat was backed with aluminum foil as a moisture barrier, and supported in place by a piece of archival-quality corrugated cardboard nailed to the frame. The frame chosen was a medium-brown walnut, lightly oiled and otherwise plain. Before closing the frame the subject's name and address, and the vintage of the photograph, were noted on the back, lightly, in pencil.

WOMAN IN CLASSICAL GARB

Original Condition

The initial examination showed that this print was in a wood and gesso-gilded frame with stamped metal corners (Figures 20.8–20.12). One corner fell off during examination to reveal that it had been attached with small brads that had corroded, probably be-

Figure 20.8
Woman in classical garb. Initial examination did not disclose any problems with the print other than some minor frame damage.

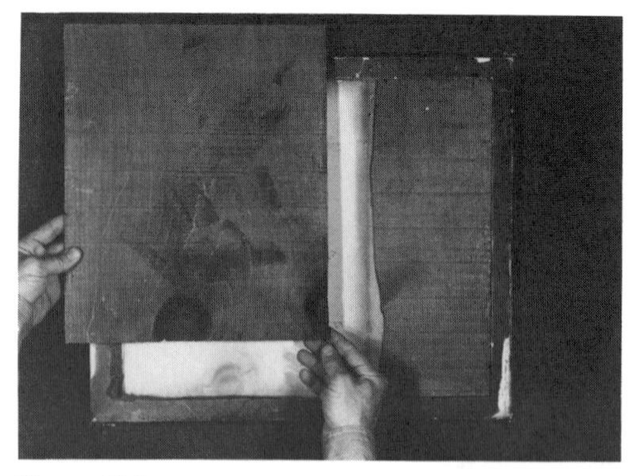

Figure 20.9
When the wooden back was taken off the frame along the lines where it had been fractured, the first signs of problems became evident.

Figure 20.10
A strongly developed stain had resulted on the back of the print's mount board from the resins exuded by a knot in the pine backing.

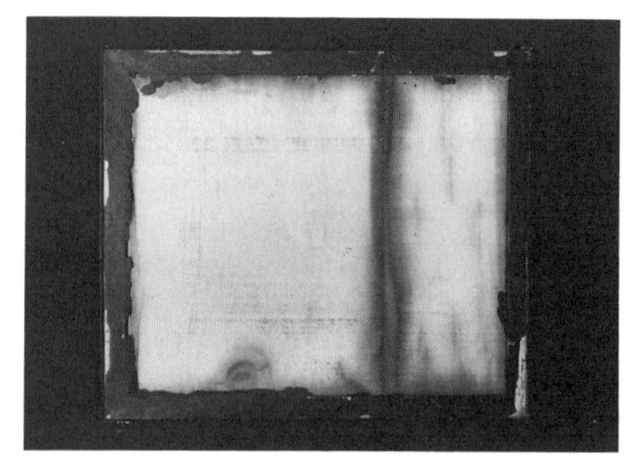

Figure 20.11
When the entire back was removed, a pattern of stain was visible from both the wood and from dirt that had entered through the fracture in the back.

cause of electrolytic action between the two different kinds of metal. The gesso-plaster underlayment on the gilding was cracked at several places on the top of the frame, and a large chip had fallen off the bottom edge. Exterior dimensions were 16½ × 17¾ inches, and the frame appeared to be the original one, although there were no identifying marks. The back of the frame consisted of two rough-sawn pieces of pine, approximately ⅛ inch thick, with several large knots and a large crack between the pieces.

Examination did not show any damage visible with the print still in the frame. There was no sign of water marks or adhesion. (This was rather surprising, for reasons discussed under "Interior Examination.") The subject of the print was a young woman in classical garb, seated on a carved marble bench of vaguely Medi-

Figure 20.12
A conservation-quality mat preserved the spirit of the original while protecting the finished pieces from the danger of adhering to the glass.

terranean pretensions. Behind the bench was an obviously painted backdrop with a misty motif of leaves and a tree trunk, which indicated that the print was a studio setup.

Interior Examination

When the nails holding the pine backing had been removed and the wood moved slightly apart, it was immediately apparent that resins in the wood had contributed to the deterioration of a none-too-stable backing. The imprint of several large knots was clearly seen as an almost black positive image, and the crack between the two boards had allowed pollution from the outside environment to mark the backing in similar fashion.

When the print had been removed from the frame and the glass lifted carefully off after testing for possible adhesion, the surprisingly good condition of the print appeared even more remarkable. It was an albumen print, hand-colored and mounted by overall adhesion to a piece of highly acidic mount board. An edge, beveled outward, had been cut around the entire mounted piece, and it had then been glued onto another sheet of gold-painted board. This painted backing had been the only protection against chemical intrusion from the wood at the rear of the frame. Furthermore, the print had been in close contact with the glass of the frame, but it had never been seriously threatened by water.

With the dirty glass of the frame removed, it became possible to read a small inscription written originally on the negative; it said "Copyright 1898, Br Baker's Art Gallery, Col. O." This made it possible to date the picture to around the turn of the century, although it was obviously a mass-produced piece that might actually have been made somewhat later than the ostensible date.

It was decided to remove the painted back matting rather than try to preserve it. In addition to being brittle, it contained residues of contamination by the wood boards; further, the value of the print did not justify the extra effort of building spacers to hold the print away from the glass while reframing. Although the board to which the print had been joined also showed signs of deterioration, the danger of trying to soak it off was great enough to warrant caution in proceeding further, especially since there was no sign of deterioration on the print. There were no signs of mold or insect damage, and the surface of the print was pristine, free of any mechanical scrapes, cuts, or abrasions.

Evaluation

The photograph is an excellent example of Victorian "kitsch" with pseudoclassical pretensions, a genre meant to hang in the drawing rooms of the well-to-do middle class. Despite its excellent condition, it has a low historic and artistic value in itself, although it could hang very well in a period environment.

No restoration techniques were called for. This was due, in part, to the probably excellent processing the print received in the hands of its manufacturer. Despite the highly acidic nature of the board on which the print was mounted, it did not seem to be suffering noticeably from this contact. Finally, any contact with moisture might have damaged the hand-coloring pigments or removed them entirely, and in the process removed whatever artistic merit the image possessed.

Once the potential for damage from the wood backing boards had been removed it appeared that the only other significant problem to solve was the direct contact of the print with glass in the frame.

Reframing

Despite its somewhat garish appearance, the original frame was retained as appropriate to the distinctly ostentatious appeal of the print. Chips off the gesso were refilled. Gesso would have been appropriate, but for reasons of convenience we used joint compound of the kind made for drywall construction. It worked quite well. The entire frame was then refinished in gold color. The loose corner was reaffixed with brass brads, and where needed the other corners were also tightened.

Next, the picture was rematted in conventional archival fashion, with one important exception. Because the thickness of the backing board on the print would have caused the window mat to lift in the center and settle toward the back along the edges, strips of acid-free board were cut to nearly the width of the borders and stuck to the backing board of the mat with two-sided transfer tape. These strips provided a concealed spacer of identical depth to that of the print, and so will prevent buckling.

The rabbet of the refinished frame was sealed by painting with gesso. After the glass had been cleaned thoroughly and with the print placed in the frame, the back was covered with acid-free corrugated cardboard and sealed with a kraft paper dust cover.

MEDICAL TRAINING COMPANY GROUP PORTRAIT

Original Condition

This panoramic group portrait from World War I showed signs of serious deterioration when first examined. It was framed in a half-round black painted wood molding that measured 7½ × 25½ inches on the exterior. The glass had cracked from top to bottom approximately 3 inches from the right side of the frame. Dirt had penetrated through the crack in the glass, and could be seen inside on the surface of the print as an accumulation of coarse black grains. In addition, serious stains had appeared at the top and bottom of the crack, as well as approximately halfway down its length. Closer examination of these areas revealed several greenish blotches, one about ¼ inch in diameter and another ¾ inch in diameter. These were correctly assumed to be adhesions (Figure 20.13). It was considered a good sign that both adhesions had taken place on the smaller piece of glass, which would facilitate their removal. A strip of masking tape was placed along the length of the crack to prevent the glass from shifting when the print was removed from the frame. The print had been framed without any kind of mat that would have separated it from the glass.

Figure 20.13
Medical Training Company Group. A close examination of the spot where the glass was cracked showed stains and possible adhesions in the lower corner.

Figure 20.14
The dust cover on the back of the frame showed wrinkling due to water damage. Most water had been on the bottom and on one end, where the glass was also broken.

Figure 20.15
When the dust cover came off, it was evident that two separate pieces of backing board, butted end to end, were used. Brads are being removed with pliers.

Figure 20.16
A side-by-side comparison of the dust cover, backing boards, and the back of the print allows assessment of the pattern of damage.

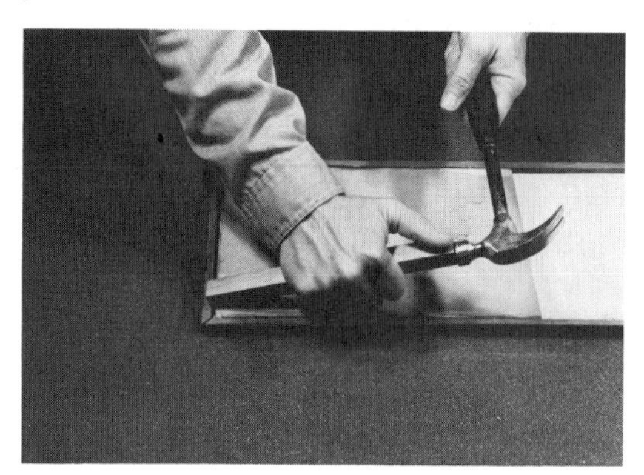

Figure 20.17
The frame was knocked apart (very gently) to free the print, so that print and glass could be removed together.

Figure 20.18
The print and glass end face down on a sheet of cardboard after the frame has been taken from around the edges.

It was possible during initial examination to decide the medium of the print (silver gelatin) and the date and place of its origin from the inscriptions. A note on the lower left side read "Medical Training Dept. Co. E. Fort Ethan Allen Vt. Sept. 1917." The other, lower right side inscription read, "Photo by McAllister/Burlington Vt." This kind of information inscribed on the negative made it unnecessary to attempt identification from the distinctive period uniforms of the soldiers in the company. The panoramic format of the photograph makes it clear that McAllister was a commercial photographer specializing, probably among other things, in group portraiture. After making his exposure, he would have sold prints to individual soldiers; this was probably one of these prints. Given the rapid rate of mobilization and training during this early period of World War I, the prints were in all likelihood cranked out at great speed and without much regard for such niceties as fixing and washing.

The back of the frame showed further signs of deterioration. The dust cover of dark brown kraft paper had become brittle and was crumbling. It showed signs all along the bottom of water staining, as though the frame had rested in water an inch or more deep (Figure 20.14). Attached to the back was a small label bearing

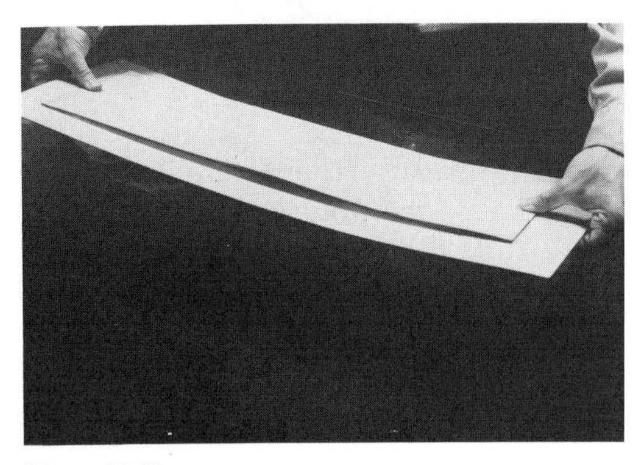

Figure 20.19
The entire piece is turned over to allow work to proceed from the front.

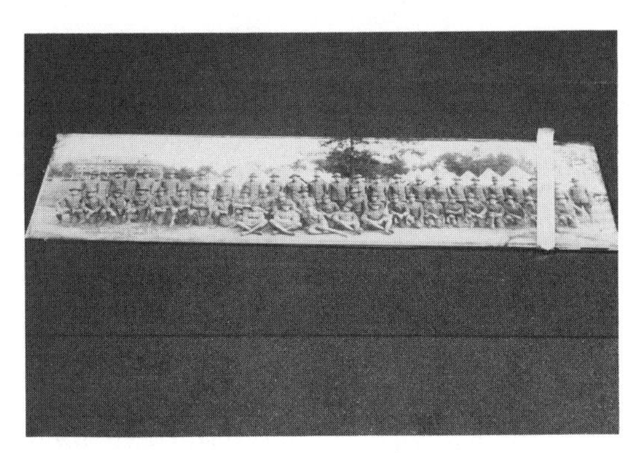

Figure 20.20
At this stage, the print still has the glass on top, held together with a strip of masking tape to prevent it from moving. The glass is now ready to be taken off; the piece on the left will come off easily, while that on the right has some pieces of emulsion adhered.

Figure 20.21
Once the glass has come off, a closer examination of the print is possible. Notice how it is darker in the center than at the edges.

Figure 20.22
Careful cleaning with a dry brush removes small bits of embedded dirt that had penetrated along the crack line through the cover glass.

Figure 20.23
A conservation-quality mat and a new frame cover the damage and protect against further deterioration, while hiding some of the problems that have already affected the photograph.

the slogan, "The Sign of Quality/'Savage'/The Picture Frame Specialist/Batavia, New York." Because the print was acquired at an antique shop in upstate New York, the assumption was made that the print had been framed later than its making, probably on behalf of one of its subjects.

Interior Examination

Removal of the print from the frame was judged a highly critical procedure, as in fact it turned out to be. Once the backing paper had been torn away, it was found that the print had been backed with two pieces of chipboard of unequal size, separated by a large crack and nailed into the frame with small brads.

Removal of the brads and careful attempts to lift the contents out of the frame showed that the glass was wedged (but, happily, not otherwise secured) in the frame. After looking to see which way the corner nails were set, we gently knocked apart all four corners until the contents could be removed, and we put them face down onto a sheet of archival corrugated cardboard.

Lifting off the chipboard revealed severe wrinkling of the back of the right side, caused again by the water. The most severe wrinkling was found behind the cracked piece of glass. Dark brown stains covered many wrinkled areas that had contacted the chipboard backing directly; these had a glossy finish, which indicated that some kind of adhesion had occurred on the back as well. There was a dark brown stain along all four edges, probably caused by the action of resinous vapors from the wood frame. No other markings or identification could be found.

The exoskeleton of a small insect, shed in molting, was found, but no signs of insect damage were visible, nor did we see any signs of mold.

The print and glass were then backed with a sheet of glass and inverted. After the masking tape was taken off, it was possible to remove the larger piece of cracked glass (Figure 20.20). Hot compresses against the front of the glass failed to release the adhesion to the smaller piece.

The print and attached glass were again turned over, and after blot testing, a small amount of moisture was applied to the back of the print itself. This water reduced the brittleness of the paper and softened the gelatin emulsion so that it was possible to carefully peel the print away from the glass by very slowly lifting the corner. Had this failed, we would have laid a moist, acid-free blotter on the back of the print, and touched it with a hot tacking iron to drive moisture into the paper.

At this time it was possible to see that there was also a stain in the upper left corner that had been largely hidden by the molding, and a light yellow-brown stain along the bottom. In addition, once out from behind the grimy glass, the print showed very faint signs of fading gradually out from the center.

Evaluation
The photograph is a not particularly significant example of popular commercial photography. Mass produced and mass marketed, it is a larger, more sophisticated example of the same kind of photographic endeavor represented during an earlier war by the tintype. If it survives, however, it will achieve a certain kind of distinction merely because of its age; already its period qualities look primitive, almost tribal, to us.

This photographic print is an example of almost all of the many dangers to the silver gelatin print that can result from improper processing, inexpensive framing, and careless storage.

- ☐ *Improper processing*. The black stains on the right showed that the print was underfixed in an exhausted fixing bath. Residual silver halides turned black by themselves without the need for a chemical developer. The fading toward the edges is also symptomatic of too much residual hypo — that is, of a lack of sufficient washing.
- ☐ *Inexpensive framing*. The sins of the Savage framing specialist (aptly named!) are too numerous to list, but they include, among others, the omission of a mat, putting the print into direct contact with the wood molding, and backing directly with chipboard.
- ☐ *Careless storage*. The cracked glass and the extensive water damage resulted directly from rough handling and keeping the print in an area, probably a basement, where water could reach it. In addition, the print was evidently not removed from the frame when the damage was uncovered.

A final question arose. Should the photograph be reprocessed in order to compensate for the obviously inadequate fixing and washing it had received initially? In accordance with the philosophy that the best conservation treatment is the most conservative, we decided to forgo reprocessing. The reasons were several, but the decisive one was that overall the picture had undergone only mild deterioration due to these causes. Protection from further contact with moisture and high humidity would do much to prevent any further degradation of the image quality, and since its slow fading — the print is already 63 years old — is not likely to accelerate abruptly, occasional checking will indicate whether it needs reprocessing at a future date. Another reason is that further fixing might actually remove some detail by bleaching out areas where the silver had been contaminated by thiosulfate residues. Finally, there was the uncertain question as to what effect further moisture might have on the areas that had already been joined to the glass by water; it was thought that the emulsion might actually peel away from the paper.

Reframing

The original frame was discarded because of size and because it had no historical or aesthetic value. Further, a window mat was desired because it made it possible to conceal some of the edge damage.

A mat of 4-ply ivory conservation board was cut with borders 2¼ inches at top and sides and 2½ inches at bottom; it was backed with another sheet of 4-ply conservation board. Overlap on the picture was sufficient to conceal most damage while leaving the identifying label visible.

A square-edged, very dark brown wood frame was chosen because it would not clash with the obvious historical period of the image. Glass in the frame was sealed along the edges with Scotch Magic Transparent Tape, and the back of the mat was covered with a sheet of aluminum foil; both these measures will protect against further moisture damage, from which this print had already suffered. Finally, a sheet of acid-free corrugated cardboard was put behind the mat in the frame, and the back sealed with a kraft paper dust cover.

REFERENCES

1. Siegfried Rempel. *The Care of Black and White Photographic Collections: Identification of Processes*. Ottawa: National Museums of Canada, 1980.
2. James M. Reilly. *Care and Identification of 19th-Century Photographic Prints*. Rochester, NY: Eastman Kodak Company, 1986.

A P P E N D I X A

Suppliers

This list includes vendors from whom you can purchase materials for the conservation of photographs. Some specialize in supplying only materials for conservation, while others supply a more diverse product line. We would be surprised if anyone could get seriously engaged in preservation without drawing on at least some of the resources offered by the people and companies on these pages.

Any list such as this must end up being incomplete, and we apologize in advance to suppliers we inadvertently left out, especially those engaged in the difficult business of making tools and materials for the conservation trade.

Aabitt Adhesives, Inc.
2403 North Oakley Avenue
Chicago, Illinois 60647

Jade 403 adhesive

Alto's EZ/Mat
818 16th Avenue West
Kirkland, Washington 98033

mat cutters

American National Standards Institute
Sales Department
1430 Broadway
New York, New York 10018

catalog of ANSI standards

Andrews Nelson Whitehead
31-10 48th Avenue
Long Island City, New York 11101

rag and conservation board

C&H Bainbridge
40 Eisenhower Drive
Paramus, New Jersey 07653

mat cutters

Bainbridge/Letraset U.S.A.
40 Eisenhower Drive
Paramus, New Jersey 07653

mat board rag and conservation

Berg Color-Tone, Inc.
P.O. Box 16
East Amherst, New York 14051

suppliers of gold protective toner

Dick Blick
P.O. Box 1267
Galesburg, Illinois 61401

mail order supplier of artists' materials and tools

Calumet Photographic
890 Supreme Drive
Bensenville, Illinois 60106

mail order supplier of darkroom equipment and archival washers

Charrette
31 Olympia Avenue
Woburn, Massachusetts 01888

mail order supplier of artists' materials

Clark Moulding Company, Inc.
11567 Hillguard Road
Dallas, Texas 75243

metal frame moulding

Clear Mount Industries
585 E. 49th Street, Suite 18-406
Hialeah, Florida 33013

face mounting services

CODA, Inc.
196 Greenwood Avenue
Midland Park, New Jersey 07432

mounting supplies/laminating film

Columbia Corporation
Route 295
Chatham, New York 12037

mat board manufacturer

Conservation Resources International, Inc.
800 H-Forbes Place
Springfield, Virginia 22151

conservation board, print enclosures

Crescent Cardboard Company
100 W. Willow Road
Wheeling, Illinois 60090

mat board manufacturer

Dax Manufacturers, Inc.
955 Midland Avenue
Yonkers, New York 10704

manufacturer of plastic box frames

Designer Moulding
6910 Preston Highway
Louisville, Kentucky 40219

metal frame moulding

E.I. du Pont de Nemours & Company, Inc.
Product Information
1007 North Market Street
Wilmington, Delaware 19898

polyester film and related plastics

Edmund Scientific Company
Edscorp Building
Barrington, New Jersey 08007

mail order supplier of laboratory apparatus

Eastman Kodak Company
Kodak Laboratory Chemicals
Rochester, New York 14650

specialty photographic lab chemicals

Eubank Frame, Inc.
P.O. Box 425
Salisbury, Maryland 21801

clip frames

Filmolux, Inc.
39 Comet Avenue
Buffalo, New York 14216

mounting supplies/laminating film

Fisher Scientific Company
711 Forbes Avenue
Pittsburgh, Pennsylvania 15219

laboratory equipment

Fletcher-Terry Company
Spring Lane
Farmington, Connecticut 06032

tools for fitting frames

Fotographika, Inc.
6735 I.H. 10 West
San Antonio, Texas 78201

face mounting services

Frame Tek
2134 Old Middlefield
Mountain View, California 94043

Framespace

Franklin Distributors Corporation
P.O. Box 320
Denville, New Jersey 07834

Saf-T-Stor

Gallery Clips
P.O. Box 2388
Boston, Massachusetts 02107

clip frames

Gold Leaf and Metallic Powders, Inc.
2 Barclay Street
New York, New York 10007

gold leaf

Graphic Laminating, Inc.
5122 St. Clair Avenue
Cleveland, Ohio 44103

laminating film

H.P. Marketing Corporation
98 Commerce Road
Cedar Grove, New Jersey 07009

Gepe mounts

Hollinger Corporation
P.O. Box 6185
3812 South Four Mile Run Drive
Arlington, Virginia 22206

boxes and storage enclosures

Howard Paper Mills, Inc.
115 Columbia Street
P.O. Box 982
Dayton, Ohio 45401

Permalife paper

Interior Steel Equipment Company
2352 East 69th Street
Cleveland, Ohio 44104

museum storage cases

JBC Imports and Marketing
326 Palomar
Shell Beach, California 93449

clip frames

Key Color Lab (Canada)
Pannier Graphics
John Fitch Industrial Park
Warminster, Pennsylvania 18974

embedment services

Kinex Corporation
34 Elton Street
Rochester, New York 14607

copy stands

K-S-H Inc.
10091 Manchester Road
St. Louis, Missouri 63122

UV-filtered glazing

Knox Manufacturing
111 Spruce Street
Wood Dale, Illinois 60191

slide viewing, slide storage

Kostiner Photographic Products, Inc.
P.O. Box 94
204 Main Street
Haydenville, Massachusetts 01039

archival print and film washers

Kulicke Frames
601 West 26 Street
New York, New York 10024

frames

Light Impressions Corporation
439 Monroe Avenue
Rochester, New York 14607

mail order supplier of matting and framing materials and archival supplies

Logan Graphic Products, Inc.
1100 Brown Street
Wauconda, Illinois 60084

mat cutters

Luxor Corporation
2245 Delany Road
Waukegan, Illinois 60085

slide viewing, slide storage

MACtac
4560 Darrow Road
Stow, Ohio 44224

laminating film

MC/B Manufacturing Chemists, Inc.
2909 Highland Avenue
Cincinnati, Ohio 45212

wholesale distributors of colorpHast pH indicator strips

Mat Magic Products
95 Mitchell Boulevard
San Rafael, California 94903

powdered colors for French mats

Museum Systems
Archival Products
4129 Sepulueda Boulevard
Culver City, California 90230

Insta Hinge

National Draeger, Inc.
P.O. Box 120
101 Technology Drive
Pittsburgh, Pennsylvania 15230

air pollution detection kits

Neumade Products Corporation
720 White Plains Road
Scarsdale, New York 10583

slide storage

Nielsen/Bainbridge
40 Eisenhower Drive
Paramus, New Jersey 07653

metal frame moulding

North American Enclosures, Inc.
35 Drexel Drive
Bay Shore, New York 11706

metal frames

Photofile
P.O. Box 123
Zion, Illinois 60099

negative storage

Photo Plastic Products, Inc.
Box 17638
Orlando, Florida 32860

print file, negative storage

Process Materials Corporation
301 Veterans Boulevard
Rutherford, New Jersey 07070

rag and conservation board

Reeves Photo Sales, Inc.
9000 Sovereign Row
Dallas, Texas 75247

polyethylene negative envelopes

Rising Paper Company
Housatonic, Massachusetts 01236

rag and conservation board

Rogers Anti-Static Chemicals, Inc.
22 West Madison Street
Chicago, Illinois 60602

suppliers of AR-8 anti-static polish for Plexiglas

Rohm & Haas Company
Independence Mall West
Philadelphia, Pennsylvania 19105

UF-3 Plexiglas

Russell Harrington Cutlery, Inc.
44 River Street
Southbridge, Massachusetts 01550

Dexter mat cutters

S&W Framing Supplies, Inc.
120 Broadway
Garden City Park, New York 11040

mail order supplier of framing materials

Seal, Inc.
550 Spring Street
Naugatuck, Connecticut 06770

dry mount press and materials

Solar Screen Company
1032 Whitestone Parkway
Whitestone, New York 11357

UV filtering materials

Structural Industries
96 New South Road
Hicksville, New York 11801

metal frames

TALAS
213 West 35th Street
New York, New York 10001

mail order supplier of conservation and matting supplies

Thermoplastic Processes, Inc.
Valley Road
Stirling, New Jersey 07980

UV filtering tubes, Arm-A-Lite filter ray

Arthur H. Thomas Company
Vine Street at Third
P.O. Box 779
Philadelphia, Pennsylvania 19105

laboratory apparatus and chemicals

3M Company
3M Center
Building 220-7E
St. Paul, Minnesota 55144

Scotch brand tapes and dry mount materials

United Mfrs. Supplies, Inc.
3 Commercial Street
Hicksville, New York 11801

framing supplies

University Products, Inc.
P.O. Box 101
Holyoke, Massachusetts 01041

mail order supplier of board and matting supplies

Victor Moulding Company
P.O. Box 2206
Oakland, California 94621

framing supplies

Vue-All, Inc.
Box 1994
Ocala, Florida 32678

negative and slide envelopes

Watco ID Systems
P.O. Box 707
Industrial Parkway
Shelbyville, Tennessee 37160

embedment services

Wei To Associates, Inc.
P.O. Box 419
224 Early Street
Park Forest, Illinois 60466

deacidification solutions and sprays

Westlake Plastic Company
West Lenni Road
Lenni Mills, Pennsylvania 19052

UV filtering materials

Windsor Graphics
P.O. Box 1287
Galveston, Texas 77553

museum mounting kits

Yale Picture Frame and Moulding Corporation
770 Fifth Avenue
New York, New York 11232

wood frames

Zone VI Studios, Inc.
RFD 1
Putney, Vermont 05346

archival film and print washers

APPENDIX B

How to Order Framing and Conservation Supplies Through the Mail

For people and institutions actively engaged in the display and storage of important prints, it is a fact of life that many tools and supplies have to be bought sight unseen (at least for the first time) through the mail. A market for these materials has not yet grown to where retail outlets can support themselves by selling and stocking these materials, and only an occasional large city has a local supplier for them.

Since both of us have had extensive experience in mail order sales, we include this section on how to order top-quality materials through the good offices of the U.S. Postal Service. A well-informed purchaser can buy materials by mail in a way that ensures satisfaction with the goods received.

RESEARCHING THE MARKET

The most important single rule in mail ordering is to know what you need. Before you order, or even consider ordering, know what product you need, what it will do for you, who makes it, and who sells it. Use the library extensively. Advertisements in trade journals and catalogues published by mail order houses provide most of the information needed for satisfactory purchasing — *if* they are read with discrimination.

Write away for mail order catalogues, technical data sheets, and price lists, and save them all for comparison of descriptions and prices. This gives necessary information about specific products available on the market. If you are ignorant when you send your money away, you will be disappointed when your order is filled.

FILLING OUT THE FORMS

On almost every order form enclosed with a mail order catalogue, there appear the words, "Print or type clearly." Do it. A thousand times over, do it. Even with the best intentions in the world, the secretary who opens your order and the shipping clerk who tries to fill it cannot tell what your desires are except by what appears on that piece of paper. Make it as complete as possible, as precise and clear as you can, and follow all of the instructions. This is not just to make easier the jobs of the people who work at the mail order house; it is also so you get what you want, quickly.

Use common sense in filling out the form. For example, supply a phone number where you can be reached during the day. Do not just put down your home phone, and assume that the mail order house will call until someone finds you at home — especially when you are at home only in the evenings.

METHODS OF PAYMENT

Basically, you have six options for payment. One of them, if selected wisely, should make parting with the money as painless as it is ever likely to be.

1. Send a *check or money order* with the order for the full amount of the order, with sales tax and shipping calculated. Never send cash; then you have no record that you ever paid your bill.
2. Fill in the order blank, and include (a) a *credit card account number and expiration date,** (b) your signature, and (c) your phone number. You can either fill in the full amount or let the mail order house total the purchases and compute tax and shipping costs. In this way, if there has been a price increase, you will not have to send them a new order, which saves much time. Also, it is easier not to have to figure out the shipping costs. *Never* mail your credit card with an order.
3. *Phone* an order into the mail order house. Again, this requires a credit card they will accept. It saves time and informs you of the current price and full amount of purchase *before* committing yourself.
4. Send your order for the goods to be delivered *COD* (cash on delivery). Few mail order houses accept this kind of order without a deposit, which is often the full amount of the purchase price, plus tax, with the shipper collecting freight charges on delivery. It is unrealistic to expect the company to send out goods COD without prior arrangement unless it states specifically that it will ship in this fashion. This method used to be much more commonly accepted than it is today.
5. Goods can be paid for by *invoice* — which is to say, *billed* to institutions or other businesses by a mail order house. Usually terms of payment are "net 30," which means paid in full within 30 days after the date on the bill. In most cases your business or institution will be invoiced only after an account (with credit references) has been opened with the seller and after credit arrangements have been made. Often, cash has to be paid for the first order anyway; it pays a business to carry an invoice on account only for customers who buy regularly and in volume. A purchase order or requisition system of ordering may be necessary. Order by mail or phone.
6. Pay *cash in person* at the mail order house, and have the goods shipped to a residence or receiving department. This is not exactly "mail order," but it gets your material delivered to where it is needed.

* Check the catalogues and advertisements to see which credit cards are accepted; most mail order businesses will take at least MasterCard and Visa.

SALES TAX

A sales tax can add considerably to the cost of a purchase, particularly when you spend several hundred dollars. It pays to be aware of several ways to legally avoid the added cost.

A sales tax is collected either by a state or local government, or both, but never by the federal government. This means that orders through the mail from one state to another may not have to include sales tax; for example, goods ordered from a New York state firm by a customer in Texas are effectively exempt from sales tax.

In general, an institution or firm registered with the Internal Revenue Service in the category of "not-for-profit" also escapes sales tax. Usually the presentation of some kind of certification will be necessary; check with the local sales tax office.

In some states, like New York, a special provision in the sales tax law exempts businesses that purchase material for resale. For example, a framing business that buys frames to sell its customers does not pay tax when buying the frames. It must, however, act as a collecting agent for the state when the frames are sold to the end user. Instructions on tax exempt certificates can be had from the local sales tax office.

SHIPPING

One way or another, the customer pays for shipping. If a catalogue says that materials are shipped *postpaid* (the shipper pays shipping charges), examination usually shows that the prices tend to run somewhat higher than those of competitors who sell the same material. After all, shipping is a legitimate cost of business, and the customer must expect to pay for the convenience and expense of individual shipping of packages, which costs a great deal more per unit of weight than do bulk shipments sent to wholesale buyers.

Specify your preferred method of shipment. Since you are paying for it, you might as well choose, if possible. Some firms, like the ones with concealed shipping charges or fixed "handling" costs, will not ship via your method unless extra is paid; check with these firms before ordering.

The different methods of shipment:

Parcel Post

This is via the U.S. Postal Service. It is not necessarily the cheapest, fastest, or most convenient way, although service has of late been improving slightly. In general it is not really competitive with other ways. "Insured registered" designation requires the carrier to get a signature, but it might be your neighbor's. There are also size and weight limits, and some smaller post offices have limits on the weight and parcel size they will deliver. Check with your post office before ordering.

United Parcel Service (UPS)

This is the best of the private parcel delivery services — usually fast, safe, and reasonably priced, even though it pays its drivers Teamster wages. The carrier gets a signature but, again, it might be your neighbor's. UPS also has size and weight limits. It delivers only inside the United States and parts of Canada, and *never* to Post Office or APO box numbers.

Bus

Most bus companies offer high speed at modest prices, and tend to have larger size and weight limits. The package must be picked up at the station nearest its destination, and the mail order house may charge additional to deliver to the bus station at its end.

Truck

This can be very tricky. Trucking firms are geared for commercial deliveries to businesses that know what they are doing. Somebody has to be there to receive the shipment, because all a trucking company is legally required to do is to put the load on the sidewalk — and they may do just that, although many drivers will be quite helpful if you are there and need a little assistance. The terminology FOB, as used in the expression "FOB Tucson," for example, means that the shipper will put the load "free on board" the truck at its plant in Tucson; from there, the recipient pays the shipping charges. The "free" means that, generally, the recipient does not pay crating or packaging charges.

If you decide to ship by truck because your load exceeds size or weight specifications for other methods, do some research. Specify a *direct shipper* when ordering; it may be necessary to call around, using the Yellow Pages as a guide. Make sure that your order avoids transshipment by regional carriers; every time a package gets taken off one truck and put onto another, time is wasted and the risk of damage or loss goes up.

Air Express

This is very expensive and very fast. Firms like Emery, Federal Express, Purolator, and REA, and most airlines, offer quick delivery of small packages. A courier often picks up the package and delivers it from the airline, or shipment can go from the terminal. It is hard to beat, but make sure that you want to pay the price.

Rail

Forget it. It is too expensive, too slow, and not geared for the small shipper.

RECEIVING

Look it over as soon as your package comes in. If your inspection shows that it is broken or damaged on the outside, note this on the receiving slip when signing for delivery. Save the original packing if damage is found after the package is opened. Inspectors will want to see it.

Contact the company from which you ordered the goods to get instructions. You (or it) have 30 days to file a claim with the shipper; in the meantime, it is a good idea to be hopping mad, let the company know it, and tell it that you want to know what it plans to do to remedy your problem. Now!

The law and the federal government are very clearly on your side if your goods are damaged, so you will get your money back or a satisfactory replacement. Do not be shy about your rights. Check applicable duties before ordering from abroad, and be ready to pay them when the customs office tells you that the shipment has come in. If you live in Canada, it may save you some money to know that there is no duty on types of conservation supplies not made in Canada. Check with your nearest customs office.

REASONS TO ORDER THROUGH THE MAIL

In general, we favor the practice of ordering by mail or phone. Institutional purchasers, of course, do not go into a retail store and browse for items because of the accounting requirements of purchase orders and so forth, but private individuals can also find many benefits from ordering from catalogues.

As mentioned, there is the practical necessity of mail ordering many items used in archival framing and storage. A larger selection of related items is available by mail.

Many times it is possible to compare prices most accurately from catalogues because all the facts and figures are right in front of you. Shopping for quality can sometimes be easier, if you take the time to compare prices. A great discrepancy in prices may give you a clue that items that seem identical are in fact the genuine article and a cheap imitation. Beware of bargains that seem too good, and look at the photographs and descriptions very carefully. A close scrutiny often discloses important differences.

Catalogues and price lists serve for a small reference library if collected and used assiduously. If read from cover to cover they often disclose a great deal more information than seems possible at first.

Look for quantity discounts offered through catalogues. Large mail order houses stock in quantity, which means that they buy in quantity. Often it pays them to pass along a discount to increase the size of their sales. Minimum purchases may be a hassle to the small buyer, but often several people can get together to earn a quantity discount or to order enough to make a minimum purchase on large-scale items.

Federal antifraud laws give the mail order buyer better protection in many cases than the average retail buyer gets for purchases of similar items.

SELECTING FROM CATALOGUES

By carefully perusing a catalogue, you can gain a sense of the integrity of the entire operation. Sometimes the most revealing part of a catalogue is material not directly related to goods offered; the letter from the president at the beginning, for example, or disclaimers on shipping policy.

Remember that a catalogue's production and wording are entirely under the control of the company that puts it out. This company controls the image it wants you to have of it. Certainly, this means that an anonymous empty barn in Utah can pose as a center for the sale of all kinds of wonderful goods, since nobody ever sees it. However, such a place rarely gets into the field of fine arts and framing materials. More to the point, one can gather whether the approach of the company is strictly businesslike, somewhat folksy, fanatically dedicated to the finer points of its chosen area of business, or just plain loony. Then you can judge its claims one way or the other.

After picking up on the company's philosophy, take a hard look at its price structure and ordering policies. It may have minimum orders, minimum quantities on specific items, surcharges for shipping, handling charges, or very high prices on small amounts. These details can vary from company to company, and often they make a vast difference on the bottom line on the order form. When you consider a large purchase like a dry mount press or a storage system, shop around and use a calculator. A great many hours go into figuring out the price schedules that get published, and every mail order house scrutinizes the catalogues of its competitors very carefully — but it does not always consider everybody who sells the same item a competitor. For example, a house that sells to libraries and educational institutions often carries hardware at the manufacturer's list price, while a discount photographic house in New York City in the retail market sells the same gear at discounts of 30% or more.

Price and quality, as one may gather, do not always go hand in hand. It is important to be able to discern exactly what you are buying. There are a number of ways to get top quality for your money. An easy way is to buy brand names. If you order, for example, a Nikon or Olympus camera, Windsor & Newton brushes, Nielsen frames, or similar products, it is quite easy to compare prices and pick the lowest, because you know what you will get.

A second determining factor, when you cannot buy brand names, is the technical description. Look for test results, dimensions, and chemical or material specifications. At the same time, be wary of exaggerated pseudotechnical claims, like "Our mat board is the only one with a neutral pH!" — especially when they are backed up by little or dubious documentation. Photographs or the descriptive writing may give cheap substitutes away; if you are still uncertain, write and ask for specific details.

Finally, the reputation and general integrity of the company with which you are dealing can help make up your mind. But even here, remember that it is your responsibility to know exactly what you want. Returns at best are time-consuming, and at worst they are chancy unless you have a very good reason for them. A mail order house is in business to make money. Shipping the wrong item costs them time, money, and possibly lost business. So give them the exact specifications of what you want to buy.

APPENDIX C
Plastics Used for Photographic Enclosures

There are hundreds — perhaps thousands — of types of plastic. All of them can contain various kinds of plasticizers, slip coats, and antioxidants in different combinations, which can interact with one another and with their environment in complex and surprising ways. Because their chemistry is so fiendishly complex, the use of plastics in photographic enclosures is problematic. Let's look at some guidelines to minimize the risk.

To safely use plastics to hold photographs, you need to make a very conservative choice of known materials. To that end, we discuss the six most common plastics used by manufacturers of photographic enclosures such as slide pages, print envelopes, and the like.

Tests can rule out certain plastics for photographic enclosures. For example, you can test whether a film will abrade a negative or print when they come in contact; and it's possible to use conditions of high temperature, pressure, and humidity to check whether a particular plastic causes ferrotyping (the "blocking" test). Published research into particular cases where plastic has caused damage is also useful, and we have some of that.

Unfortunately, we don't have an established protocol for testing plastics to be used in photographic enclosures. It would help both manufacturer and consumer if we could put something like the Underwriter's Laboratory label on, for example, a slide box. That would tell you it had passed a test at a recognized professional laboratory. But both the current Photographic Activities Test and another (silver tarnish) test have been criticized as giving problematic results with plastics. For now, we must rely on field trials, experience, and published results to determine the safety of a particular material.

You may find yourself with a plastic enclosure of unknown composition. Refer to "Identification of Plastic Films" for tests to identify the plastic. When you can determine the material of a particular housing, it becomes easier to decide whether to continue to use the housing or to replace it.

These are the plastics you're likely to find for holding photographs:

Cellulose triacetate (Recommended)

This is the base for modern "safety" film; it is used for Kodak negative enclosures. It is highly stable and does not interact with film or paper. Only uncoated film without plasticizers is suitable.

Polyester (polyethylene terephthalate) (Recommended)

Also sold under the trademark names Mylar, Melinex, and Estar, this material has not been extensively exploited for photographic enclosures. It is becoming more widely used by suppliers of conservation-grade materials. The Estar version is used by Kodak for a film base. Polyester is hard, semirigid, and has a crystalline clarity that makes it suitable for housing photographic materials when you wish to perform a close inspection without removing the subject from its housing.

Polyvinyl chloride (PVC) (Not recommended)
Frequently used in slide pages, album pages, and sheet protectors. PVC is cheap and easy to fabricate, which must explain its popularity with makers of inexpensive, low-grade albums and slide pages. It exudes plasticizers, salts, and hydrochloric acid, and sticks to the emulsions of prints, causing physical damage. Horrible stuff around photographs.

Polyethylene (Recommended)
Widely accepted as safe when it does not contain plasticizers or coatings. A smooth, nonrigid plastic with a slightly milky appearance, polyethylene is used to make archival-grade negative enclosures (and Zip-Loc bags).

Polypropylene (Recommended)
Chemically quite similar to polyethylene, polypropylene has slightly less light stability. Excellent for storing slides. Many binder pages are made of it.

Polystyrene (Not proved safe — conditionally recommended)
Approved by ANSI but questioned by Polaroid Corporation on theoretical grounds, polystyrene seems to be generally safe. It stands up to radiation and oxidation better than other plastics. It has been used for many years in Kodak Carousel slide trays, slide storage cabinet dividers, and microfilm spools without any reported problems.

REFERENCE

1. For an excellent description of the chemistry of plastics used for photographic enclosures, see R. Scott Williams. "Commercial Storage and Filing Enclosures for Processed Photographic Materials," *Second International Symposium: The Stability and Preservation of Photographic Materials*. Springfield, VA: SPSE, 1985, pp. 16–30.

IDENTIFICATION OF PLASTIC FILMS

This table of characteristics can be used to identify many plastic films used in photographic enclosures. Because "plastics" constitute such a large class of materials, only a fraction of all the possible films are included here. In case you experience difficulty in identifying a particular piece of plastic, we suggest that you consult a competent authority.

Plastic	*Physical characteristics*	*Burning test*	*Copper wire test*[1]*: look for color change*	*Heating test: check odor and acidity/alkalinity*	*Solubility test: dissolves and may swell in . . .*	*Stretch characteristics*	*Tear characteristics*
Polyolefin (Tyvek)	Waxlike feel	Burns freely; smokeless, bluish flame; melts, drips like wax	None	Burnt paraffin wax aroma	Tetraline at 275°F	Easy to stretch	Extremely resistant to tearing
Polyvinyl chloride (PVC or vinyl)	Softness range is that of rubber; greasy feel; white fracturing will show on creasing line	Burns with yellowish, sooty flame; melts freely to form pearl-like drops	Flame turns green	Sharp, acrid odor; fumes turn Congo paper blue	Cyclohexane, trahydrolurane pyridine; solution turns red in alcoholic potassium hyrdroxide solution	Stretching depends on amount of plasticizer present	Moderately easy to tear (both initiating and propagating; ragged tear
Cellulose Acetate	Poor extensibility	Melts and drips	None	Acetic odor	Acetone	Hard to stretch	Easy to tear
Cellulose Diacetate	Poor extensibility	Melts and drips	None	Acetic odor	Acetone	Hard to stretch	Easy to tear
Cellulose Triacetate	Crystalline; thin films are similar in appearance to cellophane	Melts and drips	None	Acetic odor	Acetone	Hard to stretch	Easy to tear
Cellophane	Transparent; paperlike feel when dry; crumples easily	Burns like paper, with light flame; does not melt, drip, or form beads; continues to burn when with-drawn from flame	None	Burnt paper aroma	Schweitzer's reagent (cupro-ammonia)	Hard to stretch	Easy to tear
Cellulose Nitrate	Similar to cellulose	Burns rapidly with an intense white flame	None	Odor of camphor	Acetone at high nitrogen content	N/A	N/A

Polyamide (Nylon)	Opaque to clear; hard feel; poor extensibility	Burns with yellowish, sooty flame; melts to form pearl-like drops	None	Sour aroma, like burnt horn or hair; fumes turn litmus paper red	Formic acid	Hard to stretch	N/A
Polyester (polyethylene terephthalate; Mylar, Estar, Melinex)	Clear, very tough; nontearable	Ignites, but extinguishes upon removal from flame; shrinks without singeing of edges; melts to form pear-shaped drops	None	Faintly sweet; changes shape and shrinks severely above 300°F–350°F	Chlorphenol	Hard to stretch	Hard to initiate tear; easy to propagate; tears with rough edge
Polyvinylidene chloride (saran process)	Either milky or clear film; softer feel than PVC; resists crumple and white fracture	Self-extinguishing	Flame turns green	Sharp, acrid odor; fumes turn Congo paper blue	Cyclohexane; solution turns black in presence of pyridine and alcoholic potassium hydroxide solution		
Polystyrene	Clear; nonextensible; when crumpled gives metallic sound; sharp folds display mother-of-pearl effect	Burns with luminous, soot-producing flame; smoke is dense	None	Repugnantly sweet, like artificial illuminating gas; marigolds	Benzene hydrocarbons	Hard to stretch	Moderately easy to tear (both initiating and propagating); clean tear but not in straight line
Polypropylene		Melts, burns slowly, beading back without dripping	None	N/A	N/A	Easier than polyester	Hard to initiate tear; easy to propagate; tears with clean edge like cellophane
Polyethylene		Fairly rapid burning; melts and drips like wax	None	Shrinks by 10% when heated	N/A	Easy to stretch	Hard to tear

[1] **Hot Copper Wire Test for PVC, Saran, and PVDC**

Strip a 2-foot section of #8 or #10 wire, leaving an extra length because the wire gets hot. The wire must be copper. Strip off any insulation and burn off the residual plastic coating with a butane torch. This burns with a green flame, because most wire covering is PVC.

- When the flame turns orange, touch the red-hot copper to the sample of unknown composition.
- Put the wire back in the flame. If it burns with a green flame, the sample is PVC, PVDC, or a PVDC-coated film. If no green flame appears the sample is another material, and other tests will be required to determine what type it is.

If the copper wire is to be used again, make sure that it is burned clean after the test.

Glossary

BUFFERING The addition of alkaline agents such as calcium or magnesium carbonate during the papermaking process in order to counteract the effect of acidic contamination; the degree of buffering (usually 2–3%) is measured by percentage of paper weight.

CELLULOSE TRIACETATE An inert plastic film used for a film base and in making storage enclosures for photographic materials.

CONTACT PRINT A print made by exposing a sheet of photographic paper in direct contact with a film or glass negative. All the earliest photographic prints were made by contact printing; most contemporary prints are enlargements from smaller negatives.

CONTRAST The density range of a photograph. "High contrast" usually refers to a negative, print, or transparency with sharp blacks and whites and few grays.

COPY FILM The type of negative material (usually fine grain, continuous tone) for copying photographs. Film specifications need to be matched to the particular copying task.

COPY NEGATIVE A film negative made by photographing (copying) a print. It is used to make more prints.

DEFINITION The degree of sharpness available with a particular photographic emulsion and camera-lens combination.

DENSITY The relative opacity of an exposed photographic emulsion, usually referring to film. Differences in density make possible all photographic imagery, and the science of densitometry is devoted to the study of this phenomenon.

DESICCANT Any agent — particularly a silica gel — that removes gaseous water from the air and that reduces relative humidity. Desiccants can be used in sealed enclosures to protect photographs from humidity. Silica gels can be made to give up their absorbed moisture by heating, and then reused.

DIRECT POSITIVE A unique photograph made in the camera without a negative. Older types (daguerreotypes, tintypes) were laterally reversed; modern direct positives such as Polaroid SX-70s read correctly left to right.

D-MIN and D-MAX The minimum (min) and maximum (max) density (D) of a photographic emulsion, usually film. These are measured by a densitometer and indicate the greatest capacity and transparency created by a specific exposure and development.

DUPE NEGATIVE A negative that duplicates another negative. Often made by contact printing a reversal film with the original negative. When properly made, a dupe negative can give better prints than a *copy negative*.

EMULSION A gelatin containing a dispersed light-sensitive agent (usually some form of silver) that has been applied to a base of film, glass, paper, or similar material. Processing the emulsion causes the changes in density that constitute the photograph.

ENLARGEMENT A print larger than the negative, made by projecting the negative onto a sheet of photographic paper.

FERROTYPE 1. A tintype. 2. A finishing process involving heat, pressure, and a polished metal sheet while drying fiber-base black and white prints, to create a highly glossy surface. 3. An undesirable effect in which moisture causes a photographic print or negative to adhere to glass or any glossy surface. Creates blotchy changes in density and separation of the emulsion from the base.

FILM NEGATIVE A photographic image on a film base where the highlights are nearly opaque (i.e., have high density) and the shadows are nearly clear. The terms "film" and "negatives" commonly refer to the original film negative exposed in the camera.

FILM POSITIVE A photographic image on a film base in which the shadows are opaque and the highlights clear. In some respects this corresponds with a *transparency*, but this terminology is used in copy work.

FIXER Sodium thiosulfate (also called "hypo") used to remove the remaining silver halides from photographic materials after development.

FOXING Discoloration on the surface of a paper print caused by mold. Often made worse by iron particles in the paper.

GLASSINE Partly clear paper used to hold negatives; acidic and *hygroscopic.*

GRAIN Clusters of black metallic silver formed in an emulsion during development. Grain in negatives becomes more pronounced and visible during enlargement.

HALFTONE A printing process to reproduce continuous-tone photographs with black ink. The original photograph is copied through a film grid (or screen) with very small openings. Different shades of gray cause different-size dots on the copy negative, which is used to make a plate that holds the ink. Examined very closely, a halftone photograph can be seen to be composed of small dots.

HYGROSCOPIC Absorbing moisture from the air. Most paper is mildly hygroscopic. "Print flattening" agents are highly hygroscopic.

INTERPOSITIVE A film positive usually made by enlarging a negative onto negative film. An interpositive can be contact printed onto another sheet of negative working film to make a printing negative.

LATENT IMAGE An invisible image on a photographic material after exposure and before development.

NEUTRAL pH Neither acid nor alkaline, expresed on the pH scale by the value 7.0.

OXIDATION The process whereby oxygen combines with another chemical to make an oxide. Fast oxidation is fire; slower oxidation processes cause the destruction of photographs.

PLASTICIZER A chemical added to plastic resins to improve flexibility, workability, or stretching. In plastic enclosures, plasticizers tend to volatize and adversely affect photographs stored therein.

POLYESTER Flexible, transparent plastic film of polyethylene terephthalate; also sold under trademarks as Mylar and Melinex. Inert and chemically stable.

POLYETHYLENE Translucent thermoplastic with a low melting point. Inert and chemically stable.

POSITIVE PRINT A photographic print in which the values read naturally; i.e., the shadows are dark and the highlights white.

PRINTING OUT 1. The creation of a photographic image by the action of light on the emulsion. Early printing out papers (POPs) were contact printed by sunlight. When the image was judged complete, the remaining silver was removed by fixing. The cyanotype or blueprint is a POP process. 2. An undesirable effect on a print created when residual silver in the emulsion turns dark. Caused by insufficient fixation. This can be reversed by bleaching and redeveloping the print.

STAIN In photographic conservation, this refers to changes — especially in a color print — that occur in the dark. A typical example is the overall yellowing of early Kodachrome prints caused by deterioration of the magenta coupler.

TONING Treating a black and white print with a chemical to change the color and to protect the stability of the silver. Toners commonly used for photographic conservation include gold, selenium, and brown.

TRANSPARENCY A positive film image with the same arrangement of values as the original scene. A 35mm slide is a transparency, but the term is more usually applied to images made in larger formats such as 2¼ square or 4 × 5. Most often refers to a color image.

Bibliography

Adams, Ansel (with Robert Baker). *The Negative. New Ansel Adams Photography Series, Book 2*. Boston: Little, Brown and Company, 1981.

———. *The Print. New Ansel Adams Photography Series, Book 3*. Boston: Little, Brown and Company, 1983.

Banks, Paul N. *A Selective Bibliography on the Conservation of Research Library Materials*. Chicago: Newberry Library, 1981.

Barth, Miles. "Notes on Conservation and Restoration of Photographs." *PCN The Print Collectors Newsletter* 7, no. 2 (May/June 1976): 48–51.

Bowditch G. "Cataloguing Photographs: A Procedure for Small Museums." *History News* 26 (1971): 241–248.

Carroll, John S. *Photographic Lab Handbook*, 5th ed. New York: Amphoto, 1979.

Clapp, Anne F. *Curatorial Care of Works of Art on Paper: Basic Procedures for Paper Preservation*, 4th ed. New York: Nick Lyons Books, 1987.

Crabtree, J.I., G.T. Eaton, and L.E. Muehler. "The Quantitative Determination of Hypo in Photographic Prints with Silver Nitrate." *Journal of the Franklin Institute* 235 (April 1943): 351–360.

Crawford, William. *The Keepers of Light: A History and Working Guide to Early Photographic Processes*. Dobbs Ferry, New York: Morgan & Morgan, 1979.

Cummins, Jim. "Framing Pictures." *Fine Woodworking* no. 35 (July/August 1982): 61–67.

Davies, Thomas L. *Shoots: A Guide to Your Family's Photographic Heritage*. Danbury, New Hampshire: Addison House, 1977.

Denstman, Hal. "Polarized Light for High Fidelity Reproduction." *Industrial Photography* 27, no. 10 (October 1978): 23–25, 60.

Doherty, Robert J. "Editorial: Conservation and Photographs." *Image* 20, no. 3–4 (September/December 1977): 33.

Dolloff, Francis W., and Roy L. Perkinson. *How to Care for Works of Art on Paper*. Boston: Museum of Fine Arts, 1971.

Duren, Lista. *Frame It: A Complete Do-It-Yourself Guide to Picture Framing*. Boston: Houghton Mifflin, 1976.

Eastman Kodak. *Conservation of Photographs*. Kodak Publication No. F-40. Rochester, New York: Eastman Kodak Company, 1985.

———. *Kodak Ektagraphic Slide Projectors*. Kodak Publication S-74, Rochester, New York: Eastman Kodak Company, 1977.

———. *Processing Chemicals and Formulas for Black-and-White Photography*, 6th ed. Kodak Publication J-1. Rochester, New York: Eastman Kodak Company, 1963.

Eaton, George T. "Preservation, Deterioration, Restoration of Photographic Images." *The Library Quarterly* 40, no. 1 (University of Chicago Press, January 1970).

Eskind, Andrew H., and Deborah Barsel. "International Museum of Photography at George Eastman House Conventions for Cataloguing Photographs." *Image* 21, no. 4 (December 1978): 1–31.

Feldman, Larry. "Discoloration of Black and White Photographic Prints." *Journal of Applied Photographic Engineering* (February 1981): 1–9.

Gill, Arthur. "Recognition of Photographic Processes." *History of Photography* 2, no. 1 (January 1978): 34–36.

Hendriks, Klaus B. "The Preservation of Photographic Records." *Archivaria* no. 5 (1977–1978): 92–98.

Hunter, Dard. *Papermaking: The History and Technique of an Ancient Craft*. New York: Alfred A. Knopf, 1947.

Ilford. *Technical Information: Mounting and Laminating Cibachrome Display Print Materials and Films*. Paramus, New Jersey: Ilford Photo Corporation, 1988.

Jenkins, R.V. *Images and Enterprise: Technology and the American Photographic Industry, 1839–1925*. Baltimore: Johns Hopkins University Press, 1975.

Kach, David. "Photographic Dilemma: Stability and Storage of Color Materials." *Industrial Photography* 27, no. 8 (August 1978): 28–29, 46–50.

Katcher, Phillip. "How to Date an Image from Its Mat." *PSA Journal* 44, no. 8 (August 1978): 26.

Kohlbeck, Runo. "On the Measurement of Residual Thiosulfate in Photographic Paper Prints." *Journal of Applied Photographic Engineering* 4, no. 4 (Fall 1978): 205–206.

Lafontaine, Raymond H. *Environmental Norms for Canadian Museums, Art Galleries and Archives*. Technical Bulletin No. 5 (Ottawa, Ontario: Canadian Conservation Institute, November 1979).

Lafontaine, Raymond H., and Patricia A. Wood. *Fluorescent Lamps*. Technical Bulletin No. 7 (Ottawa, Ontario: Canadian Conservation Institute, January 1980).

Linville, Judi. "Fading of Color Photography: New Ways to Manage an Old Problem." *Decor* (August 1982): 62 ff.

Long, Margery S., Gerald J. Munoff, and Mary Lynn Ritzenthaler. *Archives & Manuscripts: Administration of Photographic Collections*. SAA Basic Manual Series. Chicago: Society of American Archivists, 1984.

Moor, Ian. "The Ambrotype — Research into Its Restoration and Conservation — Part 1." *The Paper Conservator* 1 (1976): 22–25.

Ostroff, Eugene. "Rescuing Nitrate Negatives." *Museum News* (September/October 1978): 34–42. Reprint purchased from American Association of Museums, Suite 428, 1055 Thomas Jefferson Street, NW, Washington, DC, 20007.

Pittaro, Ernest M., ed. *Photo-Lab-Index: The Cumulative Formulary of Standard Recommended Photographic Procedures*. Dobbs Ferry, New York: Morgan & Morgan, 1977.

Pobboravsky, Irving. "Daguerreotype Preservation: The Problems of Tarnish Removal." *Technology and Conservation* 3, no. 2 (Summer 1978): 40–45.

Reilly, James M. *Care and Identification of Nineteenth-Century Photographic Prints*. Kodak Publication G2S. Rochester, New York: Eastman Kodak Co., 1986.

———. *The Albumen and Salted Paper Book: The History and Practice of Photographic Printing, 1840–1895*. Rochester, New York: Light Impressions, 1980.

Rempel, Siegfried. *The Care of Photographs*. New York: Nick Lyons Books, 1987.

———. *The Care of Black-and-White Photographic Collections: Cleaning and Stabilization*. Technical Bulletin No. 9 (Ottawa, Ontario: Canadian Conservation Institute, December 1980).

Rowlison, Eric B. "Rules for Handling Works of Art." Reprint from *Museum News* (April 1975): 1–4.

Smith, Merrily A. *Matting and Hinging of Works of Art on Paper*. Washington, DC: Preservation Office Research Services, Library of Congress, 1981.

Society of Photographic Scientists and Engineers. *Second International Symposium: The Stability and Preservation of Photographic Images — Printing of Transcript Summaries and Advance Printing of Paper Summaries*. Springfield, Virginia: SPSE, 1985.

Steiner, Ralph. "Comparing Fixing Methods." *PhotographiConservation* 2, no. 1 (Graphic Arts Research Center, Rochester Institute of Technology, March 1980).

Stolow, Nathan. *Conservation and Exhibitions: Packing, Transport, Storage and Environmental Considerations*. London: Butterworths, 1987.

———. "Conservation Policy and the Exhibition of Museum Collections." *Journal of the American Institute of Conservation* 16, no. 2 (February 1977): 12–20.

Stroebel, Leslie, et al. *Photographic Materials and Processes*. Boston: Focal Press, 1986.

Swan, Alice. "Conservation Treatments for Photographs, a Review of Some of the Problems, Literature, and Practices." *Image* 21, no. 2 (June 1978): 24–31.

van Altena, W.F. "Envelopes for the Archival Storage of Processed Astronomical Photographs." *AAS Photo-Bulletin* 8, no. 1 (1975): 18–19.

Vestal, David. *The Craft of Photography*. New York: Harper & Row, 1975.

Weinstein, Robert A., and Larry Booth. *Collection, Use, and Care of Historical Photographs*. Nashville, Tennessee: American Association for State and Local History, 1977.

Welling, William. *Collectors' Guide to Nineteenth-Century Photographs*. New York: Collier Books, 1976.

Walker, Robert. "Black and White Slide Production: A Multitude of Means." *Industrial Photography* 27, no. 2 (February 1978): 33–36, 56, 57.

Wilhelm, Henry. "Color Print Instability: A Problem for Collectors and Photographers." *Afterimage* 6, no. 3 (October 1978): 11–13.

———. "Monitoring the Fading and Staining of Color Photographic Prints." *Journal of the American Institute for Conservation* 21 (1981): 49–64.

———. "Storing Color Materials: Frost-Free Refrigerators Offer a Low-Cost Solution." *Industrial Photography* 27, no. 10 (October 1978): 32 ff.

Zigrosser, Carl, and Christa M. Gaehde. *A Guide to the Collecting and Care of Original Prints*. New York: Crown Publishers for the Print Council of America, 1965.

Index